AF777850

Melatonin and the Mammalian Pineal Gland

Melatonin and the Mammalian Pineal Gland

Josephine Arendt

Professor of Endocrinology
School of Biological Sciences
University of Surrey
Guildford, UK

CHAPMAN & HALL

London · Glasgow · Weinheim · New York · Tokyo · Melbourne · Madras

Published by Chapman & Hall, 2–6 Boundary Row, London SE1 8HN, UK

Chapman & Hall, 2–6 Boundary Row, London SE1 8HN, UK

Blackie Academic & Professional, Wester Cleddens Road, Bishopbriggs, Glasgow G64 2NZ, UK

Chapman & Hall GmbH, Pappelallee 3, 69469 Weinheim, Germany

Chapman & Hall USA, One Penn Plaza, 41st Floor, New York NY 10119, USA

Chapman & Hall Japan, ITP-Japan, Kyowa Building, 3F, 2-2-1 Hirakawacho, Chiyoda-ku, Tokyo 102, Japan

Chapman & Hall Australia, Thomas Nelson Australia, 102 Dodds Street, South Melbourne, Victoria 3205, Australia

Chapman & Hall India, R. Seshadri, 32 Second Main Road, CIT East, Madras 600 035, India

First edition 1995

© 1995 Josephine Arendt

Typeset in 11/13 Times by ROM-Data Corporation, Falmouth, Cornwall

Printed in Great Britain at the University Press, Cambridge

ISBN 0 412 53600 5

Apart from any fair dealing for the purposes of research or private study, or criticism or review, as permitted under the UK Copyright Designs and Patents Act, 1988, this publication may not be reproduced, stored, or transmitted, in any form or by any means, without the prior permission in writing of the publishers, or in the case of reprographic reproduction only in accordance with the terms of the licences issued by the Copyright Licensing Agency in the UK, or in accordance with the terms of licences issued by the appropriate Reproduction Rights Organization outside the UK. Enquiries concerning reproduction outside the terms stated here should be sent to the publishers at the London address printed on this page.

The publisher makes no representation, express or implied, with regard to the accuracy of the information contained in this book and cannot accept any legal responsibility or liability for any errors or omissions that may be made.

A catalogue record for this book is available from the British Library

Library of Congress Catalog Card Number: 94-68244

Printed on acid-free text paper, manufactured in accordance with ANSI/NISO Z39.48-1992 (Permanence of Paper).

Contents

Foreword

A wonderful book – comprehensive, up to date and beautifully written. That is my description of Josephine Arendt's book, *Melatonin and the Mammalian Pineal Gland.* There have been good international meetings on melatonin, the pineal gland and biological rhythms, and the effect of light on these systems. The proceedings from these meetings have been excellent. Nevertheless, having a book by a single author who has played a major role in the development of these subjects makes the treatise particularly valuable.

Beginning with her development of a reliable radioimmunoassay for melatonin 20 years ago, Josephine Arendt has helped to open up an entirely new way of investigating the action of light and melatonin on metabolic processes, including biological rhythms, seasonal affective disorders – such as winter depression, reproductive capability in seasonally breeding animals, sleep patterns, insomnia, jet lag, disturbances in night-shift workers and blind people – and other topics. In addition there is her unique enthusiasm for the research – enthusiasm that is contagious. Her statement regarding research on the pineal that appears at the end of the Introduction, 'It is enormously exciting, exhausting (unsociable hours are the rule, not the exception) and will be of benefit to humanity', describes her feelings for the subject and sets the tone for her book. The nine chapters cover a wide range of topics, from basic science to clinical and commercial aspects of this new field. The book is an excellent reference for undergraduate students and scientists in general. It is a must for graduate students, postdoctoral fellows and established investigators in this field. Josephine, thank you for putting together such an interesting story.

Aaron B. Lerner MD
Yale University
New Haven, Connecticut
August 17, 1994

Preface

Biological rhythms, whilst an intrinsically fascinating subject for study and research, also form an increasing part of both undergraduate and postgraduate taught courses. An understanding of the functions of the pineal gland and its hormone melatonin is central to this area and this book was written to fulfill a perceived need. I have tried to keep it simple, to provide as much visual matter as possible and to restrict text references to a minimum. At the end of each chapter is a list of further reading, so that all material referred to in the text can be followed up. Where studies are not explicitly cited they should be retrievable through the many reviews listed, some citations appearing at the end of more than one chapter. I apologize to those colleagues and friends who may feel that their own work is not sufficiently represented.

However, I have also exercised the right to select those topics that I feel to be of particular interest. Others may not agree with the choice, in which case they should be encouraged to provide coverage of pineal topics which are either sparsely treated or absent here. I would much appreciate comments and suggestions as to how the book could be improved. Personal bias appears in some of the interpretations offered. In this fast moving field both recent data and opinions may well need revision in the near future.

Acknowledgements

Many people, and indeed all pineal research workers, have contributed to this book. I would particularly like to thank those who rummaged in their ancient and modern files to provide good illustrative material. This was much appreciated both by myself and by the publishers. It is impossible to thank them all here by name, the figures will speak for themselves. In addition to figures, David Klein and the late Bill Quay both provided very helpful comments on parts of the manuscript. Johannes Ariens-Kappers, Russ Reiter, Paul Pevet and Peretz Lavie supplied information that I would not otherwise have found.

My own work in this area has benefited greatly from wonderful collaborators. Merton Sandler, Soli Contractor, Luc Paunier and Pierre Sizonenko enabled my early work. Meeting Lennart Wetterberg in 1975 was my first encounter with a real pineal enthusiast and fueled a lifetime obsession. Andy Symons and Isabel Forsyth were ideal co-workers in the field of seasonal rhythms. Anna Wirz and Simon Folkard provided the initial indoctrination into circadian physiology, subsequent friendship, collaboration and ongoing speculative discussions. Jim Waterhouse, Dave Minors and Rutger Wever have provided both inspiration and suitable skepticism with regard to the importance of melatonin in human rhythm physiology. Alex Borbely and David Parkes have been essential to our sleep studies, Anna Wirz, Stuart Checkley and Chris Thompson continued our interest in psychiatry. Helena Illnerova's insight helps my muddled imagination. Sabbatical leaves with Greg Brown and Jean-Paul Ravault put everything into perspective, and particularly in France provided much nourishment for the body as well as for the mind.

In my laboratory over the years Judie English, Carol Laud, Steve Pryde, Chris Franey, Debra Skene, Chris Bojkowski, Margaret Aldhous, Tony Poulton, Paul Griffiths, James Broadway, Jeremy Owen, Ian Makkison, Rory O'Conor, Jonathan Ross, Sharon Deveson, Fiona Gebbie, Steve Deacon and Benita Middleton have generated the data presented here from the University of Surrey. Especial thanks go to Judie English who makes it all possible.

Vincent Marks gave essential help and encouragement both to establish my laboratory in the UK and to fund our work. John Wright enables our clinical studies in many ways not least by his ability to stay awake all night while taking blood samples. He and I first took timed melatonin in 1981 and realized its potential. Brian Follett and Ann Warner gave much needed support.

Our work has primarily been funded by the Swiss National Science Foundation, the MRC, the AFRC, the Wellcome Trust, the British Antarctic Survey, the South West Thames Regional Health Authority, the Ministry of Defence, several pharmaceutical companies and our own company Stockgrand Ltd.

The small Channel Island of Guernsey, where most of this was written, restores my sanity at frequent intervals. Finally I have to say that without the tolerance, understanding and love of my husband and family I would not have been able even to contemplate this book. I hope they feel that their domestic and other supportive efforts have allowed me to produce something worthwhile.

Melatonin and the mammalian pineal gland

1

Contents

1.1 INTRODUCTION

The pineal gland, the mysterious unpaired organ of the brain, the 'third eye', the seat of the soul, a 'calcified vestigial organ with no function', subject of medical jokes, together with its hormone melatonin is finally taken seriously during this latter half of the 20th century. This book attempts to describe how the multiple facets of pineal function integrate with mainstream physiology and medicine and indeed how melatonin itself, 'the darkness hormone' acts as a coordinator of internal physiological rhythmicity. This ancient molecule may well have the same functions in all life forms from unicellular organisms to man.

The pineal gland or epiphysis cerebri has been known for more than 2000 years. The most ancient existing written description is that of Galen (130–200 AD) who refers to Herophilus (325–280 BC). Herophilus apparently thought that the pineal was a valve regulating the flow of 'pneuma' or 'spiritus' in Latin from the 3rd to the 4th ventricle. Pneuma was considered to be derived from air and transformed into 'pneuma psychikon' or 'spiritus animalis' in the brain ventricular system. Pneuma psychikon was then the direct cause of the development of knowledge. Galen, however, considered that the pineal was a gland, at that time an organ which filled gaps between blood vessels and supported them. The first good descriptions of the pineal derive from the 15th century and particularly from Vesalius in the 16th century who still considered that spiritus animalis was found in the ventricles.

Probably the most famous of ancient texts on the pineal is that of the influential French philosopher Descartes (Figure 1.1). In 1662 Descartes considered that by movements, the pineal controlled the flow of 'animal spirits' into motor nerves and thus influenced movements of the body. He thought that the stimulus for pineal function came from visual input to the retina – the latter is a most remarkable insight, as effectively this is true today. It is usually Descartes to whom is attributed the concept of the pineal as the seat of the soul, although this idea probably derives from Herophilus. These ancient ideas persisted for many centuries and to some extent have colored 20th century perceptions of the pineal. They may have been responsible for the slow progress of early pineal research, in rendering the gland somehow not quite scientifically respectable.

Ahlborn (1884) was the first to notice the remarkable resemblance between the pineal/parietal organ (now the 'third eye') of some lower vertebrates and the structure of the lateral eyes. This was certainly the crucial observation leading to extensive interest in the evolutionary history of the pineal this century and to the statement (Studnicka, 1905) that the photosensory organ of lower vertebrates became the secretory mammalian pineal. This is extensively supported by modern comparative anatomy (Oksche *et al.*, 1971; Collin, 1972). It underpins our current knowledge of the influence of light on this secretory organ, be that direct or indirect.

The focal early work leading to our understanding of pineal endocrine function was summarized in a book published in 1954 by Kitay and Altschule. Early observations on pineal tumors by Heubner (1898) and Marburg (1907), the authors' own work and that of others provided one of their important conclusions that the gland influenced reproductive function.

Our primary understanding of pineal control mechanisms derives

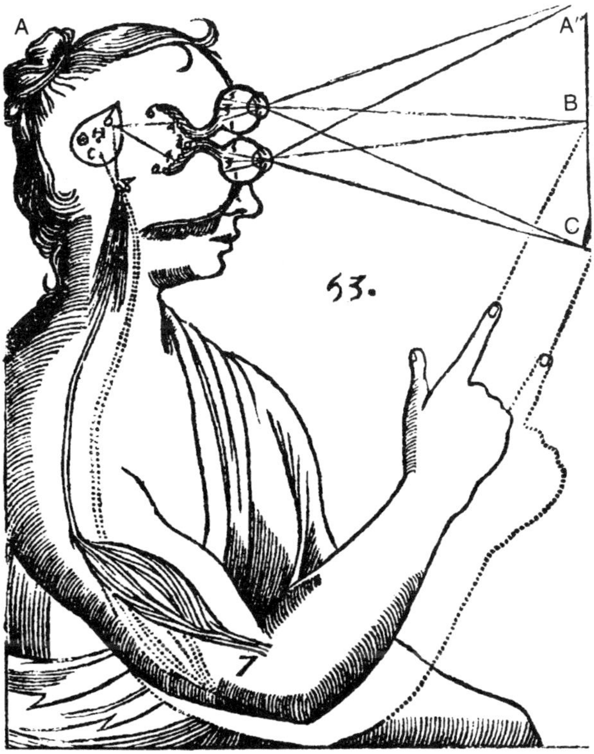

Fig. 1.1 Descartes' conception of the function and connections of the human pineal gland.

from Kappers' classical studies (e.g. 1960) on the innervation pathway of the pineal from the superior cervical ganglion of the sympathetic nervous system.

In all of the history of pineal research there is, however, nothing more extraordinary or more important than the discovery of the structure of its principle hormone melatonin by Aaron Lerner in 1958 whilst hunting for the amphibian skin-lightening factor of the pineal. Lerner named the molecule melatonin after its effects on amphibian melanophores (pigmented skin cells). Rumor has it that he originally considered the name 'yalin' as the work was accomplished at Yale University. We may be thankful, if this is true, that he changed his mind. 'Melatonin' has a certain resonance that 'yalin' undoubtedly lacks. Moreover, a world class molecule does not deserve such a parochial name! The molecule of melatonin, being lipophilic, reaches all parts of an organism and has now infiltrated into all aspects of physiology. Its primary role is that of a time-giver, or coordinator of biological rhythms. The range of periodicities affected by melatonin has yet to be fully defined, but certainly it acts as an internal calendar for the timing of seasonal events and as an internal clock for the timing of daily events. It has been referred to as the 'darkness hormone' (it is made at night), as 'time in a bottle', and as 'circadian glue' as it appears to help hold together, or synchronize, circadian or daily rhythms.

At present it could almost be considered the best candidate to date for a universal panacea. It has been proposed as the treatment of choice for an enormous range of ills, the most reasonable being natural rhythm disturbance occurring when our biology conflicts with our culture – such as shift work and jet lag, and the unnatural artificial light conditions, especially in winter, in which we live. It has essentially founded a whole new pharmacology which is yet in its early infancy.

Pineal research is truly interdisciplinary, particularly with respect to its central place in biological rhythms. It ranges from molecular biology to psychology and in this way its participants are required to keep their perspectives broad while still remaining absolutely rigorous. It is enormously exciting, exhausting (unsociable hours are the rule not the exception) and will hopefully be of considerable benefit to humanity. What more could one ask? I hope this brief account will convince, you, the reader, to join in or at least to follow our progress with interest.

ACKNOWLEDGEMENTS

For this brief history of pineal research, I am particularly indebted to Professor Johannes Ariens-Kappers, Professor J.P. Collin and to Dr Claudia Zrenner in whose reviews all the above references can be found.

REFERENCES

Kappers, J.A. (1979) Short history of pineal discovery and research, in *Progress in Brain Research, The Pineal Gland of Vertebrates including Man*, Volume 52 (eds J. Ariens-Kappers and P. Pevet), Elsevier/North Holland Biomedical Press, pp. 3–22.

Collin, J.P. (1972) Differentiation and regression of the cells of the sensory line in the epiphysis cerebri, in *The Pineal Gland* (eds G.E.W. Wolstenhome and J. Knight), Churchill Livingstone, Edinburgh, pp. 79–125.

Zrenner, C. (1985) Theories of Pineal Function from Classical Antiquity to 1900: A History. *Pineal Research Reviews*, Volume 3, pp. 1–40.

2 Structure of the pineal gland

Contents

2.1 ANATOMY

The pineal gland is a small unpaired central structure, essentially an appendage of the brain. It originates as an envagination of the diencephalon and in most species retains a 'stalk' connection to the habenular commissure (Figure 2.1). Frequently it is closely associated with the 3rd ventricle of the brain and the area of the 3rd ventricle receiving the pineal stalk is known as the pineal recess. Great variation in both size and position within the brain is shown

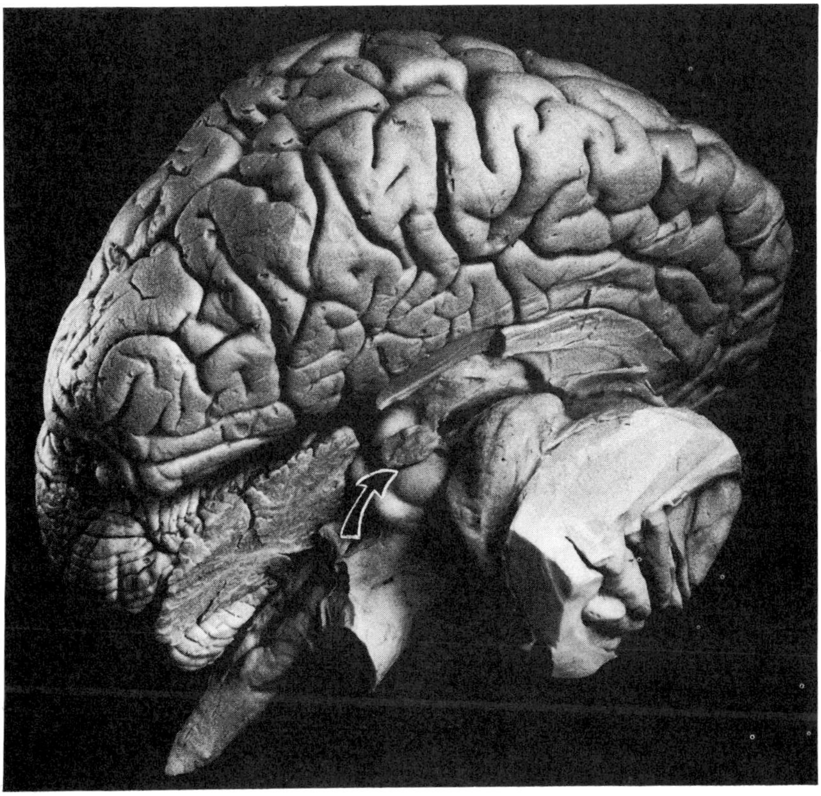

Fig. 2.1(a) Dissected human brain showing the pineal gland (epiphysis cerebri) *in situ* (arrow). Dissection and photography by Dr S. Heil. (Reproduced with permission from Vollrath, L. *The Pineal Organ*; published by Springer-Verlag, 1981.)

by the mammalian pineal gland, even within species. In humans it is about the size of a pea and assumes a shape resembling a pine cone (hence pineal). Again due to its shape, it has been referred to as the 'penis of the brain'.

The mammalian pineal is a secretory organ, whereas in fish and amphibians it is directly photoreceptive and in reptiles and birds it has a mixed photoreceptor and secretory function. During development in some, but not all, fish, amphibians and lacertilian reptiles

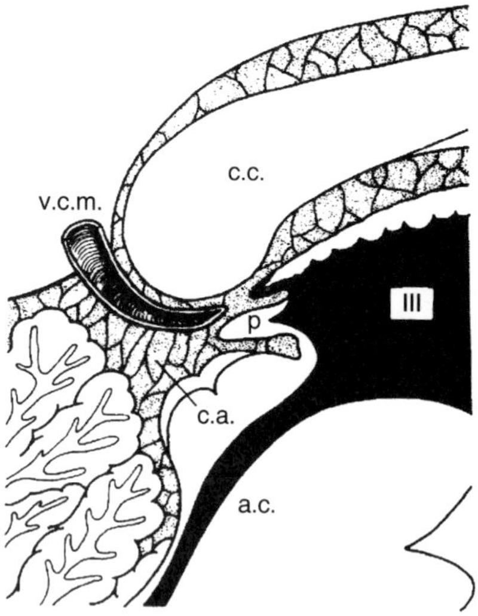

Fig. 2.1(b) Relationship of the human pineal gland to the CSF. a.c.=Aqueductus cerebri; c.a.=cisterna ambiens; c.c.=corpus callosum; p=pineal gland; v.c.m.=vena cerebri magna; III=third ventricle. (Reproduced with permission from Vollrath, L. *The Pineal Organ*; published by Springer-Verlag, 1981.)

the pineal forms two components, the extracranial parietal (parapineal, frontal) organ and the intracranial pineal organ. In some lower vertebrates the parietal organ lies directly below the skin on the dorsal surface of the cranium and is sometimes referred to as the third eye (Figure 2.2). The principal cellular components of the parietal organ are photoreceptors which in some cases may also have a secretory function.

Elements of its photoreceptive evolutionary history remain within the mammalian pineal in both structure and function. Its primary function in all species studied to date is to transduce information concerning light–dark cycles to body physiology, particularly for the organization of body rhythms, via the secretion of its major hormone melatonin. In some birds and lower vertebrates it serves as a rhythm generating system, or biological clock. In mammals it is intimately concerned with the coordination of rhythm physiology without having the capacity to act as a rhythm generator.

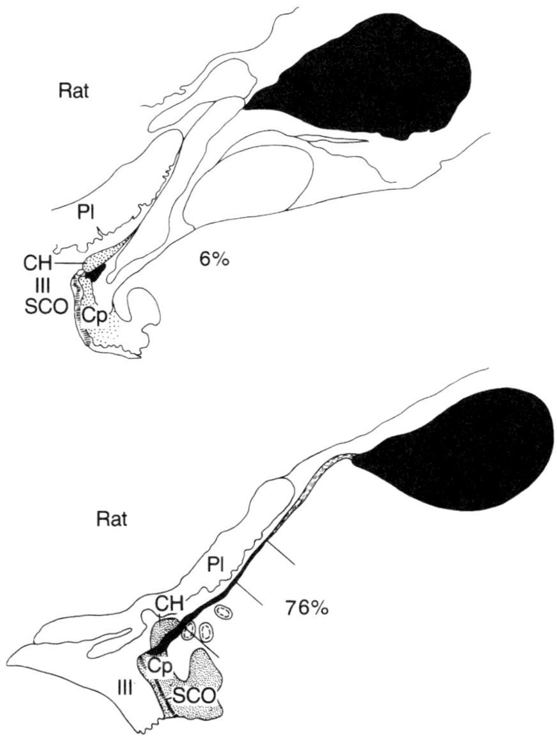

Fig. 2.1(c) Schematic drawing of midsagittal section of a rat pineal (in black). The major portion lies superficially below the skull with some deeper components. The connection to the deeper components is not always present. CH = habenular commissure; Cp = posterior commissure; Pl = choroid plexus; SCO = subcommissural organ; III = third ventricle. (Reproduced with permission from Vollrath, L. *The Pineal Organ*; published by Springer-Verlag, 1981.)

Vollrath (1981) has proposed a classification based on the form and arrangement of pineal tissue and its relationship to the 3rd ventricle. Essentially three types are distinguished: (A) where the bulk of the pineal tissue lies close to the 3rd ventricle; (AB) where the organ is elongated; (ABC) where a substantial portion of the pineal lies superficially or close to the cerebellum; and (C) where all pineal tissue is superficial. Further variations can be accommodated, e.g. if one area is missing, the appropriate letter is discarded, or if an

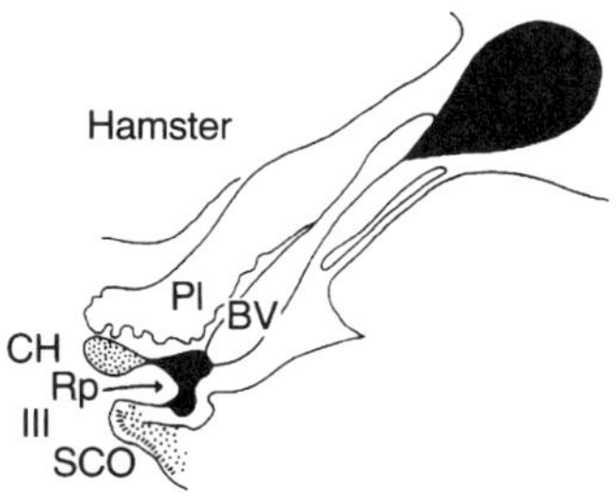

Fig. 2.1(d) Schematic drawing of midsagittal section of a hamster pineal complex (in black). The major portion lies superficially below the skull (caudal diencephalic roof) with a separate, distinct deep pineal. CH = habenular commissure; Pl = choroid plexus; SCO = subcommissural organ; Rp = pineal recess of the third ventricle III; BV = blood vessel. (Reproduced with permission from Vollrath, L. *The Pineal Organ*; published by Springer-Verlag, 1981.)

area is small the corresponding Greek alphabet is used. The human pineal, deeply recessed, is type (A), the rat pineal being situated superficially is usually classified as (C) and the hamster, which shows both a superficial and a small deep pineal belongs to the type (αC) (Figure 2.1).

Pineal weight and volume also varies greatly within and between species and with respect to time, age and physiological status. Pineal volume correlates positively with body weight across species. The mean weight of the adult rat pineal is reported to vary from 0.9 mg (pigmented Long-Evans) to 1.56 mg (Sprague-Dawley) whereas the adult human pineal is on average 100–150 mg. There is a suggestion of a higher pineal to body weight ratio in young compared to adult animals.

Pineal weight has been taken by some as an index of functional activity. In this respect it is interesting to note that the Weddell seal, and indeed a number of other species inhabiting polar regions, has an exceptionally large pineal and this may be related to the extreme seasonal variations of daylength found in Antarctic regions.

In the Pholidota, the pineal is reported to be absent.

2.2 INNERVATION

2.2.1 Peripheral sympathetic innervation

Both afferent and efferent connections to the pineal are present in lower vertebrates. During phylogenesis efferent innervation is lost and in mammals the most physiologically important innervation of the pineal is from post-ganglionic sympathetic fibers arising from the superior cervical ganglion. In addition to this input, there is also strong evidence for parasympathetic, commissural and peptidergic innervation. Sympathetic innervation is clearly essential for the appropriate production of the major hormone of the pineal – melatonin. The function of other inputs remains to be determined.

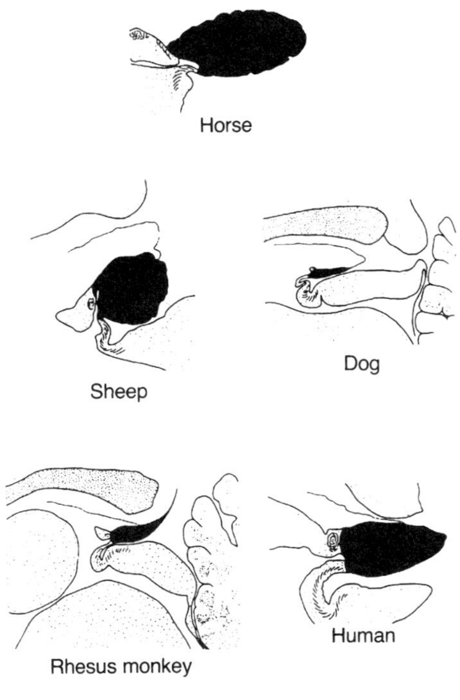

Fig. 2.1(e) Schematic representation of the pineal gland of a number of species. (Reproduced from Hulsemann (1967) cited in Vollrath, L. *The Pineal Organ*; published by Springer-Verlag, 1981.)

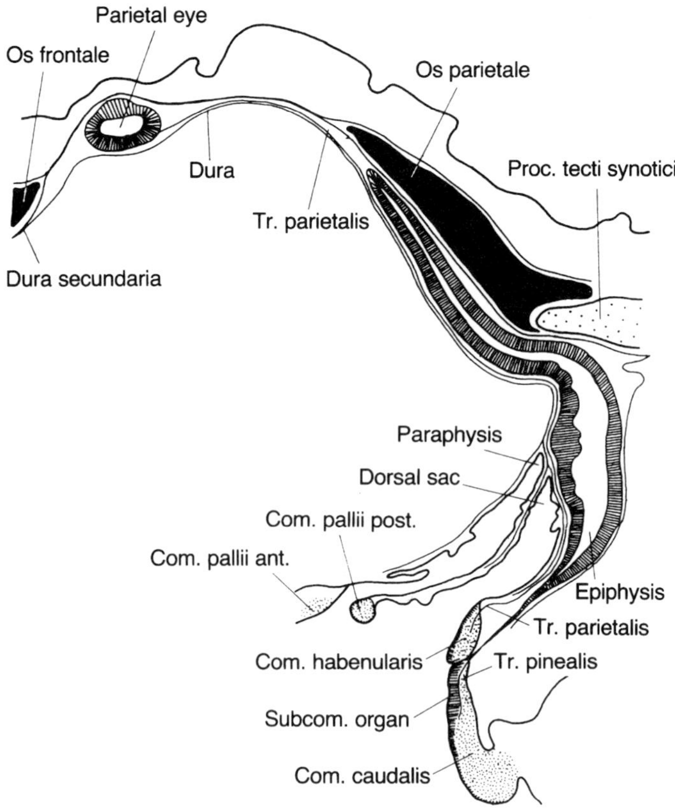

Fig. 2.2 An example of a reptile (*Phrynocephalus undulatus*) with a pineal complex consisting of a parietal eye and epiphysis cerebri. (From Kummer-Trost (1956), by permission, cited in Vollrath, L. *The Pineal Organ*; published by Springer-Verlag, 1981.)

The gland itself can show spontaneous electrical activity which may exhibit daily rhythmicity.

The nerve cells of sympathetic innervation are adrenergic, with cell bodies located in the superior cervical ganglion. Kappers (1960,1965) provided a comprehensive survey of pineal sympathetic innervation. The fibers follow the course of blood vessels – internal carotid plexus to tentorium cerebelli – and usually penetrate the apex of the pineal capsule via the nervii conarii (Figure 2.3). They appear to terminate within the pineal, usually in the perivascular space, with more or less rare synaptic contact with pinealocytes. The transmitter is thought to diffuse through the perivascular space to reach its target cells, although if pinealocyte processes lie in the perivascular space, this process would be speeded up. There is also an intracellular system of continuous microscopic channels around pinealocytes which may transport neurotransmitter.

In mammals the major neurotransmitter is noradrenalin, but the pineal adrenergic receptor type varies with species. Moreover whilst noradrenalin stimulates melatonin secretion in mammals, it is inhibitory in chickens. Whether this is true of all birds remains to be seen.

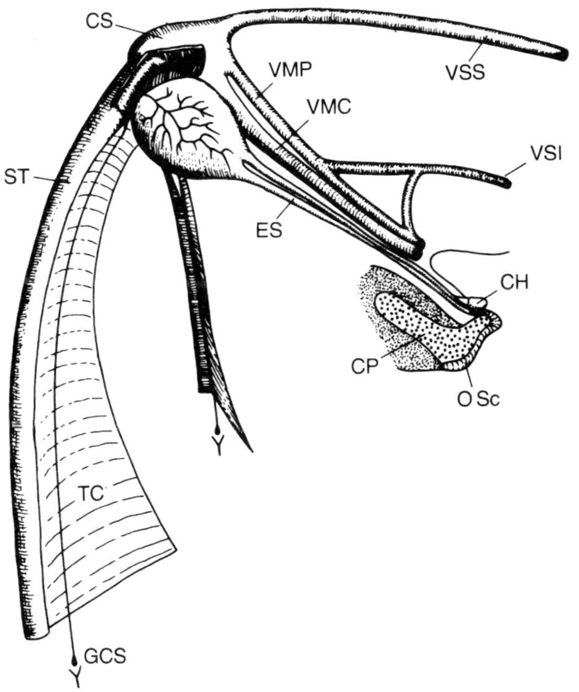

Fig. 2.3 Diagram showing the sympathetic innervation of the rat pineal gland. CH = habenular commissure; CP = posterior commissure; CS = confluens of the sinuses; ES = pineal stalk; GCS = superior cervical ganglion; OSc = subcommissural organ; ST = transverse sinus; TC = tentorium cerebelli; VMC = vena magna cerebri ; VMΠ = vena mediana prosencephali; VSI = inferior sagittal sinus; VSS = superior sagittal sinus. (Reproduced with permission from Kappers, J.A.; published by Elsevier, 1965.)

Neuropeptide Y (NPY) appears to act as a cotransmitter (possibly inhibitory) in the pineal, and NPY fibers arise in the superior cervical ganglion at least in the hamster.

Ultimately, neuronal pathways connect the pineal, via the superior cervical ganglion and the hypothalamic suprachiasmatic nucleus (SCN, the central rhythm generating system of the brain) to the retina (Figure 2.4). The paraventricular nucleus (PVN) acts as a relay station in this circuit, but the detailed pathways between the PVN and the superior cervical ganglion remain to be fully described. The PVN may also project directly to the pineal via a central innervation pathway in the rat.

A great deal of work concerning neuronal input from the retina to the SCN has succeeded in demonstrating a monosynaptic pathway – the retinohypothalamic projection – in all species studied to date including humans. This relay provides the major retinal input from environmental light–dark cycles to the SCN for the control of circadian rhythms (Chapter 6) including those of the pineal. A secondary pathway involved in the control of rhythms is from the retina via the intergeniculate leaflet (IGL) of the lateral geniculate body.

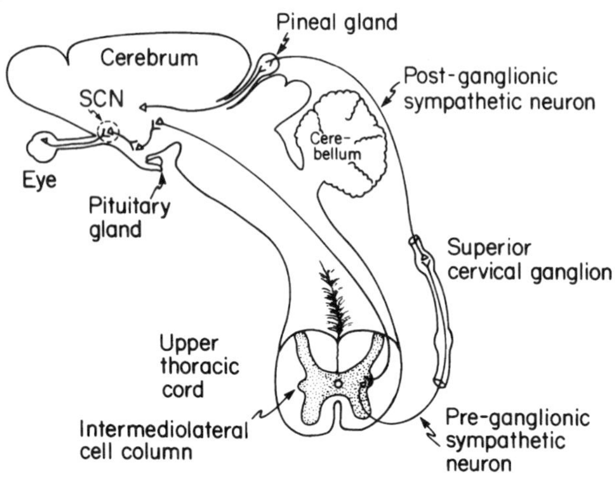

Fig. 2.4 An early diagrammatic representation of the major neural connections between the eyes and the pineal gland. Some fibers enter the gland via its connection with the posterodorsal diencephalon. (Reproduced with permission from Reiter, R.J. *The American Journal of Anatomy*, **162**, 287–313; copyright © 1981, Wiley-Liss, a division of John Wiley and Sons, Inc.)

2.2.2 Central innervation

It is of interest to note that a direct neuronal link between the IGL and the pineal of rats and gerbils, entering the pineal from the posterior commissure and pretectal area, has recently been described. A number of other direct neuronal projections from the brain to the pineal have been reported, for example from the lateral hypothalamic and pre-optic areas, the anterior hypothalamic area and the subparaventricular area (Moller, Mikkelsen and Larsen, 1991). All of these in turn receive input from the retina and the SCN. Thus there is reason to believe that multiple nervous input from peripheral sympathetic innervation and direct central innervation influence pineal function. The transmitters concerned with direct innervation may include a number of peptides such as vasoactive intestinal peptide (VIP), peptide histidine isoleucine (PHI), arginine vasopressin (AVP), arginine vasotocin (AVT), oxytocin, NPY and luteinizing hormone releasing hormone (GnRH/LHRH).

Initially, it was thought that central neural projections to the pineal formed hairpin loops in the pineal stalk, passing from the habenular to the posterior commissure and thus did not innervate the pineal. These loops certainly exist, but there is now substantial evidence for central innervation as described above.

2.2.3 Parasympathetic innervation

Parasympathetic innervation of the pineal, both centrally and peripherally derived, has been described in a few species (rabbit, macaque monkey, cow) and nicotinic cholinergic receptors are found in the rat pineal. Peripheral parasympathetic neurons have cell bodies in the superior salivatory nuclei of the brain. The tracts reach the pineal via the nervii conarii. As yet their function, as with central innervation, remains obscure.

2.2.4 The pineal 'nerve'

In the human, sheep and rabbit fetus a pineal 'nerve' is present, deriving from the subcommissural organ, with indications of both afferent and efferent fibers. This nerve has no known function at present but may be incorporated into the pineal stalk as part of adult central innervation.

2.3 VASCULAR SUPPLY OF THE PINEAL

The pineal is richly supplied with blood vessels. It has been calculated that blood flow (4 ml/min/g) is second only to that of the kidney. Arterial supply is provided by fine branches of the posterior choroidal arteries. Arising from these, arterioles may spread in the pineal capsule before entering the organ proper. In some species the gland appears to be supplied only by capillaries rather than arterioles. In most mammals the internal capillary network is extensive. The venous drainage of the pineal capillary network ultimately empties into the great cerebral vein and also in some cases into the internal cerebral veins (Figures 2.5 and 2.6).

There is doubt as to the direction of the blood flow in the region of the great cerebral vein, and this question is important when considering the vascular relationship of the pineal to the brain. According to Quay (1974) the only direct vascular link between the pineal and the brain is via the choroid plexus. If pineal secretions were to pass directly to the brain via the choroid plexus, this would require a reversed blood flow in the great cerebral vein.

With such a copious blood supply it is evident that pineal secretions can rapidly reach all regions of the body. There is evidence in some species that the pineal hormone melatonin is secreted essentially into the peripheral circulation and reaches its target areas in the brain subsequently. On the other hand it is possible that, again in selected species, there is direct secretion into the cerebrospinal fluid (CSF) via the pineal recess and the ependymal cells and tanycytes situated therein, or even by pinealocytes in contact with the CSF.

In the rat and the sheep, pronounced vasoconstriction occurs at the onset of light. This evidently suggests some degree of vascular control of pineal secretion in relation to the light–dark cycle.

Using dye injection techniques and administration of drugs, the position of the pineal within or without the blood–brain barrier has

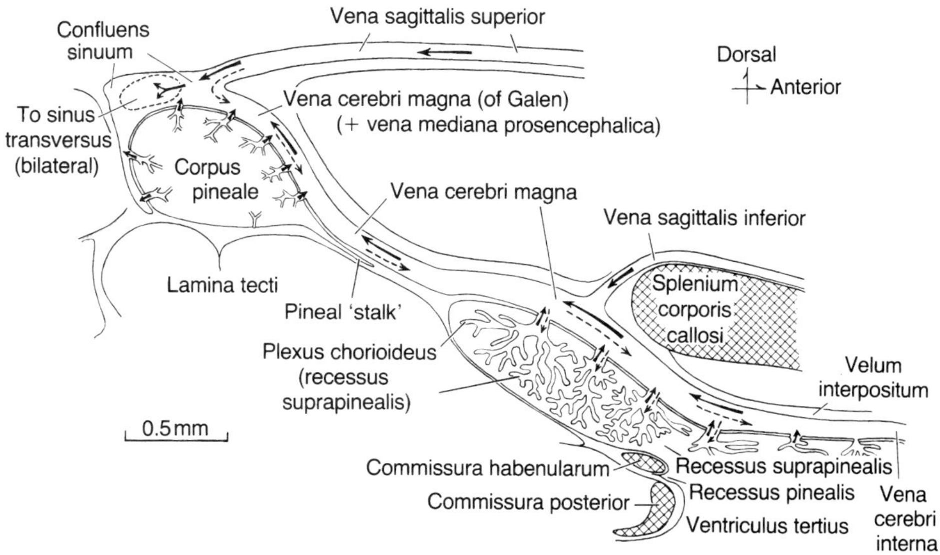

Fig. 2.5 Diagram of the blood vessels in the pineal region of the rat. Heavy arrows indicate the usually presumed direction of blood flow; broken arrows indicate sites of anatomically possible reversed blood flow on the basis of perfusion studies. (Reproduced with permission from Quay, W.B., *The American Journal of Anatomy*, **137**, 387–402; copyright © 1973, Wiley-Liss, a division of John Wiley and Sons, Inc.)

been assessed. Again this appears to be species dependent, and with some controversy in the results of dye injection. The rat and human pineal clearly lack a blood–brain barrier, as both respond to peripherally acting drugs. The ferret pineal on the other hand may be within the blood–brain barrier. The ferret is unusual in other ways, possessing a pineal ganglion.

2.4 HISTOLOGY

2.4.1 Photoreceptive remnants

As previously noted, pineal function has evolved from photoreceptive activity in lower vertebrates, through mixed photoreceptor and secretory activity in reptiles and birds to secretory function in

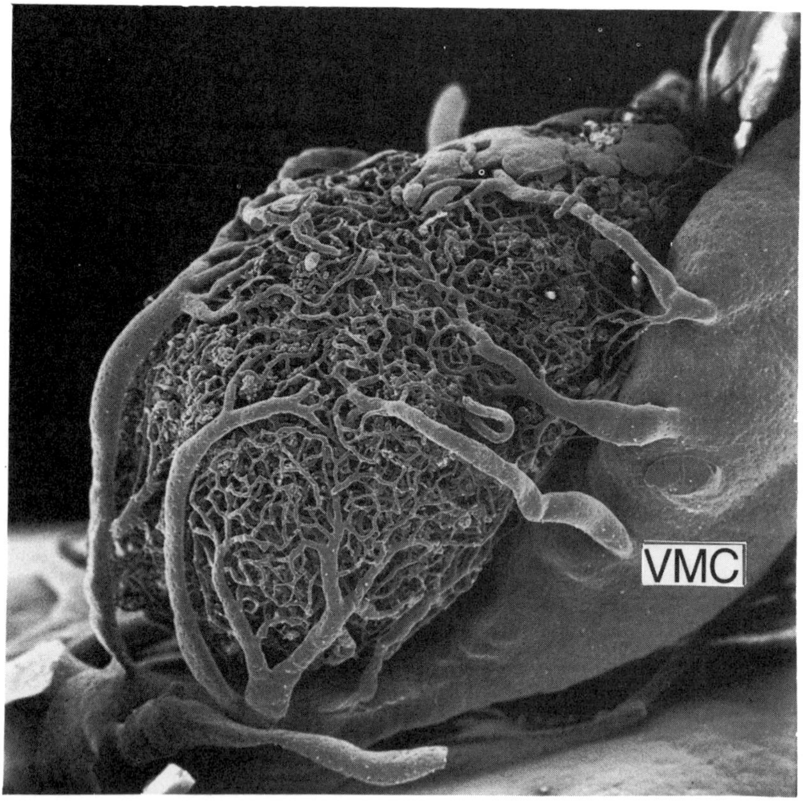

Fig. 2.6 Latex casts of the venous drainage of the rat pineal. Twelve to 16 veins usually drain the gland in this species; these vessels empty into the great cerebral vein (right). VMC = vena magna cerebri. (Reproduced from Hodde, 1979, with permission from Elsevier Science Publishers BV, Academic Publishing Division, Amsterdam.)

mammals. Until quite recently, probably the largest body of work on the pineal concerned the evolution of function through studies at light and electron microscopic level of organ and cell structure and biochemical content.

Pineal cells of the so-called sensory line are considered to be modified photoreceptor cells. The major cellular component of the mammalian pineal, the pinealocyte, is believed to have evolved from truly photoreceptive cells in lower vertebrates (Figures 2.7 and 2.8) and structural remnants of the outer segments of photoreceptor cells are reported in higher vertebrates.

This contention is reinforced by the resemblance, in many ways, of the pineal to the retina. Not only are methoxyindoles such as melatonin produced in both organs, but typically retinal proteins are found in the pineal. For example rod-opsin, the apoprotein of the visual pigment rhodopsin is present in the pineal gland of rodents. Other components of the rod phototransduction cascade such as *S*-antigen and rhodopsin kinase are also present in the rodent pineal. However 11-*cis* retinal and all-*trans* retinal are absent, suggesting that rod-opsin is not coupled to a retinaldehyde chromophore and hence phototransduction cannot take place (Korf, Kramm and de Grip, 1991).

It is beyond the scope of this book to provide a detailed account of the many photoreceptor-derived types of pineal cell and intracellular structures in various species. A simple account is given by Binkley (1988) and authoritative reviews by Collin (e.g. 1972) should be consulted. A comprehensive description of the histology of the pineal in all species is to be found in Vollrath (1981) and shorter reviews of the mammalian pineal are found in Reiter (1981) and Karasek (1983). This brief account will be confined to the mammalian pineal.

2.4.2 Pineal cell types

Pineal microscopic structure varies with time of day, time of year, physiological status and with age. In the rat, the pinealocyte (pineocyte, parenchymal cell, chief cell) accounts for 80% of the glandular components (Figure 2.8). This cell type forms processes which may terminate near blood vessels or interweave amongst other cells. Glial cells are present, some of which are clearly astrocytes. Ependymal cells from the 3rd ventricle may be found closely associated with the pineal. Efferent nerve fibers are present in quantity but with few nerve cells. Nerve fibers are primarily sympathetic with some parasympathetic and commissural fibers. Intercellular canaliculi have been found and may connect pinealocytes with blood vessels. Small numbers of other cell types may be present, even including contractile muscle cells.

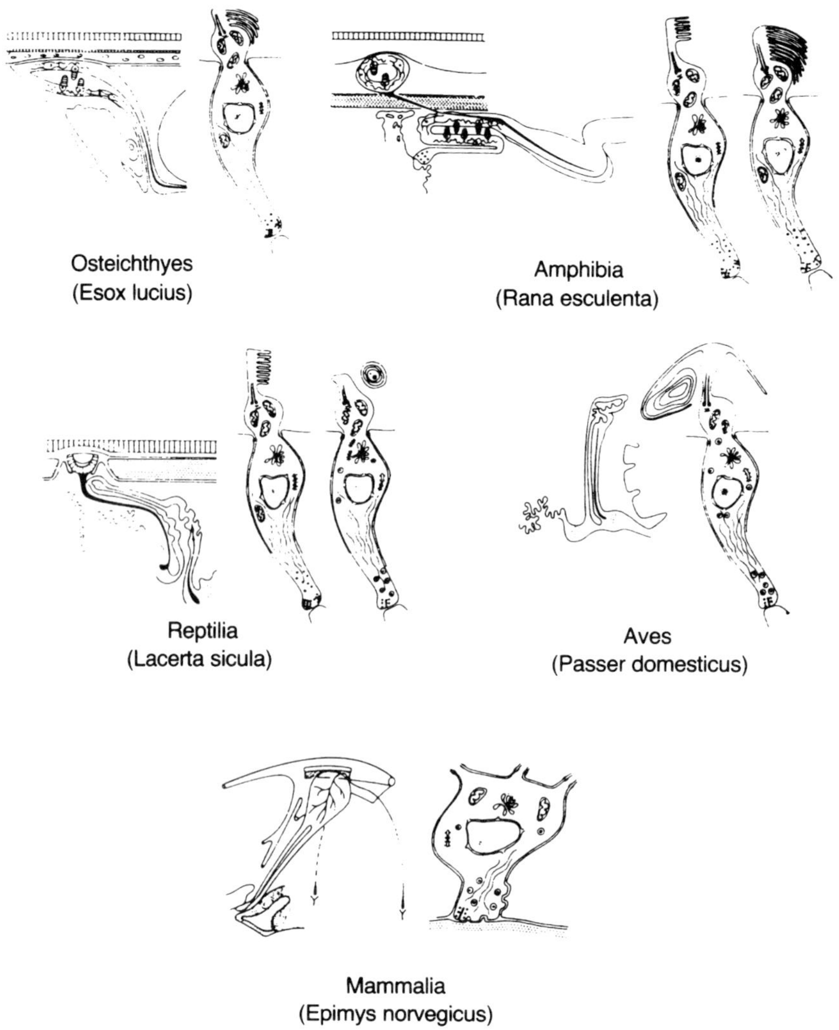

Fig. 2.7 Diagrammatic median sagittal sections of different pineal systems in relation to their characteristic types of pinealocyte. Note the patterns of innervation. The submicroscopic diagrams illustrate the ultrastructure of sensory and secretory elements. Regular outer segments (characteristic of photoreceptor cells) of pineal sensory cells are found in fishes, amphibians and lizards. Overlapping outer segmental forms appear in teleost fish and in some anurans. A considerable number of intact outer segments have been observed in the pineal organ of lizards. Pinealocytes with rudimentary or degenerated outer segments and secretory

A substructure to the gland is frequently evident, with the usual species variations. Commonly observed are lobules, follicles, rosettes, cords and clusters, with or without connective tissue septa. Acinar or pseudo-acinar arrangements are less common. Sometimes a lumen is present. Regional differences, for example an outer cortical and an inner medullary layer have been reported, but their functional significance is obscure.

The pinealocytes themselves have been the subject of a number of attempts at subclassification. In general, dark, light and clear pinealocytes are distinguished depending on their staining properties. These subtypes may be specifically distributed within the gland. Some authors have warned against subclassification however, and no functional correlates are evident.

2.4.3 Subcellular organelles (Figure 2.8)

The pinealocyte may contain more than one nucleus with one or two nucleoli. Although this is a secretory cell, secretory granules are quite infrequent. Figure 2.7 illustrates the variety of cells of the sensory line, i.e. photoreceptor-derived pinealocytes. The endoplasmic reticulum is atypical being, according to Vollrath (1981), intermediate between smooth and rough. Curious inclusions of paracrystalline structure and the formation of concentric lamellae can be related to physiological status but without establishment of cause–effect relationships. Ribosomes do not present any unusual features – most are freely present in the cytoplasm. A Golgi apparatus is regularly present and appears to participate in the formation of dense core vesicles and other vesicles.

The majority of pinealocyte mitochondria are of normal size with a few extreme variations of both shape and size. Lysosomes are present but rare. Pinealocyte centrioles may transform into microtubular 'sheaves' or 'bouquets' which may be related to antique photoreceptor function. Pinealocyte cilia are found in some mammals and again this is thought to relate to antique photoreceptor

Fig. 2.7 (contd) granules dominate in the avian pineal. Mammalian pinealocytes are exclusively secretory. (Adapted from Oksche, A., Ueck, M. and Rudeberg K.; published by Cambridge University Press, 1971.)

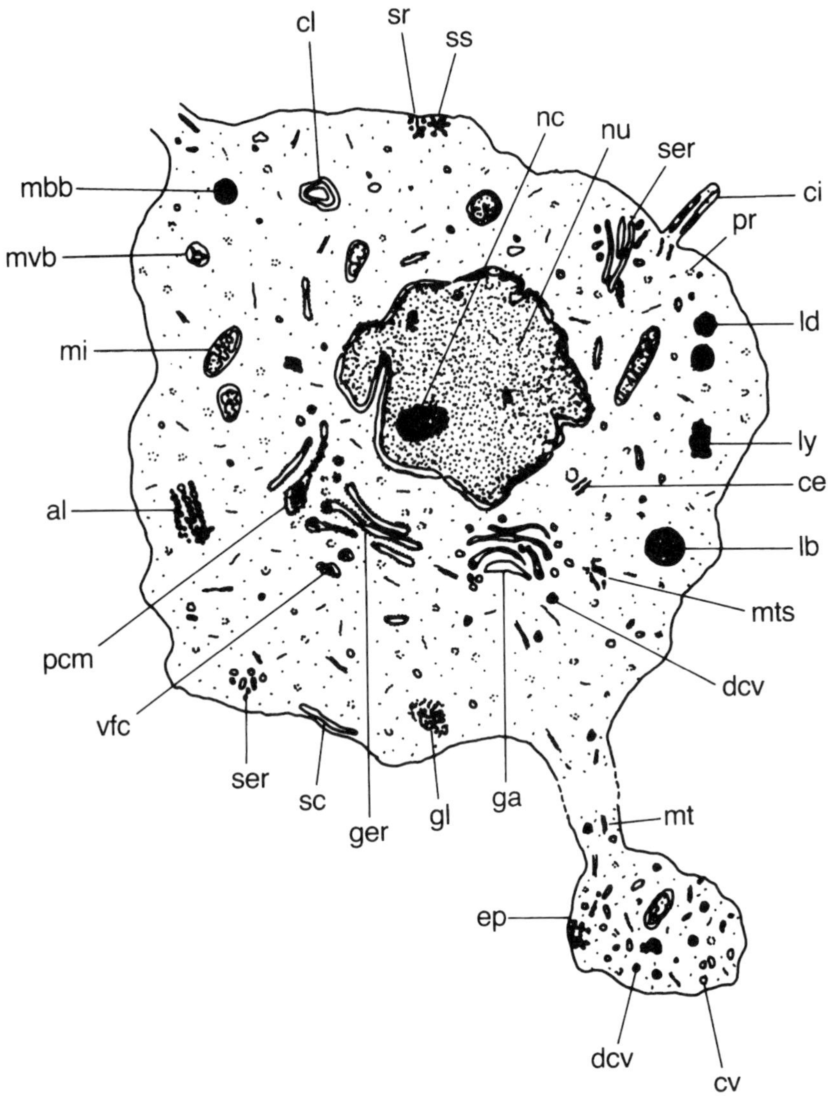

Fig. 2.8 Diagram of a generalized mammalian pinealocyte. al = annulate lamellae; ce = centriole; ci = cilium; cl = concentric lamellae; cv = clear vesicle; dcv = dense core vesicle; ep = ending of pinealocyte process; ga = Golgi apparatus; ger = granular endoplasmic reliculum; gl = glycogen particles; Ib = inclusion body; ld ;

function. Microtubules, microtubular sheaves and microfilaments are frequently observed.

2.5 CALCIFICATION

The pineals of many species contain calcified lumps or concretions. In humans it has generally been considered a feature of adult, or at least postpubertal pineals, as assessed by radiological studies. In fact, calcification is probably initiated in early life and is present in very fine granules undetectable by radiography. It appears to take place in 'ground substance' which is thought to be secreted by pinealocytes and is present in both parenchymal cells and in intercellular tissue (Tapp, 1979). It may be an index of past secretory activity: denervation of the gerbil pineal leads to decreased calcification.

Interestingly, calcified masses usually consist of concentric layers of material infiltrated by a number of minerals. It has been suggested that the rhythmic metabolic activity of the pineal is responsible for the ring formation – but whether rings are laid down daily, yearly, or with other periodicities remains unknown. There is no evidence that the presence of calcification leads to degeneration of pineal cells and metabolic activity (at least melatonin formation) within the gland of adult humans is not related to quantity of calcification. It is nevertheless possible that in the young, calcification has an important physiological role.

The pineal, the retina and the SCN together form the basic structures perceiving and transducing the non-visual effects of light. The major pineal hormone melatonin provides a closed loop to this system. The next chapter will consider the biochemistry of the pineal with particular reference to melatonin.

Fig. 2.8 (contd) = lipid droplet; ly = lysosome; mbb = membrane bound body; mi = mitchondrion mt = microtubule; mts = microtubular sheaves; mvb = multivesicular body; nc = nucleolus; nu = nucleus; pcm = paracrystalline material; pr = polyribosomes; sc = subsurface cisterna; ser = smooth endoplasmic reticulum; sr = synaptic ribbon; ss = 'synaptic spherule'; vcf = vacuole containing flocculent material. (Reproduced with permission from Karasek, M. *Pineal Research Reviews*, **1**, 1–48; copyright © 1983, Wiley-Liss, a division of John Wiley and Sons, Inc.)

REFERENCES

Binkley, S. (1988) *The Pineal: Endocrine and Nonendocrine Function,* Prentice Hall, Englewood Cliffs, New Jersey, USA.

Collin, P. (1972) Differentiation and regression of the cells of the sensory line in the epiphysis cerebri, in *The Pineal Gland* (eds G.E.W. Wolstenhome and J. Knight), Churchill Livingstone, Edinburgh, pp. 79–125.

Hodde, K.C. (1979) The vascularisation of the rat pineal organ. *Progress in Brain Research,* **52**, 39–44.

Hulsemann, M. (1967) Vergleichende histologische Untersuchungen uber das vorkommen von Glia fasern in der Epiphysis cerebri von Saugertieren. *Acta anatomy (Basel)* **66**, 249–78 (cited in Vollrath, 1981).

Kappers, J.A. (1960) Innervation of the epiphysis cerebri in the albino rat. *Anatomy Record,* **136**, 220–1.

Kappers, J.A. (1965) Survey of the innervation of the epiphysis cerebri and the accessory pineal organs of vertebrates. *Progress in Brain Research,* **10**, 87–153.

Karasek, M. (1983) Ultrastructure of the mammalian pineal gland: its comparative and functional aspects. *Pineal Research Reviews,* **1**, 1–48.

Korf, H-W., Kramm, C. and de Grip, W.J. (1991) Further analysis of photoreceptor-specific proteins in the rodent pineal organ and retina in *Advances in Pineal Research,* Volume 5 (ed. R.J. Reiter), John Libbey, London, pp. 115–22.

Moller, M., Mikkelsen, J.D. and Larsen, P.J. (1991) in *Advances in Pineal Research,* Volume 5 (eds J. Arendt and P. Pevet), John Libbey, London, pp. 13–32.

Oksche, A. (1982) Aspects of evolution of the pineal organ, in *The Pineal Gland and its Endocrine Role* (eds J. Axelrod, F. Fraschini and G.P. Velo), Plenum Press, New York.

Oksche, A., Ueck, M. and Rudeberg, C. (1971) Comparative ultra-structural studies of sensory and secretory elements in pineal organs. *Memoirs of the Society for Endocrinology,* **19**, 7–25.

Quay, W.B. (1973) Retrograde perfusions of the pineal region and the question of pineal vascular routes to brain and choroid plexuses. *American Journal of Anatomy,* **137**, 387–402.

Quay, W.B. (1974) *Pineal Chemistry in Cellular and Physiological Mechanisms,* Charles C. Thomas, Springfield, Illinois, USA.

Reiter, R.J. (1981) The mammalian pineal gland: Structure and function. *American Journal of Anatomy,* **162**, 287–313.

Tapp, E. (1979) The histology and pathology of the human pineal gland. *Progress of Brain Research,* **52**, 481–500.

Vollrath, L. (1981) *The Pineal Organ.* Springer-Verlag, Heidelberg.

FURTHER READING

Bargmann, W. (1943) Die Epiphysis cerebri, in *Handbook mikrosk. Anatomy Mensch.,* Bd. VI, 4, Hrsg. W.V. Moellendorf, Springer, Berlin, pp. 309–502.

Collin, J.P., Arendt, J. and Gern, W.A. (1988) Le 'Troisieme Oeil'. *La Recherche,* **203**, 1154–65.

David, G.F.X. and Herbert, J. (1973) Experimental evidence for a synaptic connection between habenula and pineal ganglion in the ferret. *Brain Research,* **64**, 327–43.

Korf, H.-W., Moller, M., Gery, I. *et al.* (1985) Immunocytochemical demonstration of retinal S antigen in the pineal organ of four mammalian species. *Cell Tissue Research,* **239**, 81–5.

Kummer-Trost, E. (1956) Die Bildungen des Zwischenhirndaches der Agamidae, nebst Bemerkungen uber die Lagebeziehungen des Vorderhirns. *Morphologisches Jahrbuch,* **97**, 143–92 (cited in Vollrath, 1981).

Lukasyk, A. and Reiter, R.J. (1975) Histopathological evidence for the secretion of polypeptides by the pineal gland. *American Journal of Anatomy,* **143**, 451–64.

Moller, M. and Korf, H.-W. (1984) The innervation of the mammalian pineal gland with special reference to central pinealopetal projections. *Pineal Research Reviews,* **2**, 41–86.

Moller, M., Mollgard, K. and Kimble, J.E. (1975) Presence of a pineal nerve in sheep and rabbit fetuses. *Cell and Tissue Research,* **158**, 451–9.

Pevet, P. (1981) Ultrastructure of the mammalian pinealocyte, in *Pineal Gland, Volume 1, Anatomy and Biochemistry* (ed. R.J. Reiter), CRC Press Inc. Boca Raton, Florida, pp. 122–54.

Welsh, M.G. (1985) Pineal calcification: structural and functional aspects. *Pineal Research Reviews,* **3**, 41–68.

Biochemistry of the pineal

3

Contents

3.1 EARLY WORK

During the 1950s, interest in the indoleamine serotonin (5-hydroxytryptamine, 5HT) was just beginning to bear scientific fruit, although its fundamental role as a neurotransmitter had yet to be determined. With the relatively crude measurement methods of that time it rapidly became apparent that serotonin was present at high concentrations in the blood, the gut and the brain in particular.

The peripheral and central localization of this and other biogenic amines led to the concept of the 'amine precursor uptake and decarboxylation system' (APUD), a widely distributed cell system with the ability to synthesize biogenic amines such as serotonin, noradrenalin, adrenalin and dopamine. It was considered that embryologically these cells derived from the same primordial germ cells. The pineal is perhaps a good example of such an organ, containing large amounts of biogenic amines, the metabolic machinery for their synthesis and catabolism, and many associated peptides.

Intriguingly the pineal was found to contain the highest concentration of serotonin per unit weight in the body, and it was present in both parenchymal cells and nerve endings. At this time the function of the gland was unknown; indeed it was still thought by most to be a rudimentary vestige (albeit a good radiological marker in view of its frequent calcification).

The important review by Kitay and Altschule in 1954 had concluded that if the pineal had a role in physiology it was primarily concerned with reproduction, with light–pigmentation response and possibly with behavior. The association of pineal tumors with precocious or delayed puberty provided the main fuel for the association with reproduction. The authors also conducted experimental work showing that pinealectomy increased ovarian weight and that the effect was suppressed by pineal extracts.

Further clues to pineal function came from the pioneering work of Fiske (1941; Fiske, Bryant and Putman, 1960). She succeeded in showing that environmental lighting, which stimulated the growth of rats, simultaneously depressed the weight of the pineal. Persistent estrus induced in rats by constant light could be reduced by administering pineal extracts (Wurtman, Roth and Altschule, 1961). These observations led to the initial hypothesis that the pineal releases a substance which inhibits the gonads and is under the control of environmental lighting (Wurtman *et al.*, 1961, 1963). In parallel, critically important work by Quay (1963) demonstrated that serotonin in the pineal varied with time of day, being highest during daytime, and that this effect was dependent on the light–dark cycle.

3.2 DISCOVERY OF MELATONIN

The discovery of melatonin (*N*-acetyl-5-methoxytryptamine) in 1958 by Lerner and colleagues provided the unifying theme to all these threads of information. It was the major stimulus for an enormous amount of work. Research on this small molecule has led to the acquisition of basic physiological information on fundamental rhythmic function from micro-organisms to man.

Lerner was searching for an amphibian skin lightening factor known to be present in beef pineals. Using a classic approach, he and colleagues collected many hundreds of thousands of beef pineal glands from abattoir material and processed the tissue through a series of extraction and other separation procedures.

The skin lightening activity was monitored by a bio-assay. Amphibian skin contains pigment cells known as melanophores, which in the intact animal respond by a change of pigment distribution to the ambient light conditions. In the whole animal this response involves central mechanisms. *In vitro*, however, it is possible to look directly at cell response to stimulation or inhibition in isolated pieces of skin or cultured melanophores. Alternatively, entire live tadpoles can be used, by scoring the 'melanophore index' after animals are exposed to different agents added to their aquarium water (Figure 3.1). Melanophore pigment migrates in such a way as to change the apparent cell shape from stellate to round under the influence of skin lightening factors and vice versa in response to darkening factors (e.g. melanocyte- stimulating hormone, MSH).

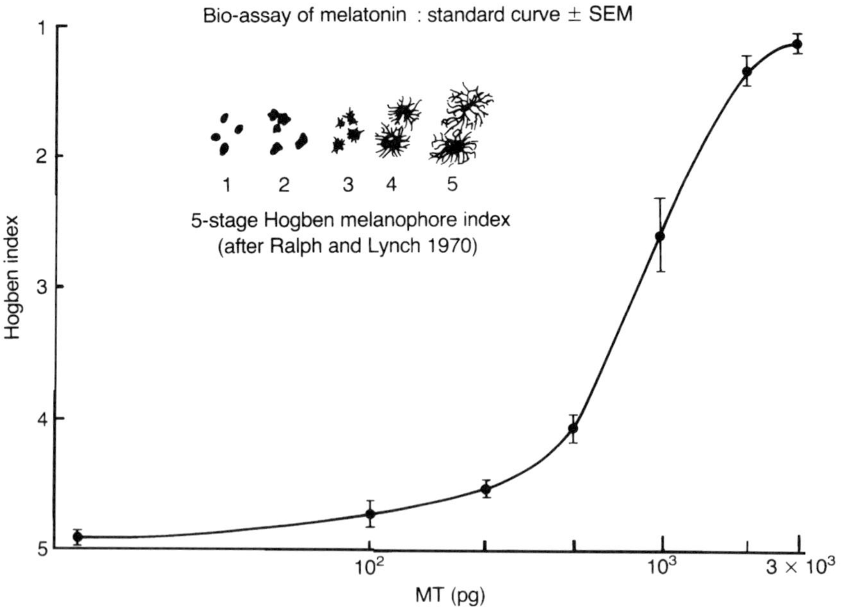

Fig. 3.1 Diagrammatic representation of dermal melanophores of *Xenopus laevis*, the African clawed toad, with increasing concentrations of melatonin (MT). The melanophore 'index' is indicated for each stage together with the calibration curve that can be thus derived.

The procedure is extremely tedious and requires great care if reproducible quantitative results are required. Two years of effort was necessary before Lerner, Case and Takahashi had a few micrograms of pure material and knew that the compound of interest was an uncharged indole. Chemical synthesis of candidate molecules led to *N*-acetyl-5-methoxytryptamine (Figure 3.2) which proved to be the most potent skin lightening substance ever tested.

Lerner is reported as saying that during the isolation process, he was conscious that each side fraction that had to be thrown away due to lack of storage space probably contained fascinating material of use to generations of pineal researchers. Some of his unfractionated material was reportedly used in the next step of the melatonin saga.

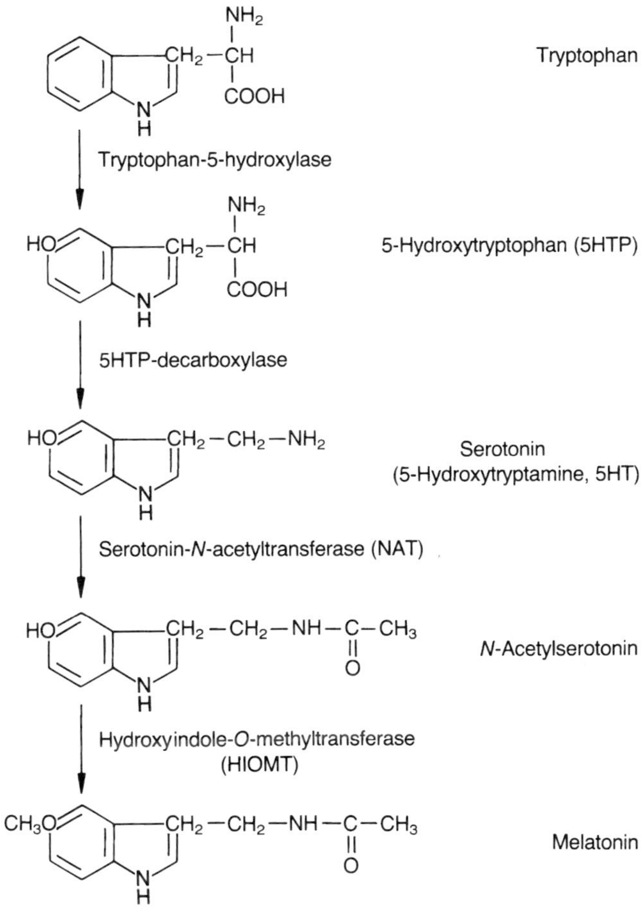

Fig. 3.2 The chemical structure of melatonin and its synthetic pathway from tryptophan.

3.3 SYNTHESIS OF MELATONIN

It is said that, on reading of the isolation and characterization of melatonin, Julius Axelrod at the National Institutes of Health, Washington, considered that the enzymic machinery required for such a synthesis within the body did not exist. He and coworkers then proceeded to demonstrate for the first time the existence of all elements of the pathway within the pineal gland, summarized in a classic review in *Science*, 1974 (Figures 3.2 and 3.3). Other reported sites of melatonin synthesis are the retina, the Harderian gland, the gut and blood platelets. These, in mammals, do not make a major contribution to melatonin levels in blood but may be of local importance.

3.3.1 Tryptophan

The first step in the formation of melatonin is the uptake of the

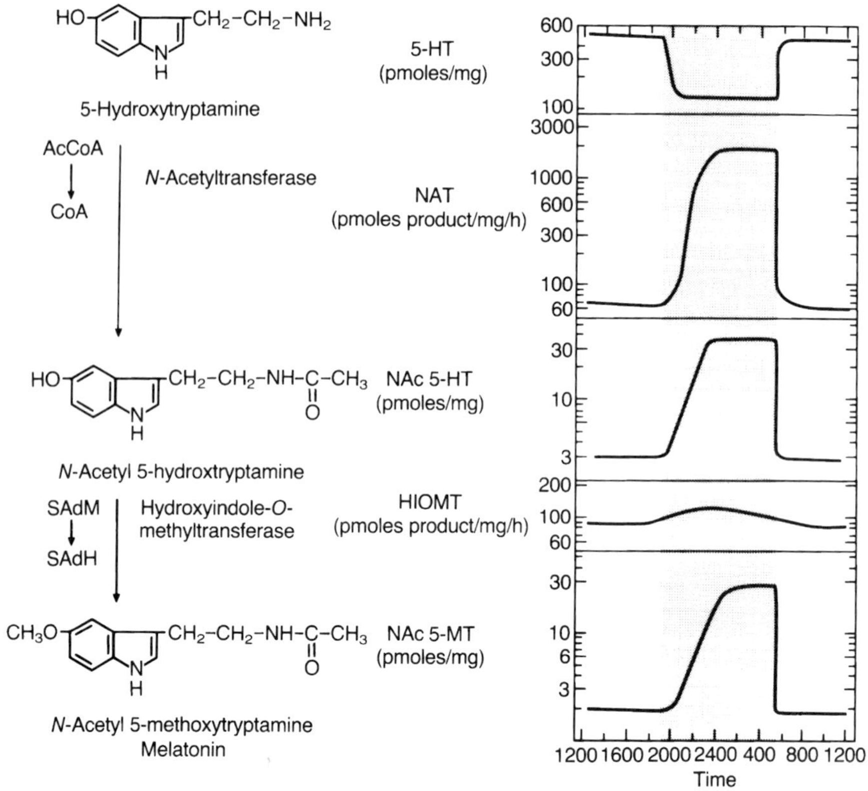

Fig. 3.3 Synthesis of melatonin from 5-hydroxytryptophan (data from the rat) with an indication of the rhythmic changes in the course of 24 hours. The stippled area represents the dark phase. (Adapted from Klein, 1974 and 1979, by permission.)

dietary amino acid tryptophan from the circulation into the gland. Uptake of tryptophan into the brain itself is dependent on transport mechanisms to cross the blood–brain barrier. These in turn are strongly influenced by the presence of other large neutral amino acids which compete for the transport system. Dietary conditions may thus affect tryptophan availability within the brain although gross manipulations such as lengthy starvation followed by specific refeeding are necessary to demonstrate this.

In most mammals studied, the pineal is without the blood–brain barrier, and one might expect that tryptophan levels with the gland reflect circulating tryptophan, or at least that fraction which is not bound to albumin. However serum-free tryptophan shows a reverse pattern to pineal tryptophan in the rat implying an active transport mechanism into the pineal against a concentration gradient. To date no dietary manipulations have been shown to modify pineal tryptophan, serotonin or melatonin secretion.

3.3.2 5-Hydroxytryptophan

Tryptophan is transformed to 5-hydroxytryptophan by tryptophan-5-hydroxylase, a mitochondrial enzyme requiring a pteridine cofactor, possibly tetrahydrobiopterin (BH4). This important reaction is rate-limiting in the formation of serotonin wherever it is produced, and the specific localization of the enzyme is restricted to serotonin-forming tissues. As in other localizations it can be inhibited in the pineal by *para*-chlorophenylalanine. Thiol reactions and reducing agents may be involved in physiological regulation. The enzyme is present in high amounts in the pineal and it is the availability of tryptophan which determines the rate of reaction, as it is in the brain. The activity is two-to-three fold higher at night in the rat.

Administration of large amounts of tryptophan will increase the formation of serotonin and there is recent evidence that melatonin formation is also increased by exogenous tryptophan. The pineal enzyme is highly homologous with phenylalanine and tyrosine hydroxylase.

3.3.3 Serotonin (5-hydroxytryptamine)

5-Hydroxytryptophan is decarboxylated to serotonin by the

cytoplasmic enzyme, aromatic amino acid decarboxylase. As its name suggests this enzyme has a broad specificity, being also an essential step in the synthesis of catecholamine neurotransmitters, and is widely distributed in tissues. It is very active with little daily variation in the pineal. It requires pyridoxal phosphate as a cofactor and may be stimulated by constant light.

A number of medications designed to increase intercerebral dopamine in Parkinson patients combine a peripheral decarboxylase inhibitor such as benserazide with the dopamine precursor dihydroxyphenylalanine. Patients receiving such medication do not show an effect on melatonin synthesis. If, however, very large doses of benserazide are given to rats, the doses can be calculated to act either peripherally or both peripherally and centrally. In this case a dose capable of acting centrally will substantially inhibit melatonin production with less but still significant inhibition using a peripheral dose. These observations served to show that the rat pineal is outside the blood–brain barrier.

3.3.4 *N*-Acetylserotonin (*N*-acetyl-5-hydroxytryptamine)

The enzyme which has received the most attention in the melatonin synthetic pathway completes the next step: *N*-acetylation of serotonin. *N*-Acetyltransferase (NAT) activity is present in many tissues. The enzyme present in the pineal cytoplasm and also the retina (arylalkylamine *N*-acetyltransferase) is distinguished from that in other tissues by its specificity and the remarkable variation in its activity as well as in other properties. Different authors give different figures but Klein's original publication (Klein and Weller, 1970) cites a 70 to 100 fold increase in activity at night in the rat pineal. This is illustrated in Figures 3.3 and 3.4 together with the relationship of pineal and plasma melatonin to NAT activity in the rat. In most circumstances it is likely therefore that this enzyme is rate limiting in melatonin synthesis. However, should amounts of *N*-acetylserotonin be produced such as to saturate the final enzyme in this pathway (hydroxyindole-*O*-methyltransferase), the activity of the latter becomes rate limiting.

It is possible to override the rate-limiting activity of NAT. Administration of substances designed to increase the availability of

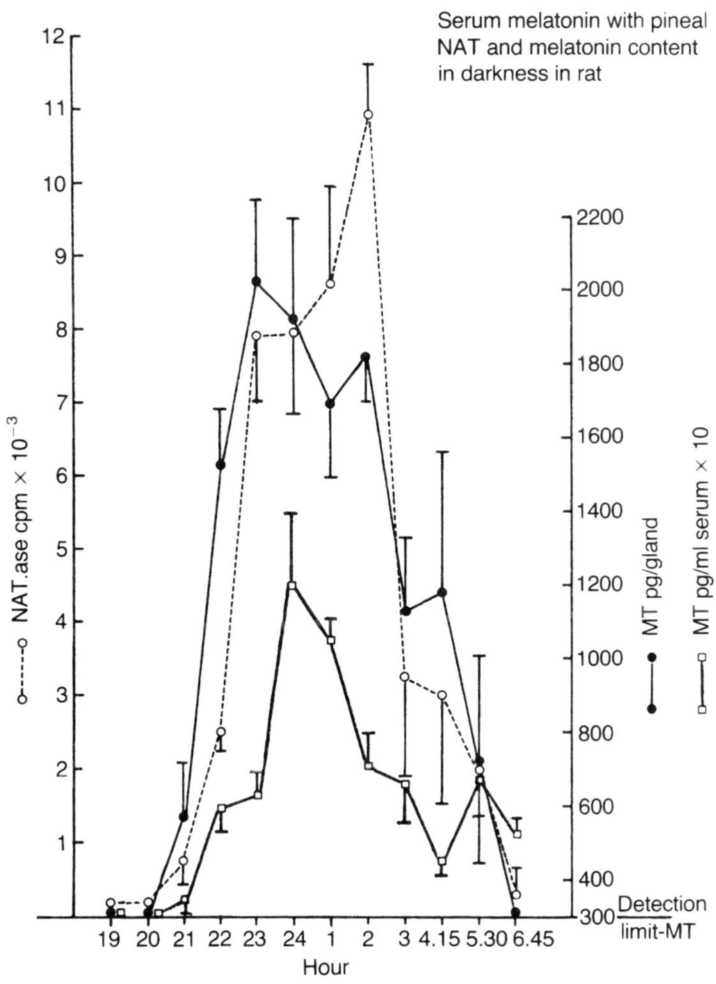

Fig. 3.4 Serum melatonin (MT) with pineal melatonin (MT) content and NAT activity (NAT.ase) during darkness (1900–0700 h) in the rat. Note the lag period from dark onset to initiation of synthesis. (From Wilkinson, M. *et al.*, 1977, reproduced by permission of the Journal of Endocrinology Ltd.)

serotonin, such as the uptake inhibitor fluvoxamine or large doses of tryptophan (see earlier) will greatly increase melatonin production, presumably by a mass action effect (Figure 3.5) although neural and even metabolic mechanisms cannot be excluded.

The enzyme uses acetyl co-enzyme A as a cofactor. It is particularly unstable and both acetyl CoA, thiol-containing compounds and peptides, will protect it against degradation. The neural control of its activity is described later.

3.3.5 Melatonin (*N*-acetyl-5-methoxytryptamine)

The final step in the pathway is *O*-methylation of *N*-acetylserotonin by hydroxyindole-*O*-methyltransferase (HIOMT) which transfers a methyl group from *S*-adenosyl methionine (SAM). Indoles other than *N*-acetylserotonin can be *O*-methylated by HIOMT as described in section 3.4. Oxaloacetate and bicarbonate are activators and pyridoxal phosphate is an inhibitor. In some species HIOMT is, of course, present in other melatonin synthesizing tissues such as the retina. The enzyme has recently been cloned.

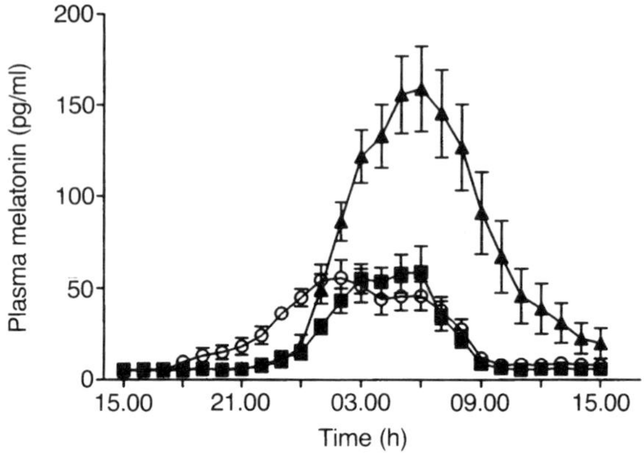

Fig. 3.5 Response of human plasma melatonin to a serotonin uptake inhibitor (fluvoxamine, 100 mg, triangles), a primarily noradrenalin uptake inhibitor (desmethylimipramine, 100 mg, circles) or placebo (boxes) at 1600 h. Fluvoxamine increases the amplitude whereas desmethylimipramine advances the onset and increases the duration of the secretory rhythm. (Reproduced with permission from Skene, D.J., Bojkowski, C. and Arendt, J.; published by Blackwell Scientific, 1994.)

Along with daily variations in pineal serotonin, this enzyme provided the first evidence of rhythmic metabolic activity within the pineal with higher values at night (attributed to an increase in enzyme protein) which could be suppressed by light. Constant dark increases activity somewhat. In fact the day–night variations of HIOMT are rather small and it is remarkable that the fundamental concept of the pineal as a photoneuroendocrine transducer could have originally been deduced from these studies and those on serotonin. Neural input to the pineal affects HIOMT as described later.

3.4 OTHER PINEAL METHOXYINDOLES (FIGURE 3.6)

It is difficult to attribute any major function to pineal indoles other than melatonin, as in most circumstances the effects of pinealectomy are reversed by melatonin. Many other indoles related to melatonin nevertheless exist in the pineal, the retina and even the Harderian gland.

In addition to the formation of serotonin, tryptophan undergoes decarboxylation to tryptamine and *O*-methylation by HIOMT to 5-methoxytryptamine. This latter indole may have a physiological role. It has a photoperiod-controlled day–night variation with higher levels during the day in the pineal, although this can only be demonstrated easily in the presence of monoamine oxidase inhibitor. Some of its effects are similar to those of melatonin. It is difficult to be sure that the observed effects are not due to *N*-acetylation of 5-methoxytryptamine to melatonin; however melatonin may be deacetylated to 5-methoxytryptamine (albeit a minor pathway) using aryl acylamidase and the reverse argument applies, that some effects of melatonin are due to the formation of 5-methoxytryptamine.

A derivative of 5-methoxytryptamine, 6-methoxytetrahydro-β-carboline, has some melatonin-like effects and may be of importance in aspects of brain function. It inhibits serotonin uptake, inhibits monoamine oxidase activity, binds to type 1 serotonin receptors and increases serotonin levels. 5-Methoxytryptamine itself is a potent agonist of presynaptic serotonin autoreceptors, and is a serotonin 1A receptor agonist. It thus may modulate serotoninergic neurotransmission.

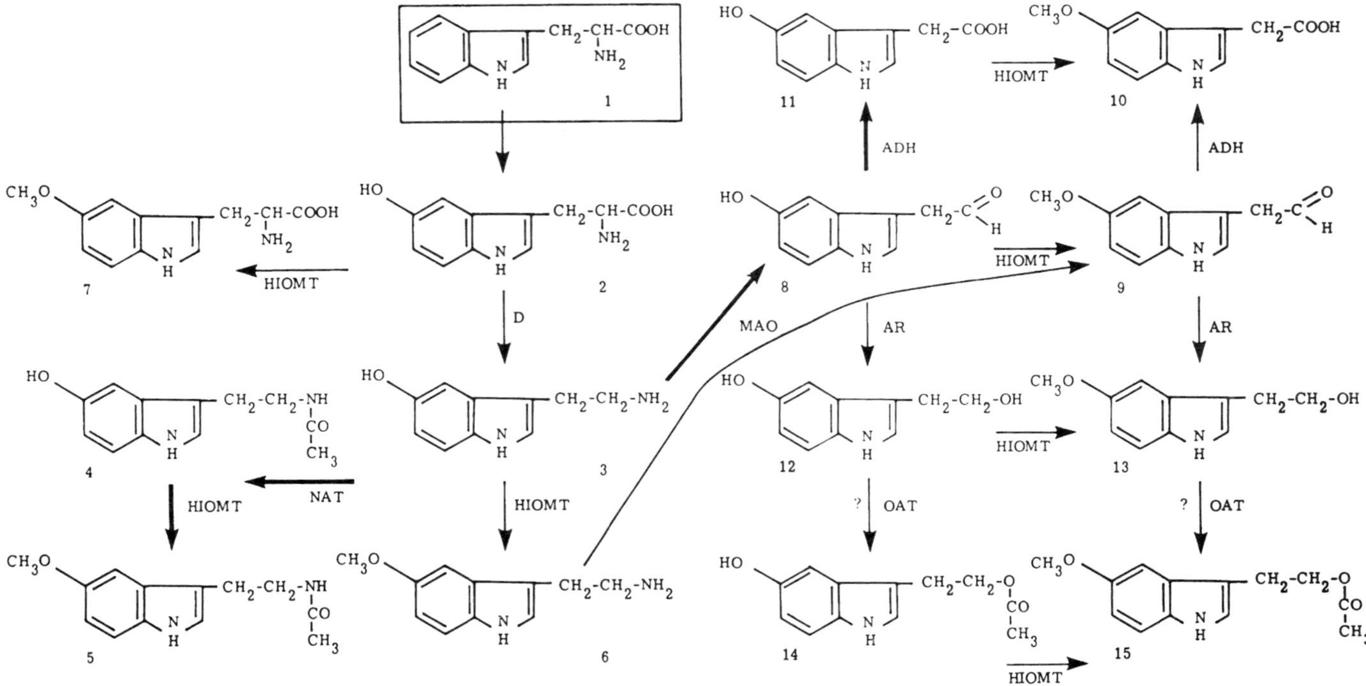

Fig. 3.6 Synthesis of known* 5-methoxyindoles in the pineal gland with common abbreviations followed by those proposed by Smith (1983) given in brackets. Of the Smith abbreviations only ML is used regularly. 1 = Tryptophan (Try, W); 2 = 5-hydroxytryptophan (5HTP, HW); 3 = 5-hydroxytryptamine (5HT, HT); 4 = *N*-acetyl-5-hydroxytryptamine/*N*-acetylserotonin (NAS, aHT); 5 = melatonin (MT, aMT); 6 = 5-methoxytryptamine (5MT, MT); 7 = 5-methoxytryptophan (5MTP, MW); 8 = 5-hydroxyindoleacetaldehyde (5HIAAld, HB); 9 = 5-methoxyindoleacetaldehyde (5MIAAld, MB); 10 = 5-methoxyindoleacetic acid (5MIAA, MA); 11 = 5-hydroxyindoleacetic acid (5HIAA, HA); 12 = 5-hydroxytryptophol (5HTol, HL); 13 = 5-methoxytryptophol (5MTol, ML); 14 = *O*-acetyl-5-hydroxytryptophol (aHL); 15 = *O*-acetyl-5-methoxytryptophol (aML). *The existence of 14 and 15 has not been confirmed. (Reproduced with permission from Arendt, J.; published by Alan R. Liss, Inc, 1985.)

3.4.1 5-Methoxytryptophol

Metabolism of 5-methoxytryptamine by monoamine oxidase gives 5-methoxyindoleacetic acid and 5-methoxytryptophol, although both of these compounds are probably produced in larger amounts from oxidative de-amination of serotonin followed by *O*-methylation (Figure 3.6). Physiological effects are not claimed for the indolic acids; however 5-methoxytryptophol, like 5-methoxytryptamine, has some melatonin-like effects, and a photoperiod-dependent day–night variation with high values during the day. If it has a function this may well reside in the retina, particularly of lower vertebrates. In the frog retina more 5-methoxytryptophol is produced than in the pineal and melatonin itself is rapidly metabolized to 5-methoxytryptophol. Retinal binding sites for 5-methoxytryptophol have been reported and it is more potent in the activation of rod disk shedding in retinal preparations than melatonin.

Both 5-methoxytryptamine and 6-methoxytetrahydro-β-carboline are present and may be synthesized in the retina. They may have important retinal effects (reviewed by Skene, 1990).

3.4.2 *O*-Acetyl-5-methoxytryptophol

The existence of *O*-acetyl-5-methoxytryptophol has been reported by one author, but this has never been confirmed.

3.4.3 Kynurenine pathway of tryptophan metabolism

It should be noted that by far the major pathway for tryptophan metabolism in other tissues is via ring-splitting, the formation of kynurenine derivatives and ultimately nicotinic acid. So far this pathway has not been studied in the pineal.

3.5 METABOLISM AND PHARMACOKINETICS OF MELATONIN

3.5.1 Metabolism

The primary site for melatonin metabolism is within the liver and secondarily the kidney. It undergoes 6-hydroxylation, followed by sulfate or glucuronide conjugation, in a classic microsomal phase 1 phase 2 reaction sequence common to the metabolism of steroids and the deactivation and detoxification of many drugs, etc. (Figure 3.7). The relative amounts of sulfate and glucuronide formed probably depend on species. In rodents and humans the sulfate predominates, indeed in humans at least 90% of a dose of melatonin may be accounted for by 6-sulfatoxymelatonin in plasma and/or urine.

The plasma and urinary profiles of 6-sulfatoxymelatonin faithfully reflect both the qualitative and quantitative aspects of melatonin secretion in humans (Figure 3.8, Chapter 7), and provide useful clinical data. In sheep the route of metabolism appears to depend on the route of administration and after acute oral administration of melatonin, 6-sulfatoxymelatonin is a minor metabolite in contrast to its production after chronic oral dosing or acute intravenous bolus.

Melatonin also forms minor metabolites. In the brain ring splitting to form *N*-gamma-acetyl-*N*-2-formyl-5-methoxykynurenamine and subsequently *N*-gamma-acetyl-5-methoxykynurenamine occurs. These metabolites may be physiologically active. It has been reported that they may have melatonin-like activity with respect to entrainment of circadian rhythms and are prostaglandin synthesis inhibitors. Cyclization of the side chain on the 3 position is another minor metabolic route. Exogenous melatonin is demethylated to *N*-acetylserotonin and the latter is hydroxylated and conjugated similarly to melatonin itself. Whether this pathway is of endogenous importance is not known with certainty (Figure 3.7).

N-Acetylserotonin may have functions of its own in addition to being the precursor of melatonin; for example, it has anesthetic activity in large doses in rats; thus the demethylation pathway may have some significance *in vivo*.

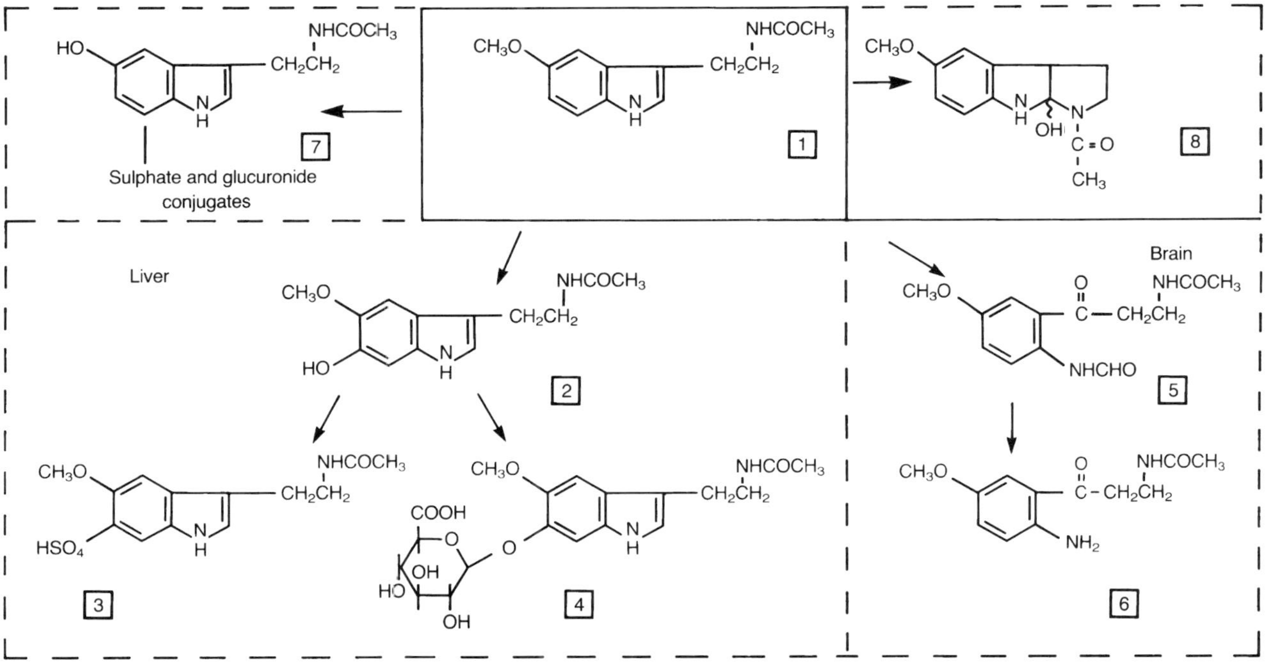

1 = Melatonin

2 = 6–Hydroxymelatonin

3 = 6-Hydroxymelatonin sulphate

4 = 6–Hydroxymelatonin glucuronide

5 = N^{γ}-Acetyl-*N* formyl-5-methoxykynurenamine

6 = N^{γ}-Acetyl-5-methoxykynurenamine

7 = *N*-Acetylserotonin

8 = Cyclic 2-hydroxymelatonin

9 = Deacetylation 5-methoxytryptamine 5-methoxyindoleacetic acid (not shown)

Fig. 3.7 The metabolism of melatonin. The primary metabolic route in humans and rodents is via 6-hydroxylation and sulfate conjugation. (See text for further details.)

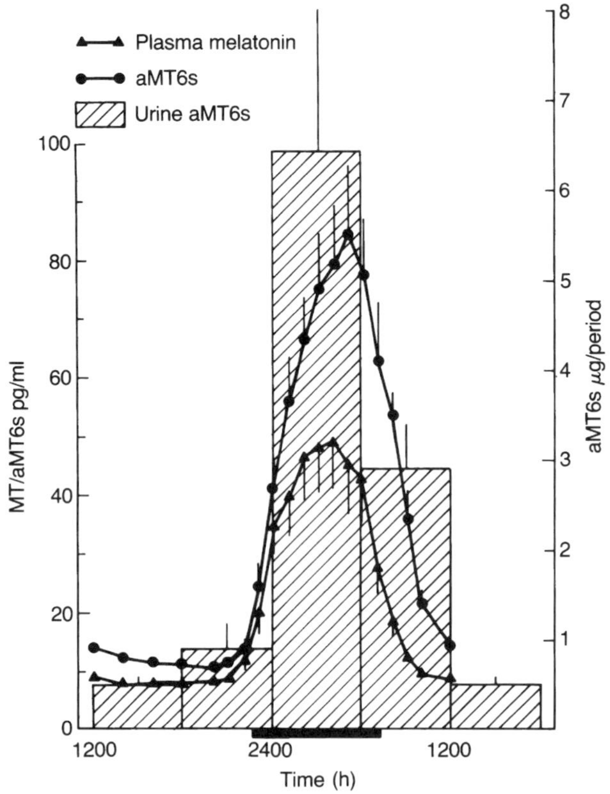

Fig. 3.8 Melatonin and 6-sulfatoxymelatonin (aMT6s) in hourly plasma samples and aMT6s in 6-hourly urine samples in normal human subjects. Note that the timing of the evening rise in both compounds is very comparable in all three fluids but that there is a delay in aMT6s clearance from plasma and urine in the morning. (Reproduced with permission from Arendt, J.; published by Blackwell Scientific Publications Ltd, 1988.)

3.5.2 Pharmacokinetics

Exogenous oral or intravenous melatonin has a short metabolic half-life with a large hepatic first pass effect and a biphasic elimination pattern in rodents and humans. The first reports by Kveder and McIsaac (1961) in mice using tritiated melatonin given intravenously indicated a two-step elimination process with half-lives of about 2 minutes and about 45 minutes. In general this has proved to be a common pattern in mammals. Humans given melatonin orally in a

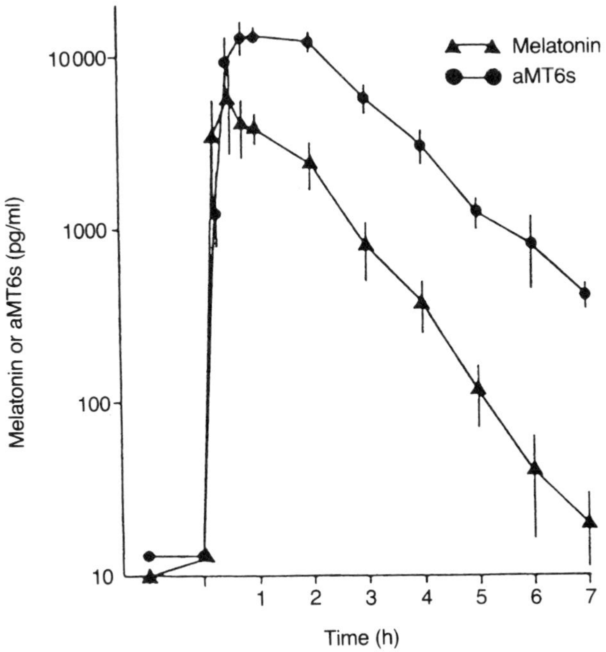

Fig. 3.9 Pharmacokinetics of melatonin and 6-sulfatoxymelatonin (aMT6s) after oral administration to humans (N=6), 2 mg in a gelatine capsule at 1000 h. Very large interindividual differences were seen in maximum plasma melatonin (37-fold) whereas only 2-fold differences were present in aMT6s. The biological half-life (mean ± SD) was 44.0 ± 6.6 min for melatonin and 58.2 ± 8.2 min for aMT6s. (From Bojkowski (1988), previously unpublished diagram. See Aldhous *et al.* (1985) for earlier work and a slow-release preparation.)

gelatine capsule show peak plasma levels after 60 minutes and a biphasic elimination pattern with half-lives of about 3 and 45 minutes (Figure 3.9). Very large individual variations are seen in peak plasma concentrations (at least 25-fold) and attributed to differences in absorption. Following attainment of the peak, plasma concentrations remain above normal for between 3 and 7 hours after administration of oral doses in the range of 2–5 mg. Intravenous infusion to humans

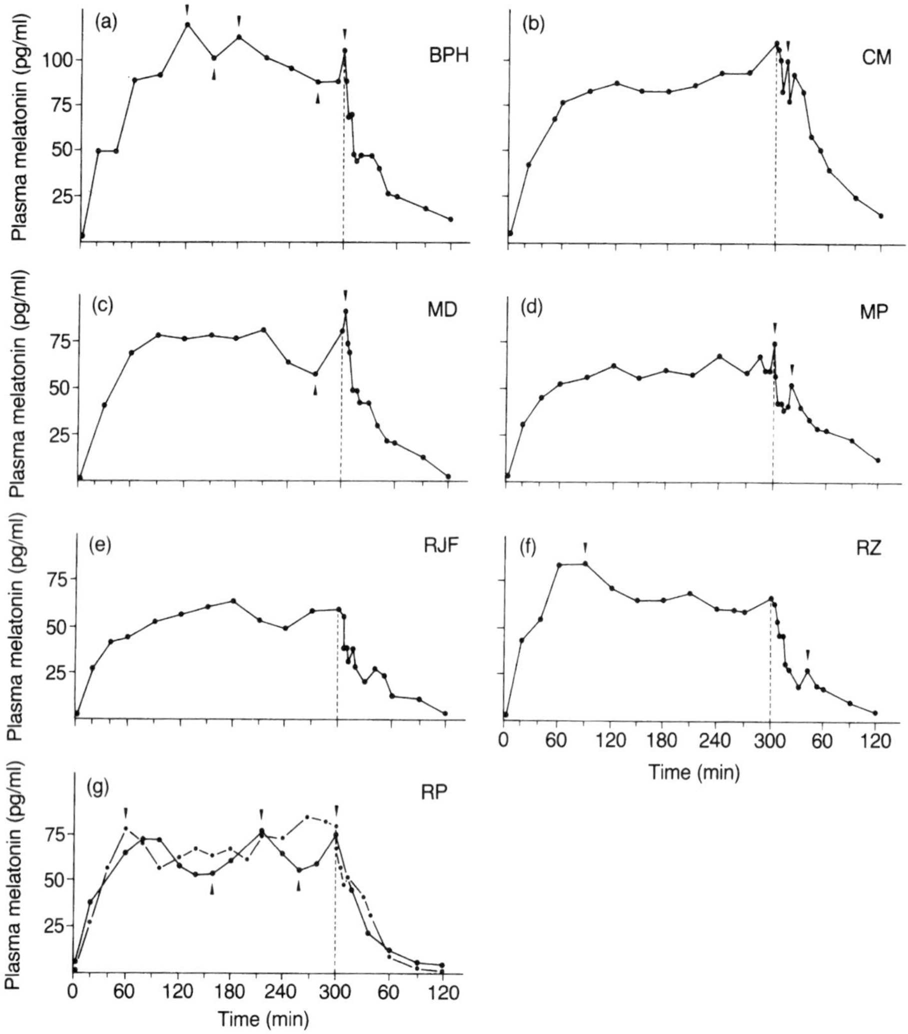

Fig. 3.10 Pharmacokinetics of melatonin during and after constant rate infusion during the day in control subjects (a–f) and in one pinealectomized patient (g) during the night time (solid line) or the day time (broken line). (Reproduced with permission from Mallo, C. *et al.*; published by Springer-Verlag, 1990.)

avoids the problem of individual variability in absorption (Figure 3.10) and intravenous bolus injection has provided useful confirmation that sulfation is the major metabolic pathway in humans (Figure 3.11).

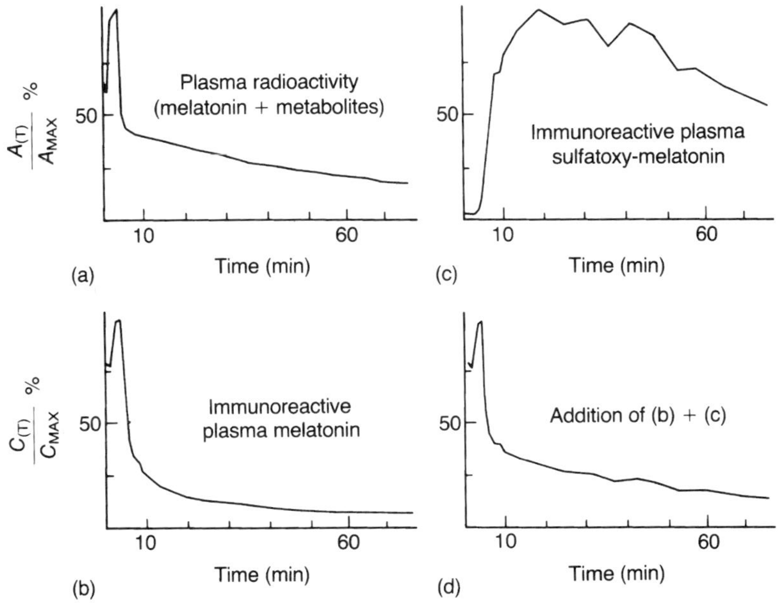

Fig. 3.11 Plasma, (a) radioactivity, (b) melatonin concentration, (c) 6-sulfatoxymelatonin (aMT6s) concentration (concentrations normalized to their maximum) after a bolus injection of [^{11}C]-melatonin, 42 μg, 700 mCi/μmol to a single male human subject. Shown in (d) is a reconstruction of the molar concentration of melatonin plus aMT6s which is closely superimposable upon the curve (a) of total radioactivity, illustrating the major sulfation route of metabolism. (Reprinted from *Nuclear Medicine and Biology*, **18**, Le Bars, D. *et al.*, PET and plasma pharmacokinetic studies offer bolus intravenous administration of [11-C]-melatonin in humans, pp. 357–62, Copyright 1991, with kind permission from Pergamon Press Ltd, Headington Hill Hall, Oxford OX3 0BN, UK.)

Lane and Moss (1985) have calculated mean endogenous melatonin production rates of 28.8 μg per day in normal adults. The bioavailability of oral melatonin is very low calculated across studies (0.03–0.76 depending on dose). Slow-release preparations are under development but are not generally available at the time of writing.

Given as a single oral dose (3–5 mg) to ruminants, plasma melatonin levels remain elevated for up to 12 hours and this phenomenon has been exploited in agriculture (Chapter 5). Melatonin has also been infused into rodents and sheep using osmotic minipumps and standard programmable infusion devices. It is possible to maintain specific steady-state plasma levels for a determined number of hours per day over periods of weeks to months. This technique has provided most elegant demonstrations of the physiological role of melatonin in many laboratories. Slow-release implants and oral bolus preparations have also been developed for commercial use in agriculture.

Early human slow-release preparations with melatonin incorporated into a beeswax–arachis oil pill were subject to major nutritional modification of plasma levels with large and inconsistent secondary peaks, presumably from gut reabsorption. Recently an efficient slow-release, transmucosal preparation has been described which is capable of maintaining steady plasma concentrations during application. This is likely to find extensive use in therapeutic developments.

3.6 PINEAL PEPTIDES

The pineal contains a multitude of biologically active peptides, some of which are identified and many that are not. When considering the biological activity of pineal extracts, workers in the field have mostly searched for activity related to reproductive function, be that anti- or progonadotropic. Evidence for the presence in the pineal of non-indolic compounds with such reproductive effects is substantial and the reader is referred to reviews by, for example, Kitay and Altschule (1954) and Benson and Ebels (1978). Pevet (1991) gives other references and also cites the commercial preparations of 'pineal factors' sold for research purposes – Epiglandol, Epiphysan, Epithalamin, Pinealin, etc. in the 1960s, none of which are taken seriously by scientists today.

Many problems dog this aspect of pineal research, the most important being that no pineal 'factor' of peptide or proteinaceous nature has so far been unequivocally identified, proved to be pineal specific and shown to be secreted and affect physiological systems in physiological amounts. Moreover it is difficult to find an important pineal function that cannot be mimicked by suitable melatonin administration. Nevertheless it is important to keep an open mind and pineal specific peptides and/or proteins may well be significant in lower vertebrates if not in mammals.

Of the identified pineal peptides, much attention was devoted to arginine vasotocin (AVT), the antidiuretic hormone of lower vertebrates. Claims were made for its presence in the mammalian pineal, for a function as a melatonin-releasing hormone and for effects on sleep *inter alia*. Whether or not AVT really exists in the mammalian pineal or indeed locally in the subcommissural organ, is still disputed. One school of thought considers that detected AVT is in fact arginine vasopressin (VP) and that artefactual identification was due to lack of methodological specificity. There is no doubt that VP occurs within the pineal. It appears to be synthesized within the gland and has most interesting season-related variations. Future work in this area is of much promise.

Pevet (1991) has proposed a classification of pineal peptides according to whether they are present in nerve fibers (class 1), taken up from the circulation (class 2) or truly of pineal origin, i.e. synthesized within pineal cells (class 3), with some peptides belonging to more than one class. The current state of knowledge does not permit a listing of pineal peptides according to this system but it provides a useful working framework for consideration of their physiology.

The pineal receives extensive peptidergic innervation, with or without co-localization of peptides and biogenic amines and with variation between species. It is within this system that most of the identified pineal peptides can be incorporated. Vasopressin and oxytocin fibers enter the pineal via the habenular and/or posterior commissures and are associated in the gland with their two neurophysins. Other peptidergic innervation includes vasoactive intestinal peptide (VIP), neuropeptide Y (NPY), peptide histidine isoleucine (PHI), calcitonin gene-related peptide (CGRP), substance P and somatostatin.

Most of these will modify (usually potentiate) stimulated melatonin synthesis or secretion *in vitro* (Figure 3.12) and the implication is that peptidergic innervation modulates the basic pattern of melatonin secretion *in vivo*. VIP exceptionally has stimulatory effects on basal melatonin secretion comparable to those of noradrenalin.

The complicated nature of peptide–amine interactions suggests that it may be many years before the precise role and functional importance of each input to overall pineal activity in the entire animal is known.

A number of hypothalamic and pituitary peptide hormones are

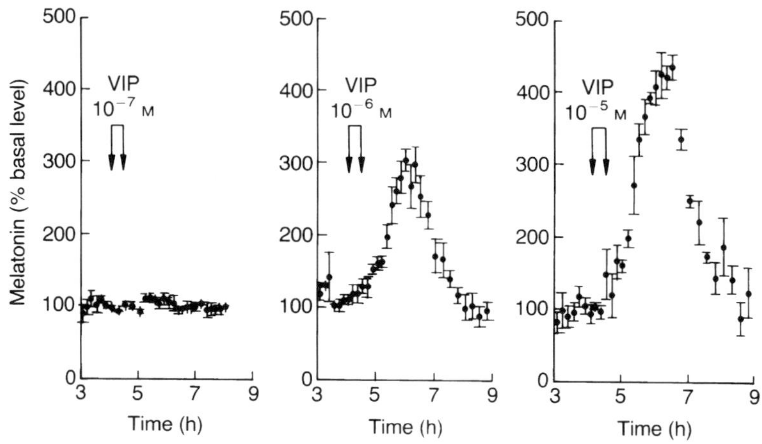

Fig. 3.12 Stimulation of melatonin release from perfusates of rat pineal glands stimulated with increasing concentrations of VIP from 10^{-7} to 10^{-5} M. (Reproduced with permission from Simonneaux, V. *et al.*; published by Springer-Verlag, 1990.)

found within the pineal including gonadotropin-releasing hormone (LHRH), thyrotropin-releasing hormone (TRH), prolactin-release-inhibiting factor (PIF), prolactin-releasing hormone (PRH), melanocyte-stimulating hormone (MSH), adrenocorticotropic hormone (ACTH), enkephalin and prolactin (PRL). All of these may be present in nerve fibers and/or accumulated by the gland but definitive information is not available. There is no evidence that the pineal is an important source of any of these factors.

Delta sleep inducing peptide (DSIP) is yet another widely distributed and interesting peptide which accumulates in the pineal and increases both basal and stimulated melatonin production. Given the close association of melatonin with sleep induction and maintenance, this relationship may prove to be of more than academic interest.

The classical neurotransmitters noradrenalin, dopamine and gamma-aminobutyric acid are all present within the pineal, probably exclusively within nerve fibers, whereas serotonin is present in both pinealocytes and nerve endings.

3.6.1 Thyroxine-5-deiodinase

Recent work suggests that the pineal contains significant amounts of this important enzyme. Day–night variations are seen in its activity with the peak of activity occurring at night. The physiological significance of its presence in the pineal is still unknown.

3.6.2 General pineal biochemistry

Most other aspects of pineal biochemistry bear no special relationship to its acknowledged physiological functions. Extensive accounts are to be found in Quay (1974) and Binkley (1988). Reference to specific compounds will be made in subsequent chapters as appropriate.

3.7 CONTROL OF MELATONIN SYNTHESIS: SYMPATHETIC NORADRENERGIC MECHANISMS

Neural control pathways leading to the pineal are considered in

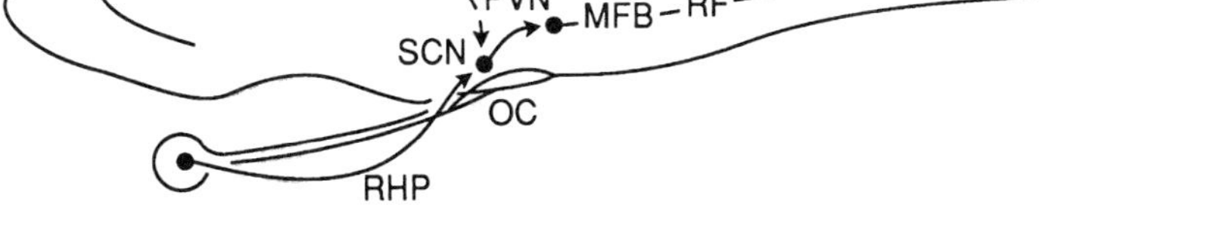

Fig. 3.13 Innervation pathways of the rat pineal gland. SCN = suprachiasmatic nuclei; P = pineal gland; RHP = retinohypothalamic projection; OC = optic chiasm; PVN = paraventricular nucleus of the hypothalamus; MFB = medial forebrain bundle; RF = reticular formation; IML = intermediolateral cell column of the spinal cord; SCG = superior cervical ganglia; ICN = internal carotid nerve; NC = nervii conarii. The dotted line represents the melatonin feedback loop. (Reproduced with permission from Klein, 1993.)

Chapter 2. Figure 3.13 provides an overview of current knowledge. This account concerns events within the pineal.

3.7.1 Noradrenalin and serotonin

By far the major control of melatonin synthesis is exerted by post-ganglionic sympathetic fibers innervating the pineal. Section of the nerves, or ganglionectomy, abolishes the rhythmic synthesis of melatonin and the light–dark control of its production. Both serotonin and noradrenalin are present in the nerve endings. Noradrenalin is clearly the major transmitter but the role of neural serotonin is probably not negligible.

Serotonin, higher in the pineal during the daytime, appears to be released from the pineal and stored in presynaptic nerve terminals during the day when sympathetic activity is low. Early work showed a day–night variation in pineal noradrenalin with highest values at night, approximately 180° out of phase with the pineal serotonin rhythm. The serotonin variations are dependent on noradrenergic input through variations in NAT as well as possible modulation of storage and release.

3.7.2 Stimulation of *N*-acetyltransferase activity

A great deal of work using *in vitro* pineal preparations developed from these initial observations, with the bulk of the information being derived from studies on NAT. There is evidence that chronic β-stimulation will elevate HIOMT but little is known about the related mechanisms. NAT is a singularly useful index of noradrenergic activity in view of the magnitude of stimulation that can be induced, and the pineal as a model system for the study of adrenergic receptor mechanisms has been enormously productive. Most work has used the rat pineal and the following account refers to the rat unless otherwise stated.

Noradrenalin, normally released at night, stimulates both α and β subtypes of catecholamine receptors. Initially the pineal was shown to possess β receptors of the β1 subtype and in general these are the essential component of noradrenergic stimulation of NAT. However α1 receptors are also present on pinealocytes and play an important role in the regulation of pineal function, potentiating the β-stimulation.

Noradrenalin binds to membrane-bound pinealocyte β-receptors, activating adenylate cyclase through GTP-binding protein in the cell membrane, and increases pineal cAMP levels 60-fold (Figure 3.14). The effect is short lived *in vitro* and continued β-stimulation is required to maintain NAT activity. Dibutyryl cyclic AMP, a stable cAMP analog, which is resistant to hydrolysis by phosphodiesterase, induces NAT activity. cAMP therefore acts as a second messenger in the pineal. cAMP subsequently stimulates NAT activity through a mechanism which involves both transcription and translation, although whether synthesis of new NAT molecules occurs is not yet clear (Figure 3.15). Klein (1985) has suggested that it might involve

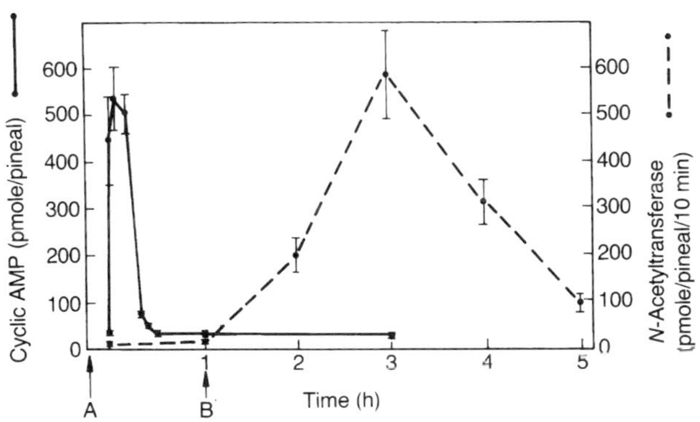

Fig. 3.14 Relationship between cAMP, *N*-acetyltransferase and β-adrenergic receptors. L-Isoproterenol (isoprenalin) hydrochloride (5 mg per kg of body weight) was given intravenously to rats (A). Cyclic AMP and *N*-acetyltransferase activities were measured at the times indicated. Arrows A and B are the times at which 1-propanolol (20 mg per kg) was injected intravenously. After injection of 1-propanolol before isoproterenol at A the rise in both cAMP and *N*-acetyltransferase is prevented, after injection at B, N-acetyltransferase is still suppressed. Vertical bars indicate SEM. (Reproduced with permission from Deguchi, T., Role of the beta-adrenergic receptor in the elevation of adenosine cyclic 3′, 5′-monophosphate and induction of serotonin *N*-acetyltransferase in rat pineal glands. *Molecular Pharmacology,* **9**, 184–90; published by Academic Press Ltd, London, 1973.)

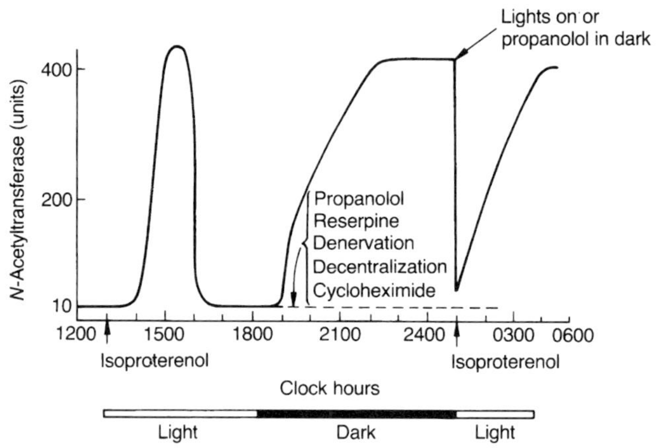

Fig. 3.15 Induction and suppression of serotonin *N*-acetyltransferase in the pineal. Isoproterenol (isoprenalin), a β-adrenergic agonist can stimulate activity during the day. Activity increases at night and can be suppressed by light, β-antagonists (propanolol), reserpine (a noradrenalin storage depleter), denervation of the gland, and protein synthesis inhibitors (cycloheximide). (Reproduced with permission from Axelrod, J. The pineal gland; a neurochemical transducer, *Science*, **184**, 1341–8; copyright 1974 by the AAAS.)

synthesis of an activating enzyme which recycles existing NAT molecules.

cAMP also serves to stabilize NAT and prevent inactivation. NAT activity and thus melatonin synthesis can be very rapidly inactivated, for example by applying a light flash during synthesis at night. The mechanism for rapid inactivation is not known but may depend on a specific inactivating factor which in turn depends on cAMP levels (see Klein, 1993, for references). Binkley (1988) lists compounds that stimulate and inhibit NAT.

3.7.3 α-Potentiation of β-stimulation

Using specific β- and α-agonists Klein (1985, 1993), Sugden (1989) and coworkers have shown that, whilst having no intrinsic stimulatory activity, α-adrenergic agonists will increase the cAMP and NAT stimulation induced by specific β-agonists 10–20 fold (Figure 3.16).

These α-receptors appear to be on the pinealocytes rather than on presynaptic terminals. The increase in cAMP is essential for the induction of NAT activity. Whilst full induction can occur with less than maximal adrenergic activation, there is good evidence that the potentiation system is important *in vivo*, where α-adrenergic stimulation enhances β-stimulation of NAT and melatonin synthesis, and α-blockade (in humans) reduces melatonin production significantly. It is likely that the additive effects of simultaneous α- and β-stimulation by endogenous noradrenalin lead to an effective

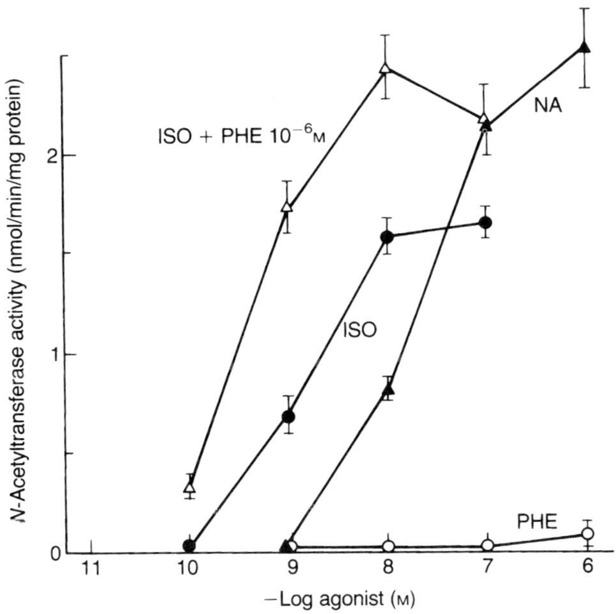

Fig. 3.16 Potentiation of β-adrenoceptor stimulation of *N*-acetyltransferase by α-adrenoceptor stimulation in cultured rat pinealocytes. Pinealocytes (2×10^5) were treated with noradrenalin, a mixed α- and β-agonist (NA), isoproterenol, a β-agonist (isoprenalin, ISO), phenylephrine, an α-agonist (PHE) or ISO plus PHE for 6 hours. Cells were collected and NAT activity measured using a radiometric assay. Each point represents the mean ± SEM of 6–9 cell aliquots. The data shown are taken from three independent experiments. PHE plus ISO clearly increases the stimulation observed with ISO alone to levels comparable to NA. (Reproduced from Sugden, D.; published by John Libbey, 1991.)

stimulus with lower noradrenalin levels than would be effective if only one receptor subtype were involved.

The mechanism through which α-adrenergic stimulation is thought to potentiate β-adrenergic induction involves an increase in intracellular calcium and stimulation of phosphatidylinositol turnover, leading to the formation of diacylglycerol as a second messenger. α-Adrenergic potentiation can be mimicked by a synthetic diacylglycerol and structurally related phorbol esters. These and the α-adrenergic agonist phenylephrine activate a calcium and phospholipid dependent protein kinase (protein kinase C) by translocation from the cytosol to the membrane, which is thought to associate the enzyme with a substrate and initiate phosphorylation of a cAMP regulatory protein located in the membrane. Protein kinase C also provides a possible feedback mechanism in that it desensitizes the α-adrenoceptor, response conceivably by an effect on the receptor

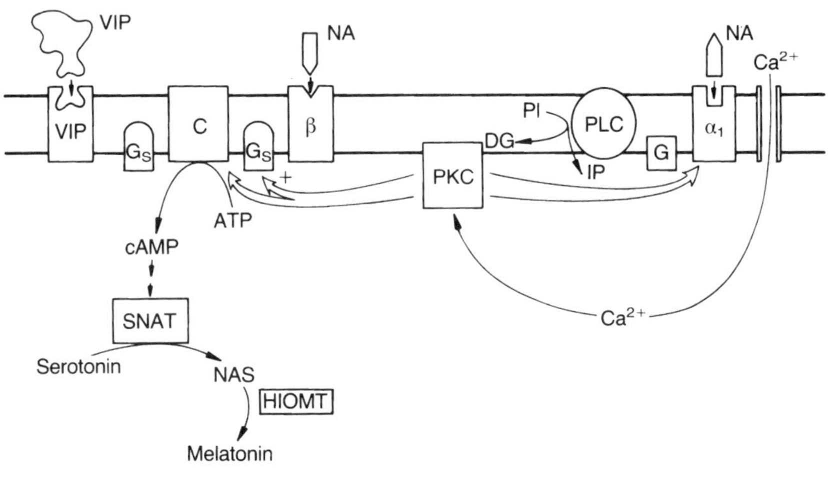

Fig. 3.17(a) Transduction mechanisms involved in the biosynthesis of melatonin. α_1 = alpha-adrenergic receptor; β = beta-adrenergic receptor; C = adenylate cyclase; ATP = adenosine triphosphate; cAMP = cyclic adenosine monophosphate; SNAT = serotonin-N-acetyltransferase; HIOMT = hydroxyindole-O-methyltransferase; PI = phosphatidylinositol; IP = inositol phosphate; DG = diacylglycerol; PKC = protein kinase C; G_s = stimulatory guanine-nucleotide binding protein; PLC = phospholipase C; VIP = vasoactive intestinal peptide. (Redrawn from Sugden, D.; published by Birkhäuser Verlag, 1989.)

itself (Figure 3.17a). Recent evidence also suggests that calcium and cAMP probably act physiologically in a co-ordinated manner to stimulate NAT activity. Figure 3.17b illustrates a proposed model for this potentiation.

The increase in NAT activity at night can show a lag period in some species (e.g. rat and hamster) and the effectiveness of β-stimulation

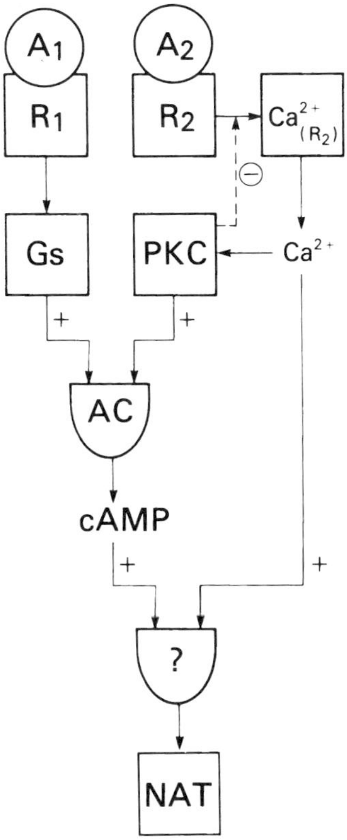

Fig. 3.17(b) Model of the role of calcium and cAMP in the regulation of NAT activity. A_1 represents an agonist that stimulates adenyl cyclase via an R_1 receptor; A_2 represents an agonist that activates R_2 receptors; R_1 represents a class of receptors that stimulates adenylate cyclase via GTP binding protein; R_2 represents a class of receptors that mediates elevation of $[Ca^{2+}]$ions; G_s represents GTP binding protein; AC represents adenylate cyclase; NAT represents *N*-acetyltransferase. (Reproduced with permission from Yu, Schaad and Klein, 1993.)

depends on the time of day when stimulation is applied. The lag increases with the time the animal is in the light and decreases with dark exposure. It is thought to reflect events in transcription and translation and that the complement of RNA accumulating with β-stimulation determines the lag.

Recently Stehle *et al.* (1993) have discovered a product of the CREM (cyclic AMP-responsive element modulator) gene, highly expressed in the pineal (and other neuroendocrine tissues), which encodes a potent repressor of cAMP-induced gene transcription. This inducible cAMP early repressor (ICER) shows a marked circadian rhythm in the pineal. ICER mRNA is induced at night under adrenergic control via cAMP. The peak of ICER expression corresponds to the declining phase of NAT and thus it is possible that it is the major component of a negative feedback loop regulating rhythmic gene expression in the pineal.

3.7.4 Sub and super sensitivity

Denervation of the pineal or depletion of catecholamine in the nerve endings leads to increased responsiveness (supersensitivity) to adrenergic stimulation. Repeated administration of β-agonists suppresses supersensitivity and induces a subsensitive state. Thus the responsiveness of the receptor is conditioned by prior exposure. This phenomenon is common to many receptor-mediated events throughout the neural and endocrine systems and the pineal served as a model for the investigation of the events concerned.

Very rapid changes in receptor sensitivity are seen in the pineal leading to a 24-hour rhythm in receptor responsiveness with pineals excised at the end of the light period showing an 8–10-fold greater response to β-stimulation than when the pineal is removed at the end of the dark phase. β-Receptor binding sites in the rat pineal vary over a 24-hour period, the lowest number being found after 12 hours in the dark, increasing shortly after lights on (Figure 3.18).

There is thus an intimate coupling between the increase in sympathetic activity and release of noradrenalin at dark onset with maximally sensitive receptors leading to onset of melatonin secretion. Tolerance to stimulation developing in the course of the night

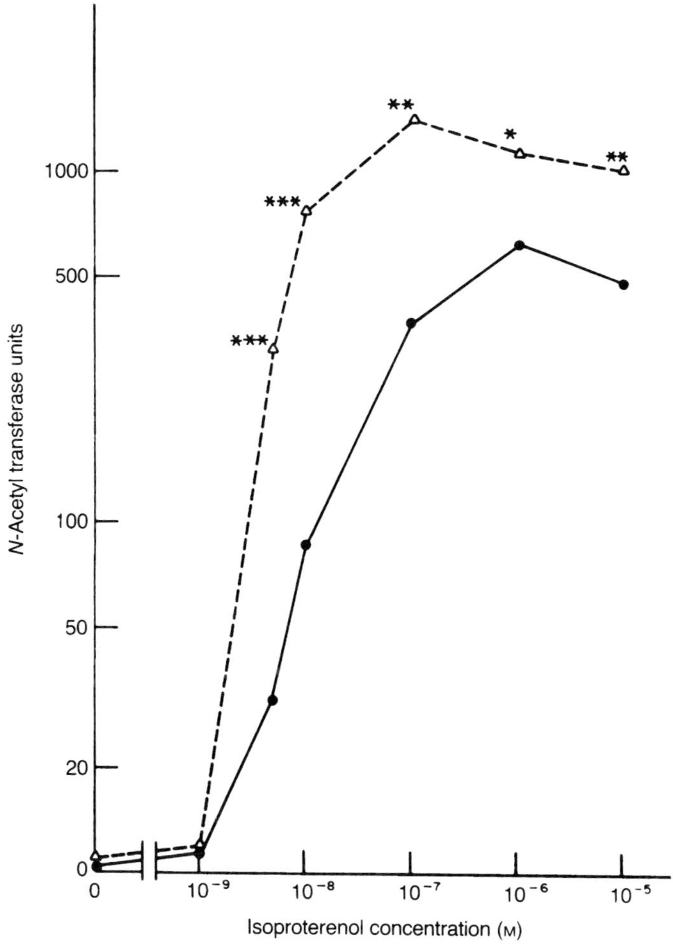

Fig. 3.18 Diurnal change in sensitivity of rat pineal serotonin *N*-acetyltransferase (NAT) to induction by L-isoproterenol (isoprenalin) in organ culture. Pineals obtained from animals killed at the end of the light period (open triangles) or at the end of the dark period (closed circles) were incubated in the presence of various concentrations of L-isoproterenol for 10 hours. Glands were homogenized and assayed for NAT; units are pmol/gland per 10 min. (Reproduced with permission from Zatz, M.; published by Springer-Verlag, 1978.)

appears to involve both decreased catecholamine sensitive adenylate cyclase activity and increased phosphodiesterase activity coupled with the ICER feedback loop described earlier. A summary of melatonin control mechanisms is shown in Figure 3.19.

It is important to note that the results in the rat are by no means necessarily true in detail for all species. Hamsters, for example, show a sharp peak of β-receptor numbers during the late dark phase only and for years it proved difficult to stimulate hamster melatonin production with β-agonists until this specific time of day was chosen

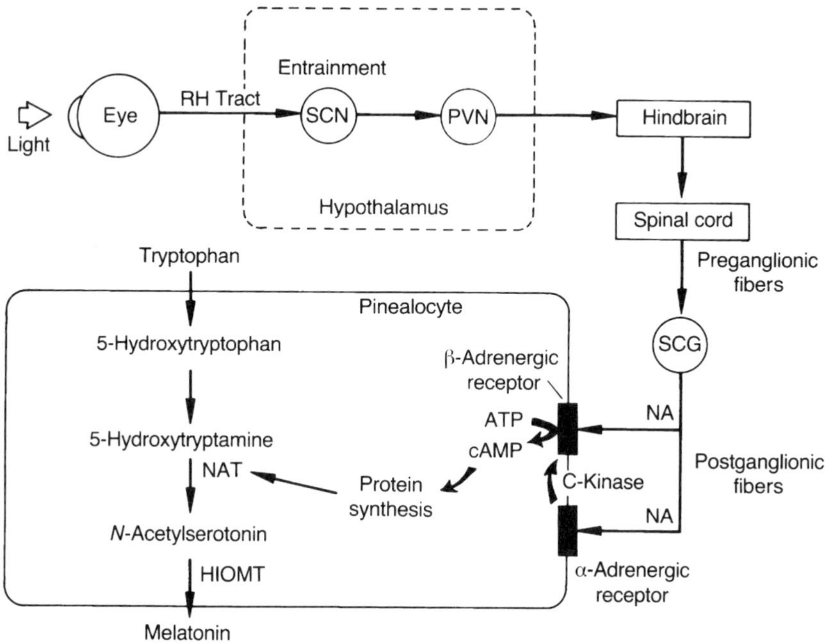

Fig. 3.19 Major control mechanisms in melatonin synthesis. A summary diagram of the noradrenergic control of melatonin production. The suprachiasmatic nucleus (SCN) generates rhythmic signals, entrained to 24 hours by light entering by the retina, relayed via the paraventricular nucleus (PVN), the hindbrain, spinal cord and superior cervical ganglion (SCG) to β- and α-adrenoceptors within the pineal. RH = retino-hypothalamic; NAT = *N*-acetyltransferase; HIOM T = hydroxyindole-*O*-methyltransferase. In mammals noradrenalin (NA) is the primary transmitter.

for experimentation. Moreover endogenous melatonin production only during the late dark phase reflects this receptor's peculiarity in the golden hamster. Sheep may use α-receptor mechanisms more than beta. In the chicken noradrenalin is inhibitory to melatonin production rather than stimulatory via an α2-adrenergic mechanism. So far, however, the only serious challenge to noradrenalin as a primary neurotransmitter is that of VIP.

cGMP has no known function within the pineal gland. It is nevertheless under adrenergic control similar in most ways to cAMP. However during induction of supersensitivity of cAMP response, the cGMP response becomes subsensitive. This phenomenon is termed 'see-saw' signal processing and may turn out to have regulatory importance in as yet undefined pineal systems.

3.8 OTHER NEURAL CONTROL MECHANISMS

The pineal contains very large numbers of other neuroreceptors and hormone receptors. Evaluation of their physiological importance is in its infancy, but most of their endogenous or pharmacological agonists have at least modulatory effects on melatonin production.

Serotonin can be released from pinealocytes by α-adrenergic activation. It amplifies the isoprenalin-induced activation of NAT *in vitro* via a receptor which does not correspond to recognized serotonin receptor subtypes (5HT1, 5HT2, 5HT3, 5HT4, etc.) but more to a 5HT receptor (5HT1P) described in the enteric nervous system.

Adenosine A2 receptors are present on pinealocyte membranes and both adenosine and its agonists are reported to stimulate cAMP production and NAT activity *in vitro*.

Pineal muscarinic and nicotinic cholinergic receptors exist, with evidence at least for muscarinic effects on indole synthesis. Dopamine D1 receptors have been described in the pineal and receptors for the vasopressin fragment (pGlu4, Cys6)VP4-9 and its agonist d(CH2)sTyr(Me)VP exist on pineal membranes. Vasopressin itself has diverse effects on pineal metabolism. *In vivo* it reduces melatonin production, but *in vitro* it enhances noradrenalin-induced stimulation of NAT. Neuropeptide Y inhibits the adrenergic stimulation of melatonin production by cultured pinealocytes.

Cytosolic receptors have been detected in the pineal for a number of steroid hormones and administered gonadal steroids undoubtedly affect pineal metabolic activity *in vitro*. The significance of steroid–pineal feedback relationships *in vivo* remains to be fully evaluated but may be of significance.

REFERENCES

Arendt, J. (1985) Mammalian pineal rhythms. *Pineal Research Reviews*, **3**, 161–213.

Arendt, J. (1988) Melatonin. *Clinical Endocrinology*, **29**, 205–29.

Arendt, J., Bojkowski, C., Franey, C. *et al.* (1985) Immunoassay of 6-hydroxymelatonin sulphate in human plasma and urine: abolition of the urinary rhythm with atenolol. *Journal of Clinical Endocrinology and Metabolism*, **60**, 1166–72.

Axelrod, J. (1974) The pineal gland: a neurochemical transducer. *Science*, **184**, 1341–8.

Bojkowski, C.J. (1988) 6-Sulphatoxymelatonin as an index of pineal function in human physiology. Thesis, University of Surrey, UK.

Benson, B. and Ebels, I. (1978) Pineal peptides. *Journal of Neural Transmission*, Suppl. **13**, 157–74.

Binkley, S. (1988) *The Pineal: Endocrine and Nonendocrine Function*, Prentice Hall, Englewood Cliffs, New Jersey, USA.

Cardinali, D.P. and Vacas, M.I. (1978) Feedback control of pineal function by reproductive hormones – a neuroendocrine paradigm. *Journal of Neural Transmission*, Suppl. **13**, 175–201.

Deguchi, T. (1973) Role of the beta-adrenergic receptor in the elevation of adenosine cyclic 3′5′-monophosphate and induction of serotonin *N*-acetyltransferase in rat pineal glands. *Molecular Pharmacology*, **9**, 612–8.

Deguchi, T. and Axelrod, J. (1973) Superinduction of serotonin *N*-acetyltransferase and supersensitivity of adenyl cyclase to catecholamines in denervated pineal gland. *Molecular Pharmacology*, **9**, 612–8.

Fiske, V.M. (1941) Effect of light on sexual maturation, oestrus cycles, and anterior pituitary of the rat. *Endocrinology,* **29**, 187–96.

Fiske, V.M., Bryant, G.K. and Putman, J. (1960) Effect of light on the weight of the pineal in the rat. *Endocrinology*, **66**, 489–91.

Kitay, J.I. and Altschule, M.D. (1954) *The Pineal Gland*, Harvard University Press, Cambridge, MA, pp. 79–95.

Klein, D.C. (1974) Circadian rhythms in indole metabolism in the rat pineal gland, in *The Neurosciences; Third Study Program* (eds F.O. Schmitt and F.G. Worden), MIT Press, Cambridge, Massachusetts, pp. 509–16.

Klein, D.C. (1979) Circadian rhythms in the pineal gland, in *Endocrine Rhythms* (ed. D.T. Krieger), Raven Press, New York.

Klein, D.C. (1985) Photoneural regulation of the mammalian pineal gland, in *Photoperiodism, Melatonin and the Pineal* (eds D. Evered and S. Clark), Ciba Foundation Symposium 17, Pitman, London, pp. 38–56.

Klein, D.C. (1993) The mammalian melatonin rhythm-generating system, in *Light and Biological Rhythms in Man*, Neuroscience (ed. L. Wetterberg), Pergamon Press (Wenner-Gren International Series Volume 63), Oxford, pp. 55–72.

Klein, D.C. and Weller, J.L. (1970) Indole metabolism in the pineal gland, a circadian rhythm in *N*-acetyltransferase. *Science*, **169**, 1093–5.

Kveder, S. and McIsaac, W.M. (1961) The metabolism of melatonin (*N*-acetyl-5-methoxytryptamine) and 5-methoxytryptamine. *Journal of Biological Chemistry*, **236**, 3214–20.

Lane, E.A. and Moss, H.B. (1985) Pharmacokinetics of melatonin in man: first pass hepatic metabolism. *Journal of Clinical Endocrinology and Metabolism*, **61**, 1214–6.

Le Bars, D., Thivolle, P., Vitte, P.A. *et al.* (1991) PET and plasma pharmacokinetic studies after bolus intravenous administration of [11-C]-melatonin in humans. *Nuclear Medicine and Biology*, **18**, 357–62.

Lerner, A.B., Case, J.D., Takahashi, Y. *et al.* (1958) Isolation of melatonin, pineal factor that lightens melanocytes. *Journal of the American Chemical Society*, **80**, 2587.

Mallo, C., Zaidan, R., Galy, G. *et al.* (1990) Pharmacokinetics of melatonin in man after intravenous infusion and bolus injection. *European Journal of Clinical Pharmacology*, **38**, 297–301.

Pevet, P. (1991) Physiological role of neuropeptides in the mammalian pineal gland. *Advanced Pineal Research*, **6**, 275–82.

Quay, W.B. (1963) Circadian rhythm in rat pineal serotonin and its modifications by estrous cycle and photoperiod. *General and Comparative Endocrinology*, **3**, 473–9.

Quay, W.B. (1974) *Pineal Chemistry in Cellular and Physiological Mechanisms*, Charles C. Thomas, Springfield, Illinois, USA.

Ralph, C.L. and Lynch, H.J. (1970) A quantitative melatonin bioassay. *General and Comparative Endocrinology*, **15**, 334–8.

Simonneaux, V., Ouichou, A. and Pevet, P. (1990) Vasoactive intestinal peptide stimulates melatonin release from perifused pineal glands of rat. *Journal of Neural Transmission*, **79**, 67–79.

Stehle, J.H., Foulkes, N.S., Molina, C.A. *et al.* (1993) Adrenergic signals direct rhythmic expression of the transcriphonal repressor CREM in the pineal gland. *Nature*, **365**, 314–20.

Skene, D.J. (1990) Alternative 5-methoxyindoles and the retina. *Advances in Pineal Research*, **5**, 129–36.

Skene, D.J., Bojkowski, C. and Arendt, J. (1994) *British Journal of Clinical Pharmacology,* **37**, 181–6.

Smith, I. (1983) Indoles of pineal origin – biochemical and physiological status. *Psychoneuroendocrinology*, **8**, 41–60.

Sugden, D. (1989) Melatonin biosynthesis in the mammalian pineal gland. *Experientia*, **45**, 922–31.

Sugden, D. (1991) Adrenergic mechanisms regulating pineal melatonin synthesis. *Advances in Pineal Research*, **5**, 33–8.

Wilkinson, M., Arendt, J., Bradtke, J. *et al.* (1977) Determination of a dark-induced increase of pineal *N*-acetyltransferase activity and simultaneous radioimmunoassay of melatonin in pineal, serum and pituitary tissue of the male rat. *Journal of Endocrinology*, **72**, 243–4.

Wurtman, R.J., Axelrod, J. and Phillips, L.S. (1963) Melatonin synthesis in the pineal gland: control by light. *Science*, **142**, 1071–3.

Wurtman, R.J., Roth, W. and Altschule, M.D. (1961) Interactions of the pineal and exposure to continuous light on organ weights of female rats. *Acta Endocrinologica (Kbh)*, **36**, 617–24.

Yu, L., Schaad, N.C. and Klein, D.C. (1993) Calcium potentiates cyclic AMP stimulation of pineal arylalkylamine *N*-acetyl transferase. *Journal of Neurochemistry*, **60**, 1436–43.

Zatz, M. (1978) Sensitivity and cyclic nucleotides in the rat pineal gland. *Journal of Neural Transmission*, (Suppl. 13), 97–114.

FURTHER READING

Aldhous, M., Franey, C., Wright J. *et al.* (1985) Plasma concentrations of melatonin in man following oral absorption of different preparations. *British Journal of Clinical Pharmacology,* **19**, 517–21.

Arendt, J., Ho, A.K., Laud, C. *et al.* (1981) Differential effect of benserazide (Ro4-4602) on the concentration of indoleamines in the rat pineal and hypothalamus. *British Journal of Pharmacology,* **72**, 257–62.

Cardinali, D.P. and Ritta, M.N. (1983) Role of prostaglandins in neuroendocrine junctions: studies in the pineal gland and the hypothalamus. *Neuroendocrinology*, **36**, 152–60.

Demisch, K., Demisch, L., Bodinik, H.J. *et al.* (1986) Melatonin and cortisol increase after fluvoxamine. *British Journal of Clinical Pharmacology*, **22**, 620–2.

Ebadi, M. and Govitrapong, P. (1986) Orphan transmitters and their receptor sites in the pineal gland. *Pineal Research Review*, **4**, 1–54.

English, J., Bojkowski, C., Poulton, A.L. *et al.* (1995) Metabolism and pharmacokinetics of melatonin in the ewe. *Journal of Pineal Research* (in press).

Fellenberg, A.J., Phillipou, G. and Seamark, R.F. (1981) Urinary 6-sulphatoxymelatonin excretion and melatonin production rate: studies in sheep and man, in *Pineal Function* (eds C.D. Matthews and R.F. Seamark), Elsevier, Amsterdam, pp. 143–9.

Franchi, A.M., Gimeno, M.F., Cardinali, D.P. *et al.* (1987) Melatonin, 5-methoxytryptamine and some of their analogues as cyclo-oxygenase inhibitors in rat medial basal hypothalamus. *Brain Research*, **405**, 384–8.

Gern, W.A. and Karn, C.M. (1983) Evolution of melatonin's functions and effects. *Pineal Research Reviews*, **1**, 49–91.

Giarman, N.J. and Day, M. (1959) Presence of biogenic amines in the bovine pineal body. *Biochemical Pharmacology*, **1**, 235.

Hirata, F., Hayaishi, O. and Tokuyama, T. (1974) *In vitro* and *in vivo* formation of two new metabolites of melatonin. *Journal of Biological Chemistry*, **249**, 1311–3.

Launay, J.M., Lemaitre, B. and Husson, H.P. (1982) Melatonin synthesis by rabbit platelets. *Life Sciences*, **31**, 1487–94.

Leone, R.M. and Silman, R.E. (1984) Melatonin can be differentially metabolised in the rat to produce *N*-acetylserotonin in addition to 6-hydroxy-melatonin. *Endocrinology*, **114**, 1825–31.

Namboodiri, M.A., Sugden, D., Klein, C.C. *et al.* (1983) 5-Hydroxytryptophan elevates serum melatonin. *Science*, **221**, 659–61.

Palazidou, E., Franey, C., Arendt, J., *et al.* (1989) Evidence for a functional role of alpha-1-adrenoreceptors in the regulation of melatonin secretion in man. *Psychoneuroendocrinology*, **14**, 131–5.

Pang, S.F., Brown, G.M., Grota, L.J. *et al.* (1977) Determination of *N*-acetylserotonin and melatonin activities in the pineal gland, retina, and Harderian gland, brain and serum of rats and chickens. *Neuroendocrinology*, **23**, 1–13.

Pearse, A.G.E. (1969) The cytochemistry and ultrastructure of polypeptide hormone producing cells of the APUD series and the embryology, physiology and pathologic implications of the concept. *Journal of Histochemistry and Cytochemistry*, **17**, 303–13.

Pevet, P. (1983) The 5-methoxyindoles different from melatonin: their effects on the sexual axis, in *The Pineal Gland and its Endocrine Role* (eds J. Axelrod, F. Fraschini F. and G.P Velo), Plenum, New York, pp. 331–48.

Raikhlin, N.T., Kvetnoy, I.M. and Tolkachev, V.N. (1975) Melatonin may be synthesised in enterochromaffin cells. *Nature*, **225**, 344–5.

Reiter, R.J. and Vaughan, M.K. (1977) Pineal antigonadotrophic substances: polypeptides and indoles. *Life Sciences*, **21**, 159–72.

Romero, J.A. and Axelrod, J. (1974) Pineal β-adrenergic receptor: diurnal variations in sensitivity. *Science*, **184**, 1091–2.

Sugden, D., Weller, J.L., Klein, D.C. *et al.* (1984) α-Adrenergic potentiation of β-adrenergic stimulation of rat pineal *N*-acetyltransferase: studies using citazoline and fluorine analogs of norepinephrine. *Biochemical Pharmacology*, **33**, 3947–50.

Vakkuri, O., Leppaluoto, J. and Kauppila, A. (1985) Oral administration and distribution of melatonin in human serum, saliva and urine. *Life Sciences*, **37**, 489–95.

Vakkuri, O., Tervo, J., Luttinen, R. *et al.* (1987) A cyclic isomer of 2-hydroxymelatonin – a novel metabolite of melatonin. *Endocrinology*, **120**, 2453–9.

Vaughan, G.M. and Reiter, R.J. (1987) The Syrian hamster pineal gland responds to isoproterenol *in vivo* at night. *Endocrinology*, **120**, 1682–4.

Voisin, P., Martin, C. and Collin, J.P. (1987) α-2-Adrenergic regulation of arylalkylamine-*N*-acetyltransferase in organ cultured chick pineal gland: characterization with agonists and modulation of experimentally stimulated enzyme activity. *Journal of Neurochemistry*, **49**, 1421–6.

Waldhauser, F., Steger, H. and Vorkapic, P. (1987) Melatonin secretion in man and the influence of exogenous melatonin on some physiological and behavioural variables. *Advances in Pineal Research*, **2**, 207–23.

Young, I.M., Francis, P.L., Leone, A.M. *et al.* (1986) Night/day urinary 6-hydroxymelatonin production as a function of age, bodymass and urinary creatinine levels: a population study in 110 subjects aged 3–80. *Journal of Endocrinology*, **111**(Suppl), abstr. 32.

Yuwiler, A. (1983) Vasoactive intestinal peptide stimulation of pineal serotonin-*N*-acetyltransferase activity: general characteristics. *Journal of Neurochemistry*, **41**, 141–56.

4 Light–dark control of melatonin synthesis

Contents

This chapter is concerned with the rhythmic aspects of melatonin secretion and the subsequent chapters deal extensively with effects of melatonin on rhythms. Rhythm theory employs certain terms and has a number of concepts which are not generally familiar. The first section therefore gives a brief overview of the general concepts and introduces terms frequently used in descriptions of pineal function. A glossary of terms used in chronobiology can be found in the Appendix.

4.1 INTRODUCTION TO BIOLOGICAL RHYTHMS

Biological rhythms of various periodicity occur in all eukaryotic organisms. The frequency displayed varies from fractions of a second (for example, the firing of neurons) to years (for example, population variations). Rhythms with a periodicity of less than a day are known as ultradian (for example, pulse rate, pulsatile secretion of hormones), and of more than a day, infradian (for example, the estrous cycle). Rhythms corresponding to major periodicities in the external environment (daily, lunar, annual) are the most prominent and by far the most information is available concerning daily rhythms. These are of particular interest with regard to human physiology and pathology (Chapters 7 and 8).

Rhythms are either externally imposed, internally generated, or more frequently a combination of these two factors. Internally generated rhythms are known as circadian (approximately a 24-hour period, from the Latin 'about a day'), circannual (about a year), circalunar, circatidal, etc. Circadian rhythms serve to temporally program the daily sequence of metabolic and behavioral changes. They may also serve as reference points for annual changes (melatonin as an internal calendar, see Chapter 5).

A rhythm displaying its endogenous period (tau or t) in the absence of time cues is said to be free-running and is desynchronized from environmental periodicity. The manifested endogenous period depends on species, individuals, experimental conditions and on previous history. It is an inheritable characteristic. Circadian tau in humans can vary from 22 hours or less to 26 hours. The environmental factors which entrain or synchronize a rhythm to a given periodicity are referred to as zeitgebers (time-givers). Some species, for example, humans, show free-running circadian rhythms which are usually longer than 24 hours: on average 25 hours (Figure 4.1), whereas others have an endogenous periodicity shorter than 24 hours, for example Soay sheep. Recently, mutant hamsters have been described with a mean tau of 20.16 hours compared to the mean tau of 24.01 hours of their original colony. Interbreeding of mutants and wild type leads to circadian characteristics ($t = 22.13$ h) intermediate between the two genotypes.

Endogenous rhythms are driven by a pacemaker (oscillator, bio-

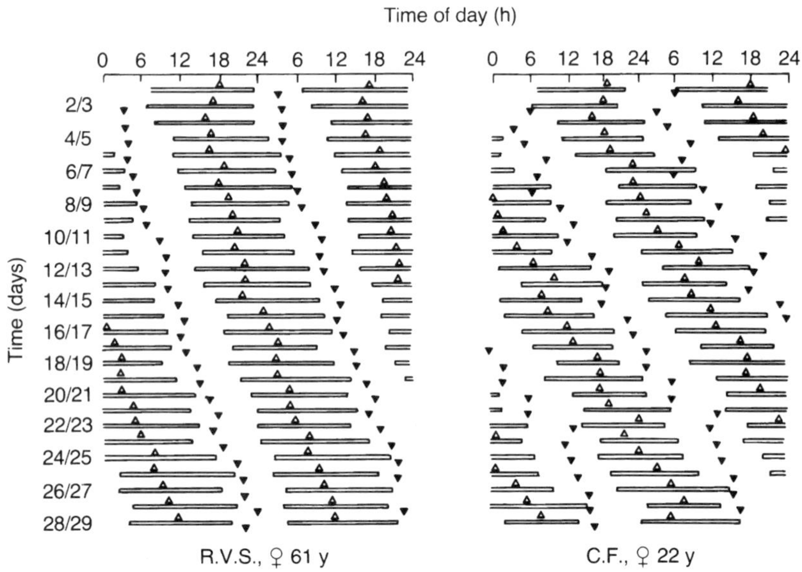

Fig. 4.1 Free-running circadian rhythms of two human subjects, living under constant environmental conditions without time cues. Sleep–wake (bars: wake episodes) and deep body temperature (upright and inverted triangles are maxima and minima of continuously recorded rectal temperature) are shown as a function of clock time. For clarity two successive days are each plotted side-by-side: a double 'raster' plot. These examples are representative of several hundred experiments estimating rhythmic biochemical, physiological and psychological variables for which a calculated free-running period is 25.00 ± 0.51 h. (Reproduced with permission from Wever, R.A.; published by Springer-Verlag, 1986.)

logical clock). The major biological clock in higher vertebrates is situated within the suprachiasmatic nucleus (SCN) of the hypothalamus whereas in some birds and lower vertebrates the pineal gland itself is the principle rhythm generator.

The phase (phi) of a rhythm or a pacemaker is an instantaneous state of an oscillation within a period. The acrophase is the calculated peak time and the bathyphase (little used) is the calculated nadir. The amplitude is the measure from nadir to acrophase. The mesor is the mathematical mean.

Zeitgebers which effect entrainment or synchronization to the environment include light–dark, temperature, social cues, rhythmic feeding and knowledge of clock time. Zeitgebers with a period T force endogenous cycles to run with a period T. This means that in each cycle the zeitgeber changes the intrinsic period (tau) by $t–T$. If the endogenous period of a circadian rhythm is greater than 24 hours synchronization requires a phase advance of the pacemaker each day by this amount. If tau is less than 24 hours the pacemaker must be phase delayed each day.

The general structure of the circadian system has three components: a pacemaker, an input pathway for entrainment of the pacemaker and an output pathway for the expression of overt rhythms (Figure 4.2). This scheme is well represented by the mammalian melatonin rhythm generating and entraining system. The pacemaker is the SCN, the input is the retina-retinohypothalamic tract, and the output is along neuronal connections to the pineal. In all other amniote species circadian rhythms are synchronized to the light–dark cycle via photoreceptors in the pineal gland, parapineal structures and deep diencephalic photoreceptors in addition to the retinal pathway.

The light–dark cycle is the major synchronizer of circadian (and seasonal) rhythms and the circadian activity (α)–rest (ρ) cycle is the most easily measured overt rhythm in small animals. Hence an enormous amount of information has accrued on the formal properties of circadian rhythms using these variables. It has been assumed that information derived from rest–activity can be extrapolated to all circadian rhythms. This may possibly be true in ideal conditions with no ‘masking’. When external factors significantly modify an endogenous rhythm in addition to synchronizing it, the rhythm is

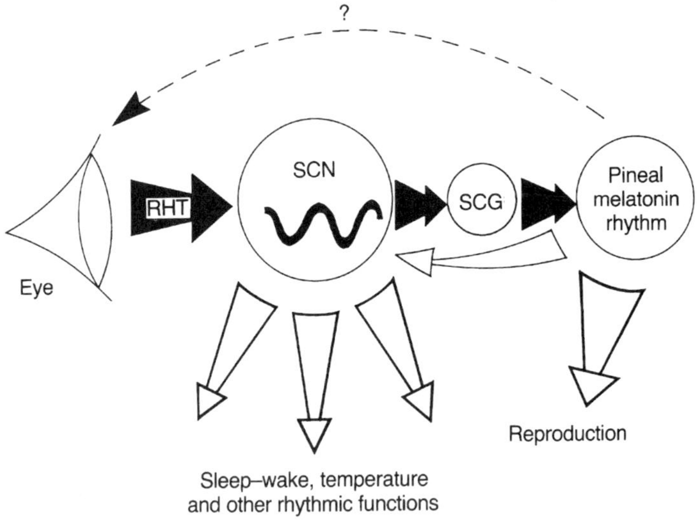

Fig. 4.2 Basic concepts of the circadian system: inputs and outputs of the central rhythm generating system in the suprachiasmatic nucleus (SCN). RHT = retino-hypothalamic tract, SCG = superior cervical ganglion. Melatonin can feed back at the level of the SCN and probably the retina (see Chapters 6, 8 and 9).

said to be masked. Examples of masking of circadian rhythms include the suppression of melatonin secretion by light, the rapid increase in body temperature with bright light and the stress-induced increase of cortisol secretion. Human studies in particular suggest that in an uncontrolled environment, for example in shift workers, and in artificially disturbed rhythms in environmental isolation, different circadian rhythms behave quite differently from the rest–activity cycle.

In continuous darkness or in dim light below a particular intensity, and in the absence of other major time cues, mammalian rhythms free-run. In constant light, disruption of the activity–rest cycle is often but not always present. In the presence of weak zeitgebers, rhythms may be synchronized to the appropriate periodicity but with an abnormal phase relationship to the driving zeitgebers. In a species with a free-running tau greater than 24 hours this implies an abnormal phase delay relative to the zeitgeber. When the light–dark cycle (standard laboratory lighting) is artificially

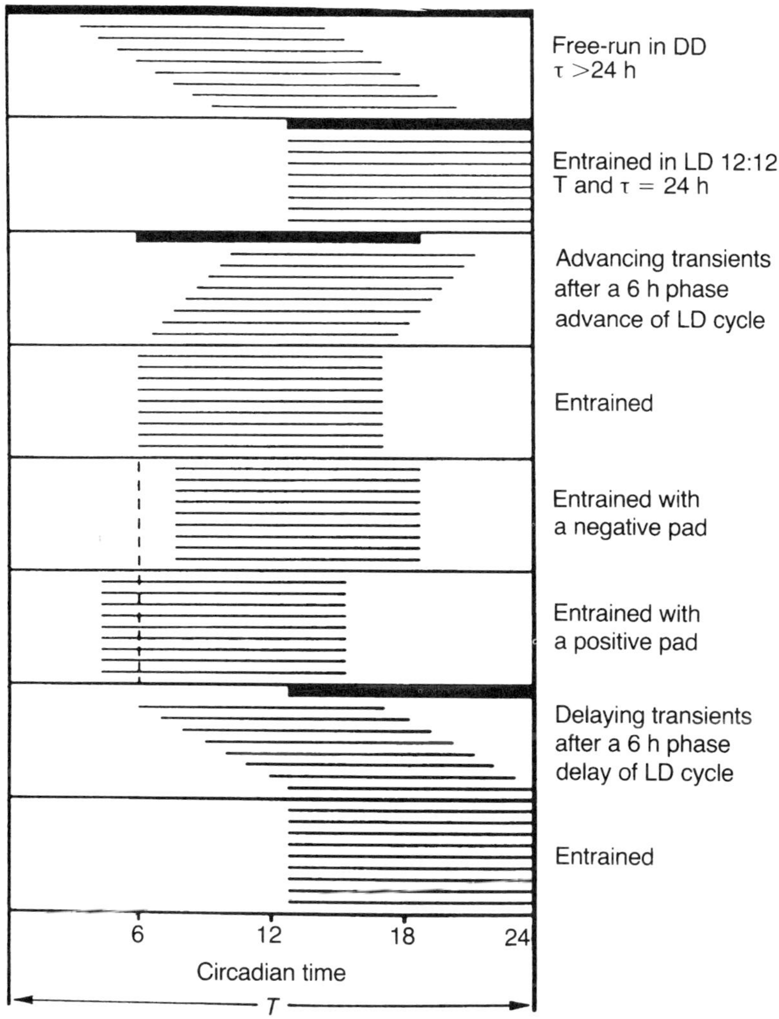

Fig. 4.3 Diagrammatic representation of various responses of circadian activity rhythms in the rat, a nocturnal species. Free-run with a period, tau (*t*) >24 h in continuous darkness (DD); entrainment by a light–dark cycle (T) of 12 hours light and 12 hours darkness (LD 12:12); advancing transients following a 6-h phase advance of the light–dark cycle; entrainment to the new light–dark cycle; entrainment with a negative and then with a positive phase angle difference (pad); delaying transients following a phase delay of the light–dark cycle. Circadian time is the time of a complete cycle of a given variable, in this case rest–activity. Circadian time (CT) 12 is usually activity onset in the rat. (Reproduced with permission from Redman, J., 1988.)

advanced or delayed, the circadian system adapts slowly to the new phase position, passing through intermediate phases known as transients (Figure 4.3). The time taken to adapt depends on the size of the phase shift, the zeitgeber strength, for example photoperiod and the light intensity, but often approximates to one hour of adaptive shift per day. During the process of adaptation endogenous rhythms are out-of-phase with the external environment (external desynchronization). They may also be out-of-phase with each other, i.e. assume a transitory abnormal phase relationship (internal desynchronization). These situations occur regularly in shift workers and in jet-lagged travelers and are considered to be responsible for many health and performance problems encountered by these groups. Blindness, old age and psychiatric disorder are all associated with disturbed circadian rhythms.

4.2 THE MELATONIN RHYTHM: A DARKNESS HORMONE

4.2.1 Introduction

Early work indicated that constant light suppressed HIOMT activity and constant dark increased it. These were minor effects compared with the very large increase in NAT activity and subsequent melatonin production at night in a 'normal' 24-hour lighting cycle. In virtually all species studied to date, whether nocturnal or diurnal, melatonin is synthesized and secreted during the dark phase of the day. This broad statement needs a little qualification as in some species, for example, in the European hamster and the penguin, a 24-hour rhythm is only evident at certain times of the year – but then melatonin is made at night. The rhythm is endogenous, i.e. internally generated. It persists in the absence of time cues, in general assuming a period deviating slightly from 24 hours and is thus a true circadian rhythm.

Remarkably, even the unicellular alga *Gonyaulax* appears to produce melatonin during the dark phase with an endogenous circadian

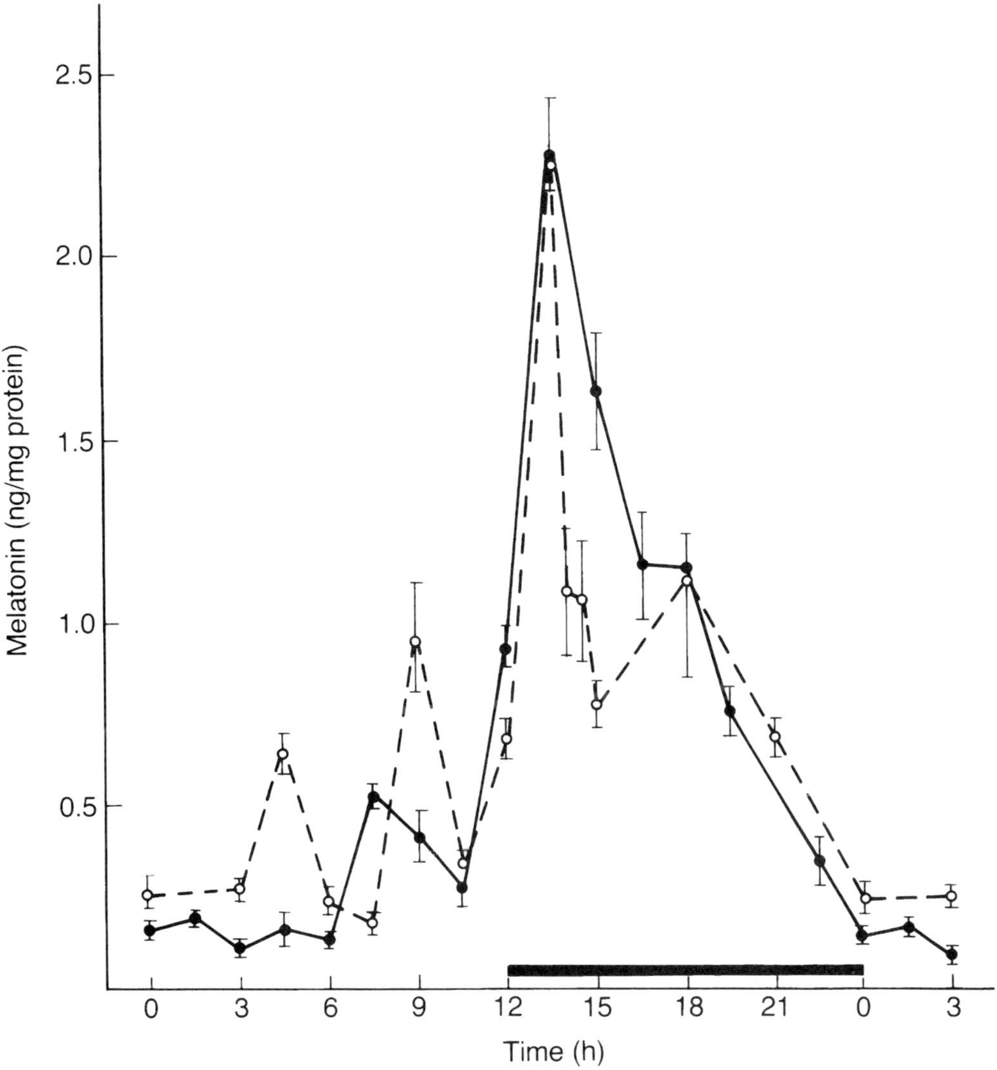

Fig. 4.4 Diurnal rhythm of melatonin in the unicellular alga *Gonyaulax polyedra* determined by (●) high-performance liquid chromatography (HPLC) and (○) radioimmunoassay (RIA). As in all other species melatonin is made largely during the dark phase. (Reproduced with permission from Poggeler, B. *et al.*; published by Springer-Verlag, 1991.)

rhythm (Poggeler *et al.*, 1991, Figure 4.4). The authors of this work have suggested that the day–night variation in *Gonyaulax* might be due to chemical destruction of melatonin during the light phase rather than through changes in enzymic activity or an endogenous clock. Melatonin production is clearly a highly evolutionarily conserved phenomenon. It is quite possible that it has served to convey light–dark information as a chemical signal from the emergence of the very first life.

4.3 RHYTHM GENERATION IN THE PINEAL

In some lower vertebrates and birds the rhythm of pineal melatonin production is clearly generated within the pineal itself. The rhythm is maintained for several days when the gland is removed and kept in culture (Figure 4.5). As these species retain a direct photoreceptor capacity in the pineal, it is perhaps not surprising that the melatonin rhythm *in vitro* can be manipulated using light–dark schedules. Furthermore the precise phasing of the *in vitro* rhythm in birds is closely related to the photoperiodic experience of the bird prior to extirpation of the gland. Thus the gland contains an internal rhythm generator precisely phased to the light–dark cycle.

Experiments in which the chicken pineal has been dissociated into its constituent cells prior to culture show that the rhythm is not dependent on the integrity of the organ structure and that indeed single pinealocytes can be shown to secrete melatonin rhythmically when cultured in isolation. Even more remarkable is the fact that it is possible to transplant a pineal taken from a donor bird kept in a specific photoperiod to a pinealectomized bird kept in a different photocycle and confer the circadian phase of the donor bird (as assessed by locomotor activity) to the recipient bird (see Chapter 6). In spite of its internal capacity to generate and phase a circadian rhythm of melatonin the pineal of lower vertebrates nevertheless appears to be at least partly controlled from light entering through the retina (for a review, see Binkley, 1988).

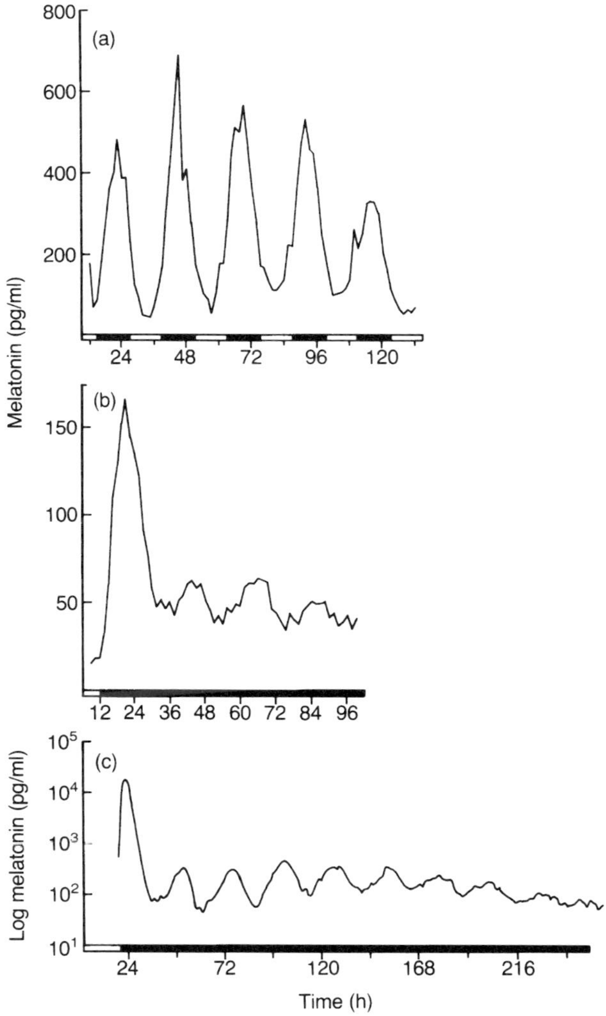

Fig. 4.5 Rhythmic production of melatonin persists *in vitro* in cultures of individual chick pineal glands (a) kept in LD cycles, (b) kept in constant darkness (DD). The rhythm in DD damps out. (c) Rhythm of melatonin release from an individual cultured lizard (*Anolis carolinensis*) pineal gland held in DD at 32 °C. The rhythm persisted for about 10 cycles. (Reprinted with permission from Underwood, H. and Goldman, B.D.; published by Guildford Press, 1987. Data from Takahashi, J. *et al.*, 1980, and Menaker, M. and Wisner, S., 1983.)

4.4 RHYTHM GENERATION IN THE SUPRACHIASMATIC NUCLEUS (SCN)

In mammals the pineal is not a self-sustaining rhythm generator. This role is taken over by the suprachiasmatic nucleus (SCN) of the hypothalamus, a small paired structure situated just above the optic chiasm. The SCN is considered to act as the major circadian rhythm generator or biological clock. Lesions which destroy this area of the brain lead to a loss of the vast majority of circadian rhythms such as locomotor activity (Figure 4.6), sleep, behavior, hormones including melatonin, urinary constituents, etc.

A few circadian variables persist after SCN ablation such as anticipatory behavior prior to feeding and visual sensitivity. It seems clear that there is more than one clock; however it is likely that in the presence of a functional SCN this structure serves to maintain coupling between subsidiary rhythmic systems.

SCN neurons show circadian rhythms in electrical activity. Electrical stimulation of the SCN, direct application of light to the retina and optic nerve stimulation lead to reduced pineal stimulation and reduction in melatonin synthesis. The electrical activity of the SCN is self-sustained and can be maintained in culture. Moreover, like the bird pineal gland, transplantation of the SCN to the 3rd ventricle in rodents restores rhythmic locomotor activity in animals rendered arrhythmic by SCN ablation.

Remarkably, a dispersed SCN cultured over a period of days or even weeks can also restore rhythmic locomotor activity. Much present discussion centers around the question as to whether in these circumstances the signal is neuronal (through regrowth of axons) or humoral. The question is not resolved. In intact animals, however the SCN communicates through an extensive neural network, in turn receiving input from a number of brain areas.

In fetal and neonatal rodents before the retinohypothalamic pathway has innervated the SCN, circadian variations in SCN metabolic activity are present and in phase with the maternal circadian activity. The maternal SCN are necessary for this entrainment to take place.

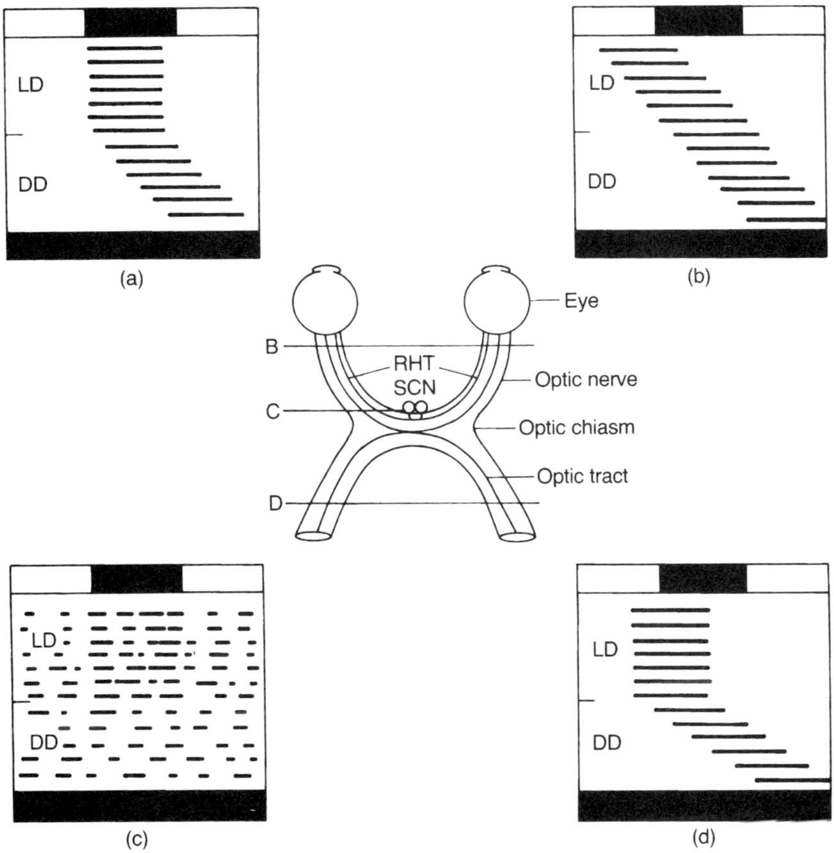

Fig. 4.6 (a) Rest–activity cycle of the rat in LD and DD, with no treatment. SCN lesion (c) but not optic tract transection or blinding (b) abolishes free-running activity–rest cycles, in DD in the rat. In 12L 12D SCN lesion shows some clustering of activity in the dark phase; with primary and accessory optic input abolished, entrainment occurs but with a phase angle difference compared to the intact animal (d). (Reprinted, by permission of the *New England Journal of Medicine*, **309**, 469–76, 1983. Moore-Ede, Czeisler and Richardson.)

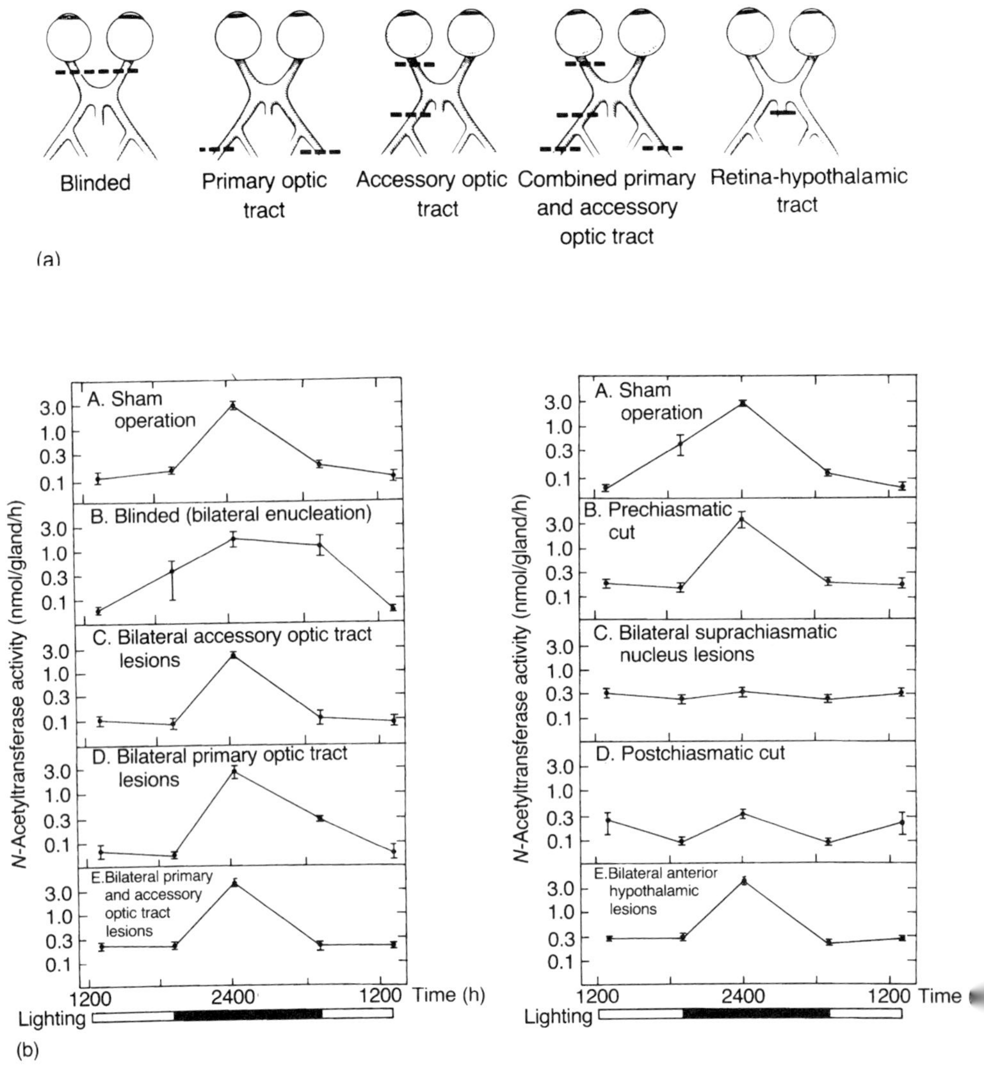

Fig. 4.7 SCN lesion and postchiasmatic section but not primary or accessory optic tract transection or blinding abolishes the rhythm of melatonin synthesis in the rat. (a) Diagram of the localization of visual pathway lesions; retinohypothalamic tract section was made by ablation of both suprachiasmatic nuclei (SCN).

4.5 ENTRAINMENT (SYNCHRONIZATION) BY LIGHT

The rhythmic activity of the SCN is entrained to the 24-hour day primarily by light–dark cycles. Retinal information is conveyed via a monosynaptic tract, the retinohypothalamic projection (RHP), directly to the SCN with a subsidiary input from the lateral geniculate. The integrity of the RHP tract is essential for appropriate synchronization of circadian rhythms to 24 hours. Sectioning of the primary and accessory optic tracts, leaving the RHP intact, while rendering the animal functionally blind does not abolish circadian synchronization, although the phase relationship of a rhythm to the zeitgeber may be modified (Figure 4.6).

Factors (zeitgebers) other than light–dark cycles which are involved in the entrainment of circadian rhythms include behavioral imposition, for example forced activity and rest, social and nutritional (rhythmic feeding) cues, temperature variations, knowledge of clock time, certain drugs and melatonin itself. In a normal environment these factors operate in concert. The extent to which any one of these can act as a principal zeitgeber in the absence of the others is species dependent, in general not well defined, and is addressed in Chapter 7. Sufficient intensity and duration of white light is, however, sufficient for entrainment to 24 hours in most species.

Lesions of the SCN abolish the melatonin rhythm (Figures 4.7a,b). Lesion of the optical input to the SCN leaves the rhythm intact but uncoupled from the 24-hour cycle. Postchiasmatic lesions also abolish the rhythm. The pathway from the SCN to the pineal

Fig. 4.7 contd (b) Pineal *N*-acetyltransferase activity rhythm was abolished only in lesions destroying the SCN and its postchiasmatic connections. Lesions of the primary and accessory optic tracts and of the anterior hypothalamus preserved the rhythm. The rhythm persists but becomes free-running in animals with prechiasmatic lesions as shown in other studies. (Reproduced with permission from Moore and Klein, 1974.) In this study some residual doubt remained concerning uncrossed visual input fibers which was resolved in a subsequent study leaving the conclusion that the intact retinohypothalamic tract controls melatonin synthesis (Klein and Moore, 1979).

passes via the paraventricular nucleus, the median forebrain bundle, the dorsal spinal cord, superior cervical ganglion and postganglionic sympathetic fibers. Destruction of any part of this pathway abolishes the melatonin rhythm. These pathways have been determined by lesioning experiments largely in the rat but evidence for their existence in other species including primates is good.

Thus the melatonin rhythm is generated in the SCN and synchronized to 24 hours largely by the light–dark cycle acting via the SCN as for many other circadian rhythms. It is therefore a 'hand of the clock' in the sense that it responds to clock (SCN) signals and indicates the status of the clock in terms of phase and possibly amplitude. It also feeds back on the clock to modify its own rhythm and other circadian variables. This subject will be treated in Chapter 7.

4.6 MELATONIN SECRETION IN RELATION TO DAYLENGTH

4.6.1 Animals

In normal entrained conditions melatonin is produced during the dark phase. In most species its secretion is related to the length of the night: the longer the night the longer the duration of secretion (Figure 4.8). This has been particularly well demonstrated in sheep where melatonin rises within a few minutes of lights off and in photoperiods of more than around 14 hours of light does not decline until lights on. In such photoperiods light serves to entrain the rhythm and to suppress secretion at the beginning and/or the end of the dark phase. This changing duration with daylength is very clearly demonstrated in ewes studied in natural light throughout the year (see Chapter 5). With longer periods of darkness secretion attains its maximum potential duration (sheep: 14–16 hours) and declines before lights on.

Rats and hamsters show a lag period before the onset of melatonin secretion even in photoperiods as long as 18L 6D. The lag period increases with increasing length of darkness. It is particularly marked in the Golden hamster (Figure 4.9). The latter secretes melatonin

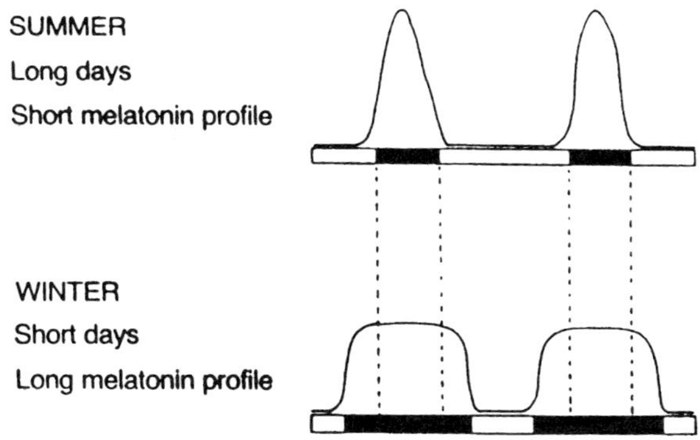

Fig. 4.8 Diagrammatic representation of melatonin secretion in long days and short days in, for example, sheep.

towards the end of the dark period but even so changes in duration are seen with changing length of the night. The duration of melatonin secretion is a critical feature of its physiological role in photoperiodic time measurement (Chapter 5).

4.6.2 Humans

In humans, for many years it has proved difficult to demonstrate a change in duration of melatonin secretion with daylength even in polar regions, although very small changes have been reported. The most consistent observation is that human melatonin profiles show a phase change from winter to summer, with earlier secretion in summer compared to winter. If however humans are kept strictly in darkness for 14 hours per day for a period of 2 months, the melatonin secretion pattern expands to cover almost the entire dark period and concomitantly, in extended periods of 16 hours of light, the rhythm contracts to less than 10 hours (Figure 4.10). Thus in appropriate experimental conditions humans show a duration change like animals.

Our behavior, particularly that of urban populations, is such that in 'normal' circumstances it is likely that we perceive only very small changes in daylength during the year. For example in the UK it is

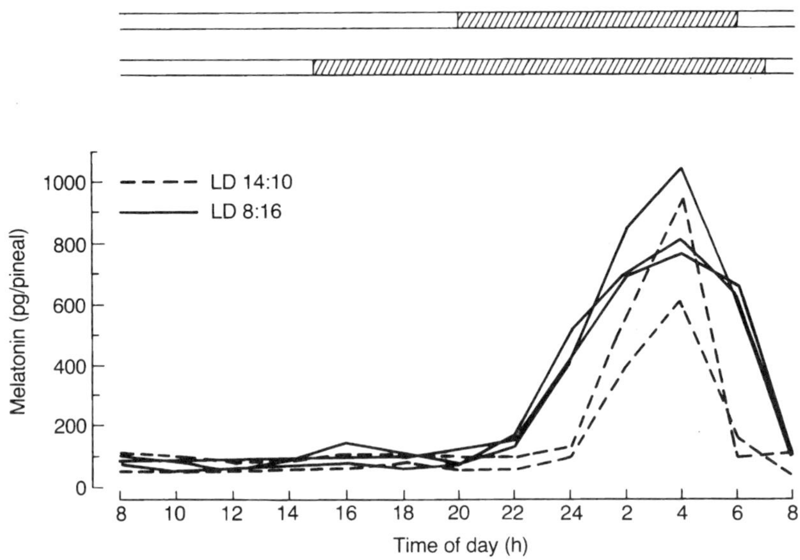

Fig. 4.9 Temporal pattern of pineal melatonin content in male Golden hamsters in two different photoperiods. Mean of five hamsters for each point. Horizontal bars on top indicate the photoperiod. Five figures combined. (Reproduced from Arendt, J. Role of the pineal gland and melatonin in seasonal reproductive function in mammals. *Oxford Reviews of Reproductive Biology*, **8**, 266–320, by permission of Oxford University Press, 1986.)

fairly typical to rise around 0700 h (well after dawn in the summer) and to be inside a dwelling from around 1900 h (well before dusk in the summer) and thus not be exposed to natural sunlight for the entire photoperiod in summer.

4.7 LIGHT SUPPRESSION OF MELATONIN SECRETION

Units of reported light exposure vary amongst authors with lux and $\mu W/cm^2$ being the most common. Human responses are usually reported in lux and animal responses in $\mu W/cm^2$. Lux are converted to $\mu W/cm^2$ by initially converting to foot-candles (divide by 10.7639)

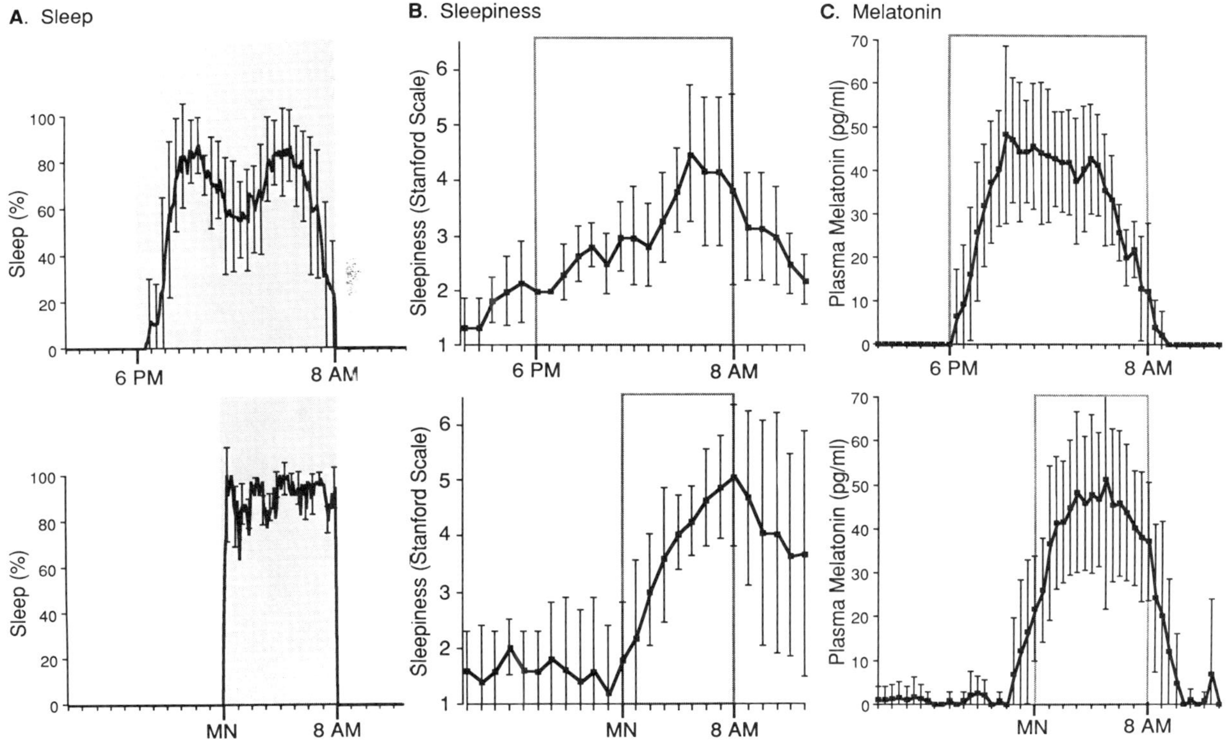

Fig. 4.10 Durations of the nightly sleep-phase (A), of the nightly rising phase of sleepiness (B), and of the nightly phase of active melatonin secretion (C) all expanded when the duration of night was lengthened from 8 (top) to 14 (bottom) hours. The average (±SD) percentage time asleep during each 6 min is shown. The dark phase of each photoperiod schedule is shown by a shaded rectangle. Sleepiness and melatonin levels were measured on the day following the last night of each photoperiod schedule during a 24-hour period (1300 h–1300 h) in which the subjects remained continuously awake in constant dim light (<1 lux). Although no light–dark cycle was applied on the day when these measurements were made, the dark phase of the photoperiod schedule that preceded the measurements is indicated by an open rectangle. Of seven men who completed both phases of the experiment, the figure shows data for the six in whom the duration of nocturnal melatonin secretion lengthened when the photoperiod was shortened. (Reproduced from Wehr, T.A. The duration of human melatonin secretion and sleep respond to changes in daylength (photoperiod). *J. Clin. Endocrinol. Metab.* **73**, 1276–80, 1991 © The Endocrine Society.)

and then multiplying foot-candles by the total spectral irradiance distribution of the light source (tables issued by Duro-Test Corporation, North Bergen, New Jersey 07047, USA).

Even a brief exposure to light of sufficient intensity at night will rapidly suppress melatonin production. In the rat, for example, one minute of 150 lux white fluorescent light is sufficient to induce a rapid decline in *N*-acetyltransferase activity with a half-life of 4 minutes, and in parallel to reduce plasma melatonin to daytime concentrations (Figure 4.11).

Various studies have assessed the spectral requirements for light suppression of melatonin production at night. In general the most effective light is in the green band (540 nm) corresponding closely to the rhodopsin absorption spectrum in both rodents and humans (Figure 4.12). Some authors report that UV light is

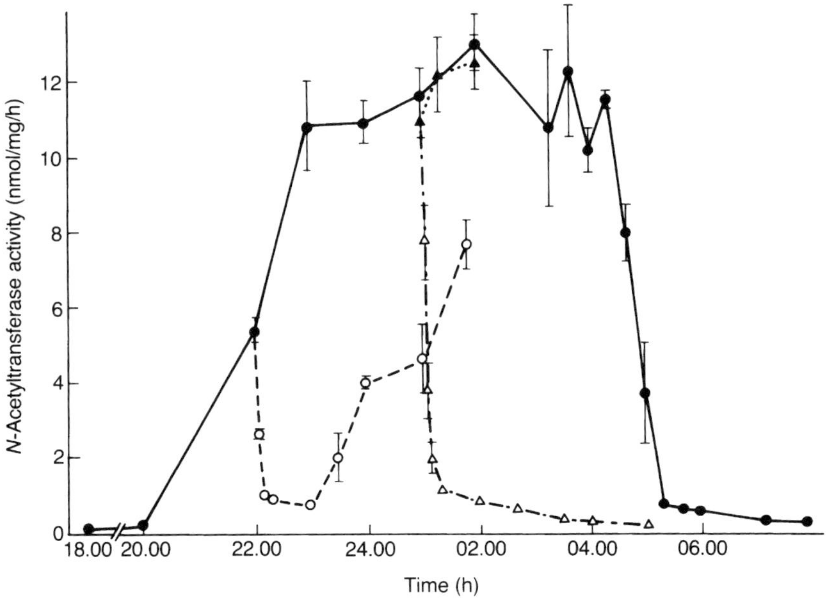

Fig. 4.11 Suppression and resumption of serotonin-*N*-acetyltransferase activity by 1 min light pulse at 2200 h (O), suppression with no resumption of activity by 1 min light pulse at 0100 h (Δ) in rats. At the end of light exposure rats were kept in darkness until killed. (Reproduced from Illnerova, H. and Vanecek, J.; published by Elsevier, 1979.)

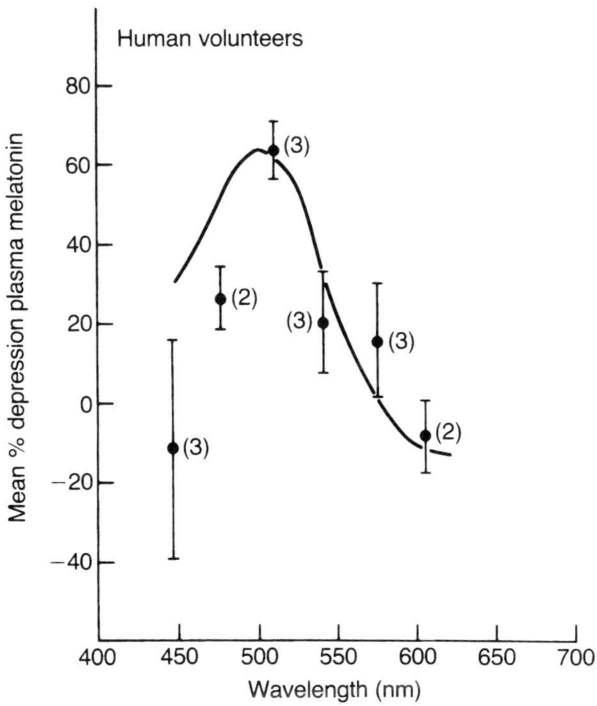

Fig. 4.12 Percentage depression (mean ± SE) of plasma melatonin in male human volunteers ($N = 3$) exposed to different wavelengths of light. 509 nm light was most inhibitory to circulating melatonin. The solid line is the spectral sensitivity of human rod cells. (Reproduced with permission from Brainard, G.C. *et al.*; published by the New York Academy of Sciences, 1985.)

also effective.There is some evidence that electromagnetic radiation can suppress melatonin production, at least in animals. The vast majority of studies have used either cool white fluorescent or incandescent light. With the recent availability of full spectrum light (Vitalite ®), and a greater awareness of the importance of controlling light conditions in pineal research (and indeed in the life sciences in general), this source tends to be used.

The amount of light required to suppress melatonin secretion either during the night or at the beginning and the end of the night varies considerably from species to species. The intensity of light applied and the duration of exposure combine to induce a given suppression. A most important factor involved in this suppression is the previous exposure to light. As little as one minute of 0.1 μW/cm^2 white light will suppress melatonin synthesis in the Syrian hamster pineal in the middle of the night whereas in the Richardsons ground squirrel 1850 μW/cm^2 (approx 6800 lux) are required. Detectable suppression can be obtained in sheep and goats with 0.15 lux (0.04 μW/cm^2) and 1–2 lux (0.27–0.54 μW/cm^2) cool white fluorescent light, respectively, and the degree of suppression bears a hyperbolic relationship to the intensity of the light source for a given duration of exposure. It is likely that light sensitivity varies acutely, increasing in the course of the night, probably through an increase in retinal sensitivity.

Seasonal variation in light sensitivity has not been reported in animals as yet. However Reiter (1985) has demonstrated that in 13-lined ground squirrels either bred and kept in laboratory conditions, or caught in the wild and subsequently maintained in the laboratory, very much less light is needed to suppress melatonin in the laboratory-bred animals than the wild-caught animals. This phenomenon is attributed to the previous exposure of animals to either natural light, which can attain 100 000 lux or laboratory light which is usually of the order of 100–200 lux. It is likely, therefore, that long-term exposure to summer photoperiod and light intensity will decrease sensitivity.

Humans also require substantial amounts of light to suppress melatonin secretion at night. Early experiments suggested that domestic intensity light (100–500 lux) did not suppress human melatonin, and the inference was made that either we are resistant to such intensities or that human melatonin could not be suppressed by light. In a classic experiment however, Lewy *et al.* (1980) showed conclusively that if sufficient intensity of white light (2500 lux for 2 hours during the night between 0200 and 0400 h) was used human melatonin could be suppressed to basal (daytime) levels (Figure 4.13). This observation has been of considerable importance, not only for an understanding of the control of human melatonin secretion but also for a general appreciation of the role of light in human physiology, in particular its importance in the control of human rhythms (see Chapter 7).

In fact partial suppression of melatonin secretion in humans can

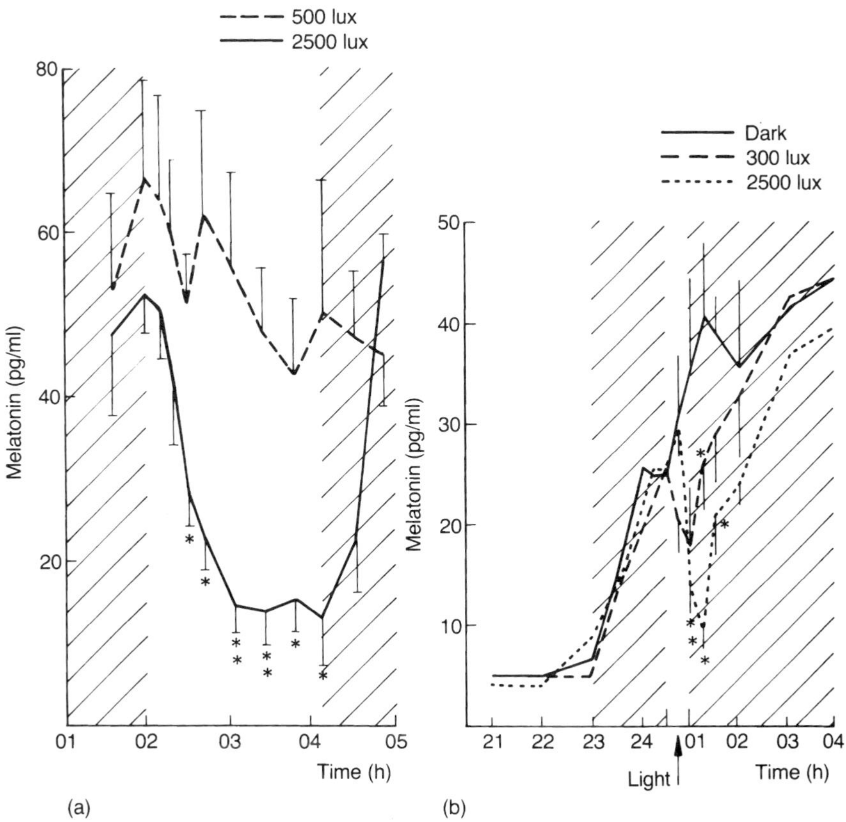

Fig. 4.13 (a) Suppression of human melatonin secretion by bright (2500 lux, solid line) and dim (500 lux, dotted line) light for 2 hours (0200–0400 h) during the night compared to darkness. (Reproduced from Lewy, A.J. *et al.* Light suppresses melatonin secretion in humans. *Science*, **210**, 1267–9; copyright 1980 by the AAAS.) (b) Partial suppression of human melatonin secretion by 300 lux compared to 2500 lux, 30 min exposure, 0030–0100 h. (From Bojkowski *et al.*, 1987, previously unpublished figure.)

be effected by lower intensity light, e.g. 200–300 lux applied for 30 minutes. Using this protocol a time-of-day and time-of-year variation in sensitivity is suggested by recent work in Antarctica – a region where long-term exposure to domestic intensity light is possible during the winter in the absence, for prolonged periods, of natural sunlight. As might be expected sensitivity increases in the course of the night and was greater in winter than in summer.

There are very substantial individual variations in sensitivity which may be both genetically and environmentally determined. For example supersensitivity is shown by bipolar (manic depressive) patients and this characteristic may be partly familial. It is possible that humans working outside, for example in agriculture, and therefore exposed to natural light much of the time, will be less sensitive than office workers. Even in California, office workers rarely encounter more than 1000 lux during a working day whereas natural sunlight can be 100-fold brighter.

Patients with seasonal affective disorder (SAD, winter depression) are conventionally treated with extended bright, full spectrum light (see Chapter 7) on a daily basis in the winter. Such treatment decreases their ability to suppress melatonin at night with a given dose of bright white light. This desensitization may relate to the therapeutic effect. These patients also show seasonal differences in sensitivity to light with a greater response in winter.

4.8 ENTRAINMENT OF THE MELATONIN RHYTHM WITH LIGHT

As previously stated, light–dark cycles are the major environmental factor involved in the entrainment of circadian rhythms in mammals. Clear evidence of the importance of light–dark for the entrainment of melatonin was demonstrated in humans where, following an inversion of the light–dark cycle, the urinary melatonin rhythm adapts to the new photoperiod over a period of several days, finally assuming the same phase relationship with the new light–dark cycle as was present originally with the old light–dark cycle (Figure 4.14).

If humans are subjected to a forced phase shift of 24 hour time cues, such as in rapid flight across several time zones, readjustment of the phase of melatonin secretion takes on average approximately 1 day per hour of shift but with substantial individual differences. Similar observations have been made in rodents. Additionally the NAT rhythm adapts much more rapidly to delaying shifts than to advancing shifts in rats. This is almost certainly true of many other species and is apparent in circadian rhythms in general.

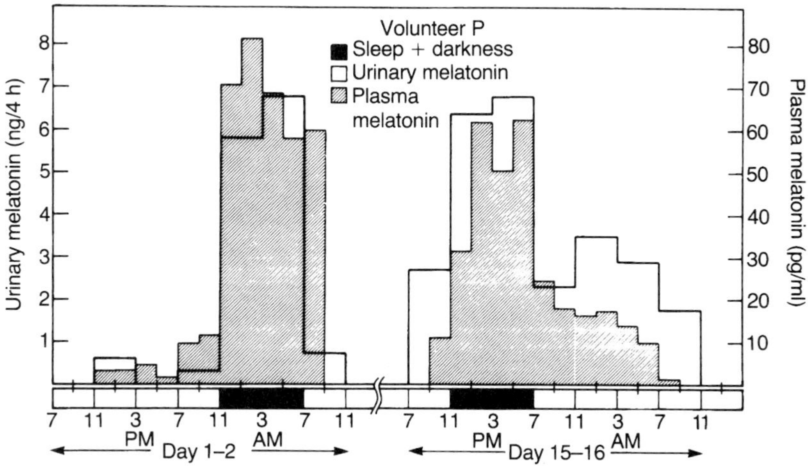

Fig. 4.14 Effect of light–dark phase shift of the human melatonin rhythm as assessed by the measurement of plasma or urinary melatonin after inversion of the light–dark cycle. Integrated 2-hourly plasma samples or sequential 4-hourly urines were obtained over 24-hour periods on days 1 and 2. The light-dark phase was reversed on days 3–4 and adaptation was complete by days 15-16 as illustrated. (Reprinted from *Life Sciences,* **23**, Lynch, H.J. *et al.,* Entrainment of rhythmic melatonin secretion in man to a 12-hour phase shift in the light–dark cycle, 1557–64, © 1978, with kind permission from Elsevier Science Ltd, The Boulevard, Langford Lane, Kidlington OX5 1GB, UK.)

If rodents are maintained in constant darkness (DD) melatonin free runs, closely coupled with the temperature rhythm. A single flash of light of 10 minutes (in hamsters) every 24 hours is sufficient to entrain the melatonin rhythm to a period of 24 hours. Similar results are seen in Soay sheep where one hour of light (160 lux) daily is sufficient for entrainment and for eliciting phase shifts. It has even been reported that 30 seconds of light will entrain activity rhythms in flying squirrels.

4.9 PHASE RESPONSE CURVES

Light pulses are conventionally used to investigate the control of circadian rhythms according to a protocol known as the phase response curve (PRC) developed from observations on activity–rest cycles. Usually animals are maintained in constant darkness, thus displaying free-running rhythms. 'Circadian time' is used as a descriptor in these circumstances, where the period (tau) of the free-running cycle is divided into 24 circadian hours each lasting tau/24 hours of real time. The phase reference point for the rhythm is the daily onset of (in rodents) nocturnal wheel running activity which is designated circadian time (CT)12 and indicates the beginning of subjective night.

For the determination of melatonin phase response curves, the phase reference points that have been used are either the onset of the evening rise in plasma melatonin, the calculated peak time or the morning and evening onset and offset of *N*-acetyltransferase. The phase of the free-running rhythm can be reset by a single light pulse, the magnitude and direction of shift being dependent on the circadian time that light is applied. Phase delays are induced by light in the late subjective day and early subjective night whereas advances are given by light in the middle to late subjective night and early subjective day. A 'dead zone' is present during the subjective day when no shifts can be induced. The direction and magnitude of shift in response to light is summarized as the phase response curve.

This approach has been used extensively in the rat by Illnerova and Vanecek (1982) to investigate the circadian control of melatonin production, but using animals pulsed with one minute of light at night under ambient light–dark cycles and then maintained in DD to assess the phase of the melatonin rhythm. In these circumstances the onset of evening activity and offset of morning activity of *N*-acetyltransferase were used as phase reference points. Quite different shaped phase response curves for the evening rise and the morning decline were obtained and Illnerova has interpreted these observations as indicating separate control of evening onset and morning offset of melatonin by two distinct oscillators – E and M (Figure 4.15).

Further evidence for two oscillators is seen in the patterns of melatonin secretion in 12L 12D in rats when two peaks of secretion are present during the night, and in sheep where a similar phenomenon is present in sufficiently long photoperiod. By shortening the night the phase relationship between the two oscillators is compressed and ultimately the rhythm is abolished in sufficiently short nights.

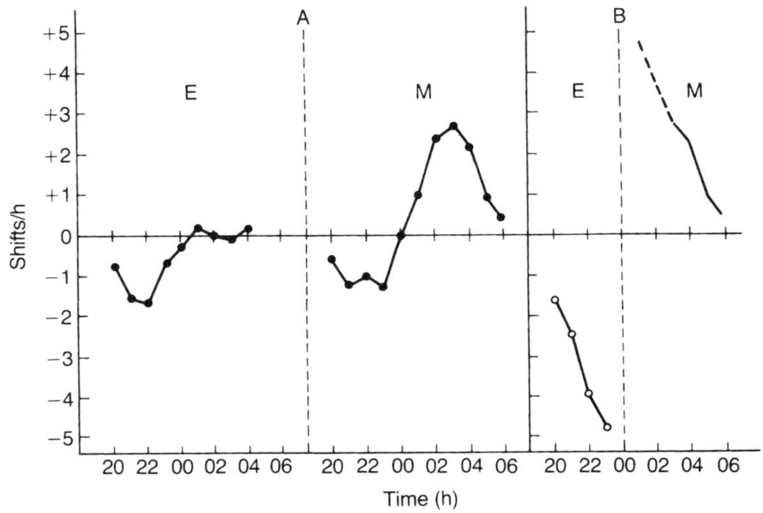

Fig. 4.15 Phase response curves (delay = −h, advance = +h as a function of the timing of the light pulse) of the evening rise (E) or the morning decline (M) of serotonin-*N*-acetyltransferase activity in the rat the following day after one minute light pulses at different times and subsequent continuous darkness. (B) Instantaneous separate phase shifts. E behaves differently from M and Illnerova has proposed that two oscillators (E and M) control the rat melatonin rhythm. (Reproduced with permission from Illnerova, H. and Vanecek, J.; published by Springer- Verlag, 1982.)

4.10 SPECIES VARIATION IN MECHANISMS OF LIGHT CONTROL OF THE MELATONIN RHYTHM

In other species the two oscillator model has not been extensively investigated; however Lewy, Sack and Singer (1984) suggest that in humans the limited amount of information available on the phase shifting effects of light can be explained on the basis of a single oscillator. Certainly there are marked differences in the response to a single light flash at night between species.

In the rat and, according to some authors, in sheep a light pulse during the second half of the night suppresses melatonin for the rest of the night in contrast to earlier times of treatment when, after suppression, melatonin secretion returns to normal in subsequent darkness. In humans, following light suppression even late in the night, levels rebound to night-time values within 40 minutes of a return to darkness, up to the point where the morning decline cannot be distinguished from induced suppression.

It is likely that in species with an endogenous circadian period greater than 24 hours, morning light is most important for entrainment whereas with a tau less than 24 hours evening light is the major effector. In the laboratory rat where tau is usually greater than 24 hours there is good evidence that the melatonin rhythm is locked to morning lights on. Lewy has provided evidence that in humans evening light serves to delay the melatonin onset by suppression whereas morning light is the primary entraining signal. In the Soay sheep, where tau is less than 24 hours, experiments with skeleton light pulses show that evening light is the primary entraining signal, with the duration of secretion being determined by light suppression in the morning in long photoperiods.

4.11 LIGHT PULSE PHASE SHIFT IN HUMANS

A single short daily pulse of light has not to the author's knowledge been investigated in free-running humans with regard to entrainment or phase shifting of melatonin. A week of daily exposure to bright light in the morning advances the onset of the evening melatonin rise (measured in dim light to prevent suppression) by up to 2.5 hours and a week of exposure to the same amount of light in the evening delays the onset by 1 hour even when the sleep–wake cycle is kept constant.

Recent work designed to clarify phase shifting mechanisms in general in humans has used constant routine conditions to evaluate phase and amplitude before and after the use of 2 to 3 daily (approximately 4 to 7 h) applications of bright (8500 lux) light to induce phase shifts. Here, to assess circadian phase before and after treatment, subjects are kept semi-recumbent in constant dim light and awake for up to 72 hours with identical hourly snacks. Such procedures abolish the 'masking' effects of light–dark alterations, sleep, activity, social contacts and intermittent meals and allow a pure assessment of circadian phase and amplitude. Bright light-induced phase shifts in the core temperature rhythm were accompanied by precisely equivalent shifts of the plasma melatonin rhythm (Figure 4.16).

Earlier work had shown equivalent shifts in a number of other

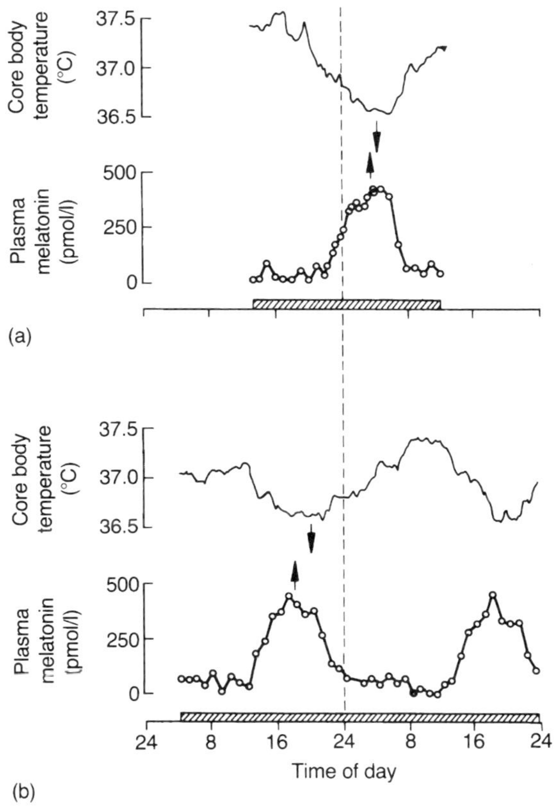

Fig. 4.16 Individual circadian profiles of melatonin and core body temperature before (a) and after (b) exposure to a 3-cycle resetting light stimulus in a 22-year-old man assessed under constant routine conditions. A phase advance shift of the same magnitude and direction in both temperature and melatonin was induced (+8.4 h and +9.8 h respectively). (Reproduced from Shanahan, T.L. and Czeisler, C.A. Light exposure induces equivalent phase shifts on the endogenous circadian rhythms of circulating plasma melatonin and core body temperature in man. *J. Clin. Endocrinol. Metab.* **73**, 227–35, 1991 © The Endocrine Society.)

strongly endogenous rhythmic variables after bright light treatment assessed in unmasked conditions; thus it would appear that resetting the human circadian clock with bright light essentially affects all clock-controlled variables simultaneously. Previous demonstration of the variability amongst different physiological rhythms in amount of shift and rates of entrainment after forced phase shift must be attributed to the prevailing environmental masking effects in different studies. Melatonin is only 'masked' to a major extent by ambient light and provides probably the purest indicator of central biological clock status provided that light exposure is controlled.

Single 3-hour light pulses have been used in free-running humans to investigate phase response curves of the temperature rhythm. Small advances and delays were obtained with appropriately timed bright (5000–9000 lux) light, demonstrating that the human circadian system behaves like other species in this protocol. The shifts were of sufficient magnitude as to entrain a rhythm free running at 25 hours on a daily basis, and presumably represent the major physiological mechanism for daily entrainment in a normal environment in humans. Melatonin was not assessed in this study but the presumption is that it would behave like temperature.

4.12 SKELETON PHOTOPERIODS

'Skeleton' light pulses at dawn and dusk, with otherwise continuous darkness, are read as a full photoperiod by animals and have been used to investigate control of human melatonin rhythms. In Antarctica, specifically on bases with an imposed structure of daily activity, human melatonin rhythms show a more marked tendency to delay in winter compared to summer, than in temperate zones, presumably due to looser coupling of the circadian system to 24 hours in the dim domestic intensity light. One hour of light (2500 lux) in the morning and in the evening for 6 weeks is sufficient to re-establish an earlier phase position comparable to summer (Figure 4.17). Pulses of 500 lux were insufficient, hence we may assume that in these particular conditions of dim light conditioning, the intensity of light experi-

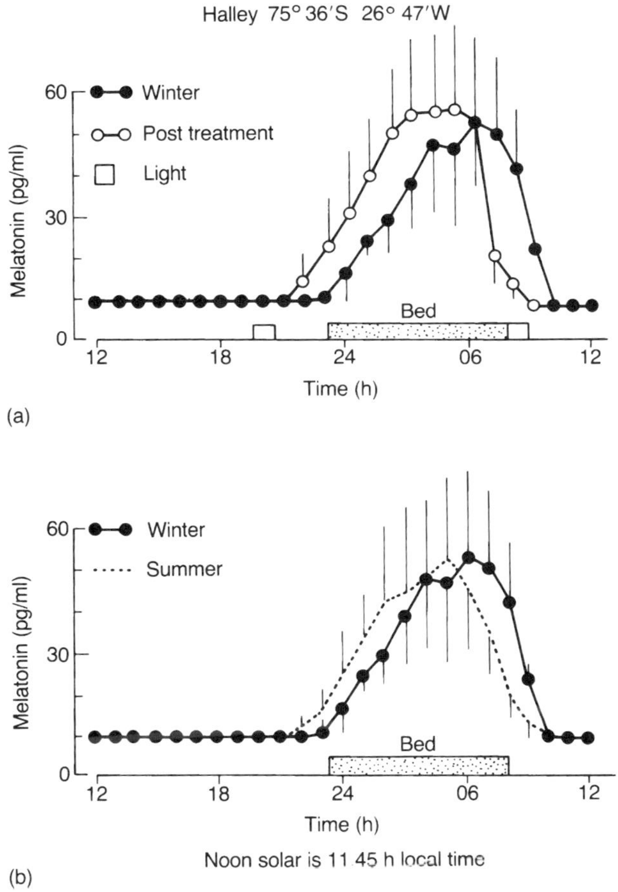

Fig. 4.17 Plasma melatonin rhythm in five normal men (a) at the winter solstice in Antarctica (75 °S) and after exposure to one hour of 2500 lux full spectrum light during the afternoon and the morning for six weeks after the solstice. Bright light induces a phase shift comparable to the phase advance seen in summer (b) compared to winter. (Reproduced with permission from Broadway *et al*; published by Elsevier Science Publishers B.V., Amsterdam, 1987.)

enced in the morning and the evening required to entrain human rhythms lies between these limits.

There is no reason to suppose that the intensity of light required for melatonin suppression and the intensity of light required for entrainment are the same in a particular individual, but they may be related. In hamsters substantially more light is required for entrainment than for suppression.

4.13 BLINDNESS

An unknown proportion of blind people have abnormal circadian rhythms. Reports of free-running sleep–wake, cortisol and melatonin rhythms exist, however to what extent these are representative of the total blind population is not known. In a small number of blind people reporting sleep disturbance and willing to provide body fluids for the assessment of hormones, the majority showed free-running melatonin with tau varying from 23.8 to 25 h. Others had a synchronized rhythm with a delayed phase relationship to the 24-hour day. In other studies normal, free-running and phase-delayed rhythms have been reported (Figure 4.18). It is important to note that the majority of blind people who have been studied were subject to all the time cues in their environment except the light–dark cycle.

Thus it would appear that for some individuals, time cues other than light will maintain synchronization albeit with the weak coupling evident from a delayed phase, whereas for others the light–dark cycle is essential. Whether these differences are genetic, environmental, related to the specific pathology of the blindness or a combination of factors remains an open question. Moreover it is possible that some visually blind people still have hypothalamic light perception via the retinohypothalamic projection, and could use this information for the entrainment of circadian rhythms.

4.14 SOCIAL ZEITGEBERS IN DIM LIGHT

Further evidence for the relative importance of different zeitgebers for the entrainment of melatonin again comes from studies in Antarctica. On the British base of Halley (75 °S) the sun does not rise for three months in the winter and does not set for three months in the summer. The maximum light intensity during the winter is around 500 lux. Sleep logs kept throughout the year (e.g. Figure 4.19) show that the vast majority of personnel on the base of Halley

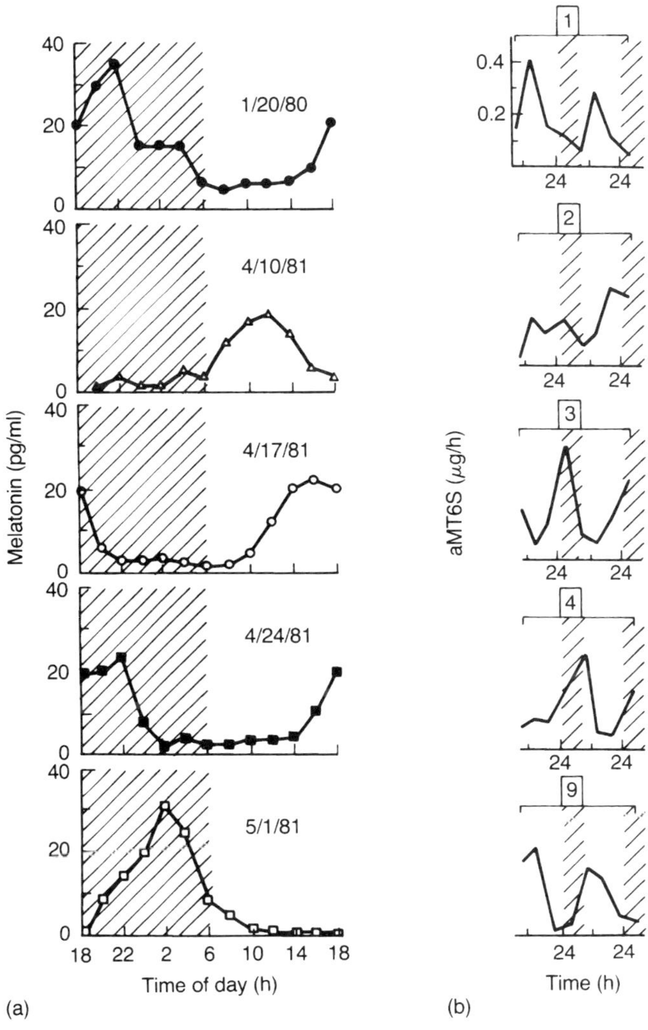

Fig. 4.18 Plasma melatonin rhythm (a) on different dates as shown and urinary 6-sulfatoxymelatonin (aMT6s) rhythm (b) at weekly intervals in two blind subjects. In many blind subjects the melatonin rhythm free-runs usually with a period slightly greater than 24 hours and hence the peak values can occur at any time during a 24-hour cycle. ((a) Reproduced from Lewy, A.J. and Newsome, D.A. Different types of melatonin circadian secretory rhythms in some blind subjects. *J. Clin. Endocrinol. Metab.* **56**, 1103–7, 1983 © The Endocrine Society. Data from Arendt and Aldhous, 1988, previously unpublished figure (b).)

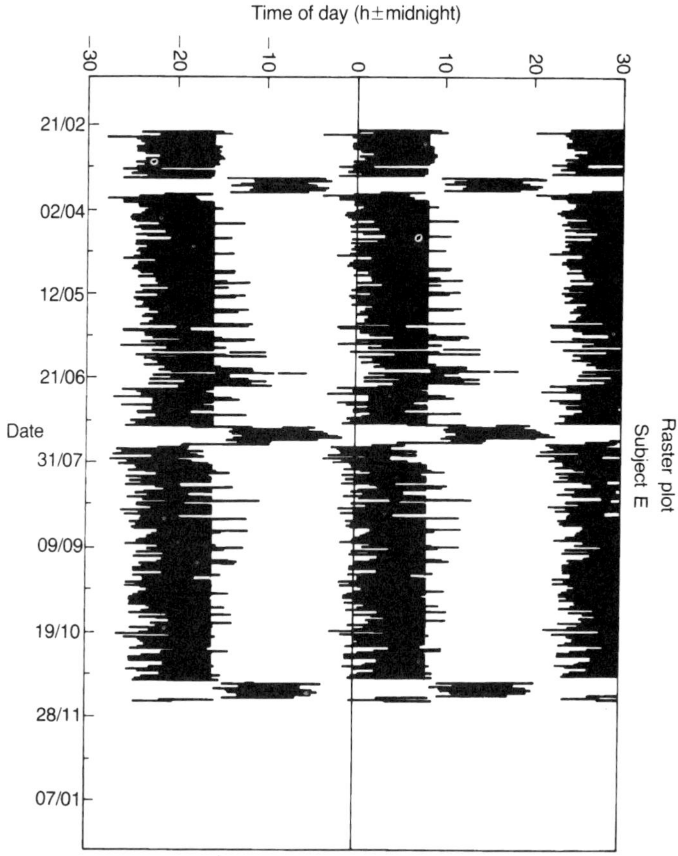

Fig. 4.19 Sleep record (sleep indicated by the horizontal lines, triple plotted, i.e. each day repeated across the page) of one subject living on the British Antarctic Survey Base of Halley (75 °S) for a complete year. Light levels do not exceed 500 lux in winter for 3 months when the sun does not rise. Three periods of night shift are evident. Note that subjects on this British Base are generally synchronized to 24 hours but with a stronger tendency to delay during the winter compared to summer. Melatonin and cortisol rhythms behave similarly. A 24-hour behavior pattern is imposed on this Base. (Data from O'Conor, 1993, O'Conor and Arendt, previously unpublished figure.)

remain synchronized to 24 hours at all times. Both melatonin and cortisol also show consistent 24-hour variations in blood and urine during the Antarctic year. In view of the low intensity light in winter the prediction was that rhythms would desynchronize from the

24-hour day and 'free-run' at their own endogenous period. In fact it is rare to observe such a phenomenon on British bases. Rhythms remain entrained at all seasons with, however, a clear delay in the timing of sleep at weekends especially in the winter, and a delay in the timing, and exceptionally a small increase in the duration, of melatonin in winter compared to summer, similar to changes found in temperate latitudes but of greater magnitude. Comparable but less robust, i.e. non-significant, seasonal variations are seen in cortisol secretion.

In contrast, Kennaway and Van Dorp (1991) have recently reported a study of four Greenpeace volunteers overwintering in Antarctica, all of whom showed free-running sleep–wake and melatonin

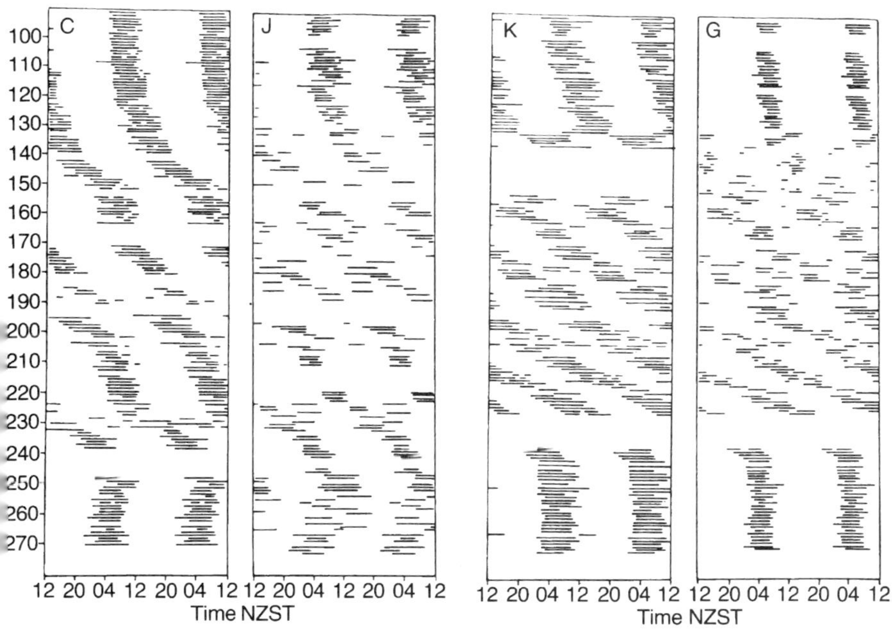

Fig. 4.20 Sleep records (horizontal lines, single plotted) for four subjects living on a Greenpeace Base in Antarctica from February to October. Light levels did not exceed 1000 lux in winter but no behavioral restraints were imposed. Note free-running sleep–wake (and melatonin and cortisol, not shown) rhythms in all subjects during the winter period (sundown April 21st to August 25th). (Reproduced with permission from Kennaway, D.J. and Van Dorp, C.F.; published by the American Physiological Society, 1991.)

rhythms throughout the period of sundown (Figure 4.20). This group were aware of clock time, maintained daily radio contact (a social zeitgeber) and even intermittent personal contact with other bases, however their daily routine was not otherwise structured. They were free to sleep and eat as they wished, in contrast to British bases where the daily routine is structured.

Some interesting conclusions can be drawn comparing these studies. Evidently in the absence of a strong light–dark cycle behavioral impositions can serve to maintain the integrity of the circadian system. Without this stricture, daily time cues and knowledge of clock time do not suffice. The delayed, but entrained, rhythms in winter reflect the weak nature of the time cues. In such circumstances coupling between the pacemaker (rhythm generating system) and the zeitgeber is loosened.

Perhaps a good analogy is towing a car with a tow rope: if the rope weakens and stretches the towed vehicle lags further behind than with a strong new rope. As human circadian rhythms have a natural tendency to delay, the weaker the coupling the greater the delay phase relationship with the 24-hour cycle up to the point when the coupling is lost altogether (i.e. the rope breaks) and free-running rhythms are displayed. The best example of free-run found on Halley was in the sleep–wake records of a base member who, by virtue of his occupation, was not obliged to live the same routine as the rest of the Base. In this case free-running rhythms were seen throughout most of the year.

For most purposes the imposed social routine enforces synchronization in combination with the dim light–dark zeitgeber and undoubtedly contributes to well-being and performance. The neural and/or biochemical mechanisms whereby imposed behavior influences the circadian system remain to be determined.

4.15 NON-24-HOUR LIGHT–DARK CYCLES

The limits of entrainment of melatonin by light–dark have been explored in humans kept in environmental isolation. Imposition of a light–dark cycle (200–300 lux) steadily increasing in length from 24 to 32 hours by 10 minutes a day leads to a phenomenon known

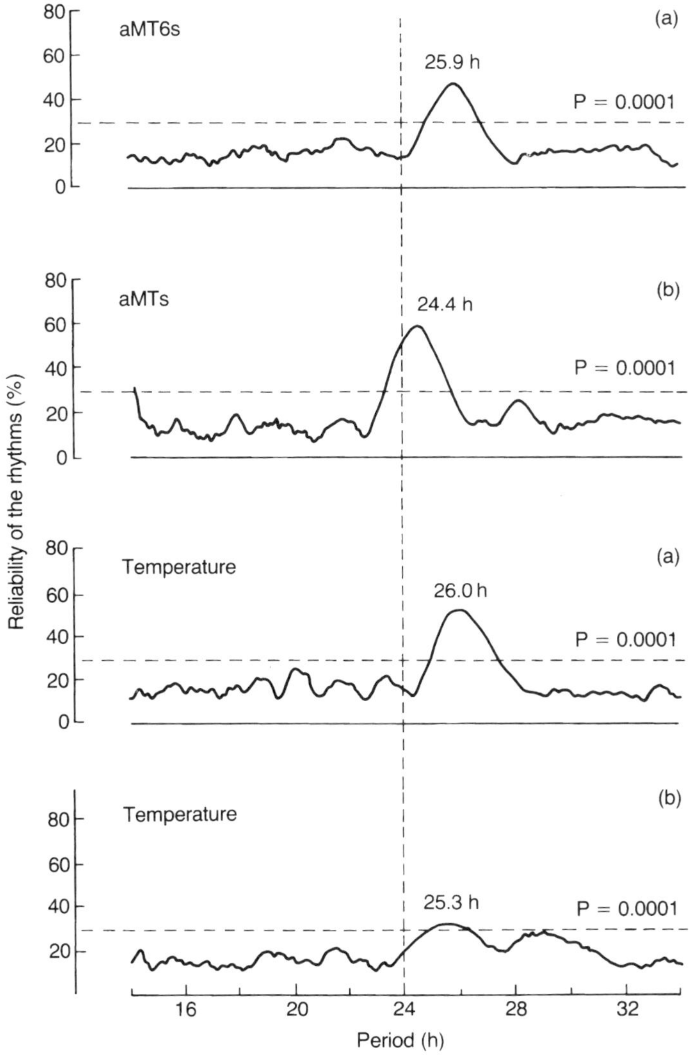

Fig. 4.21 Melatonin (assessed by urinary 6-sulfatoxymelatonin – aMT6s) free-runs with a similar period to core temperature. Fractional desynchronization of circadian rhythms in one volunteer occurs by increasing imposed light–dark (approx. 300 lux) cycle length from 26 h to 29 h. Cycle lengths up to 27 h (a) and from 27 to 29 h (b) are grouped separately. Periodogram analysis according to Wever (1979). Sleep–wake remains entrained up to 29 h (not shown) but core body temperature and melatonin free-run with imposed daylength greater than around 27 h. Note that the average period of the aMT6s rhythm (25.9 h) in (a) indicates entrainment to the light–dark cycle and that the average period (24.4 h) in (b) indicates desynchronization from the imposed 27–29-h daylength. Reliability of the rhythm >30% is equivalent to $p<0.0001$ (Wever, 1979). If the light intensity is increased to 3000 lux all variables remain entrained to the zeitgeber. (Reproduced with permission from Arendt *et al.*, 1985.)

as fractional desynchronization. The sleep–wake cycle remains entrained up to 32 hours but more strongly endogenous variables such as temperature, cortisol and melatonin uncouple from the zeitgeber at different imposed daylengths. Melatonin breaks away at approximately 27 hours and free-runs with a tau identical to that of temperature (Figure 4.21). If however, bright light (3000 lux) is used, Wever and coworkers (1983) have shown that all the variables remain coupled for up to 32 hours and the upper limit of entrainment has yet to be fully determined with this intensity of light.

Non-24-hour light–dark cycles have also been used to investigate control of melatonin secretion in animals. Soay sheep will entrain to 23-hour and 25-hour cycles (160 lux, 16 hours of light) and Suffolk sheep will entrain to 26 and 22-hour symmetrical light–dark cycles (200 lux, 13L 13D and 11L 11D).

Length of the light phase of a light–dark cycle also plays a role in entrainment for a given intensity of light. Suffolk sheep, for example, will entrain fully to 22-hour photoperiods of 18L 6D but only with a phase delay to 6L 18D.

4.16 RELATIVE NATURE OF THE LIGHT–DARK CYCLE

It is clear that light perception is intensity dependent. However, in terms of melatonin secretion, what is read as darkness and what is read as light are not absolute values. Rats kept in cycles of dim light and darkness perceive dim light as the light phase, however if the same animals are kept in dim light (of the same intensity) and bright light, the dim light is perceived as darkness (Figure 4.22). Therefore it is the relative change in intensity which is important, not the absolute intensity. Similar considerations arise in the perception of photoperiod length (see Chapter 5).

4.17 LIGHT CONTROL OF THE PRODUCTION OF OTHER PINEAL INDOLES

Serotonin rhythms in the pineal were the first to be shown to be controlled by light–dark cycles, with high concentrations during the

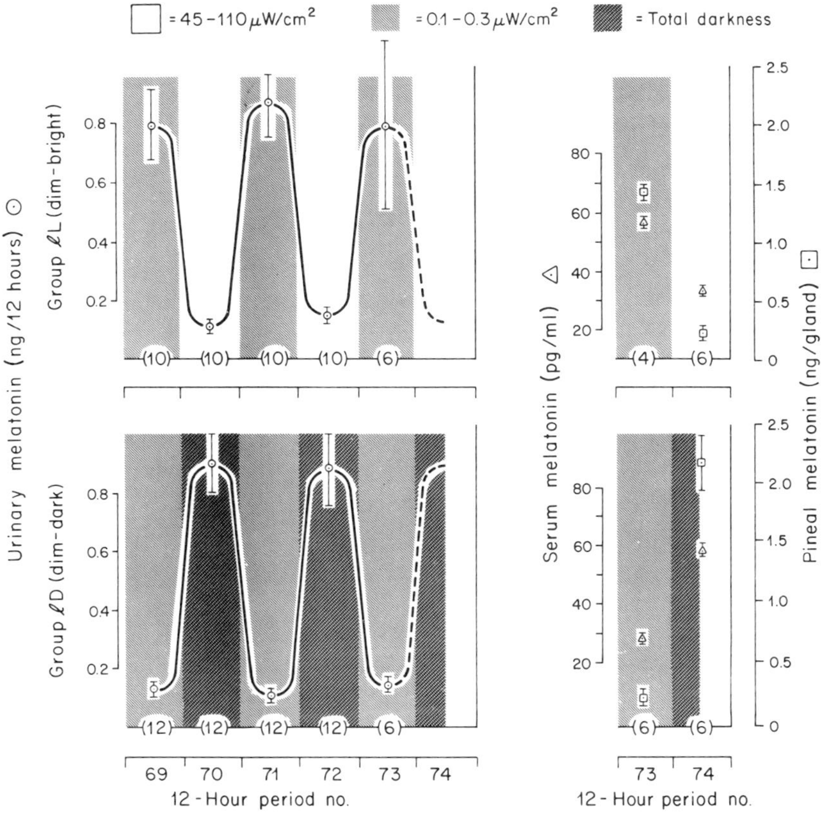

Fig. 4.22 Relative perception of light intensity as shown by urinary melatonin excretion in 10 (top panel) and 12 (bottom panel) rats. Animals kept in dim light and darkness will perceive dim light as the light phase, i.e. melatonin is high in darkness. If they are kept in the same intensity of dim light and bright light, the dim light will be perceived as darkness as shown by high melatonin excretion. (Reproduced with permission from Lynch *et al.*; published by S. Karger AG, Basel, 1981.)

day and low at night. Inversion of the lighting cycle inverts the rhythm. The rhythm is endogenous and persists in constant darkness and in blinded animals. Neural control operates through the noradrenergic receptor, but, to an as yet undetermined extent the consumption of serotonin during the night-time synthesis of melatonin undoubtedly contributes to the decline at night. Hence those factors discussed previously which act on melatonin secretion will also affect serotonin content of the pineal. Certainly light will increase its concentration rapidly (within 40 minutes) in the pineal at night. As serotonin specifically derived from the pineal cannot be measured peripherally there is lack of detailed information in this area.

5-Methoxytryptophol and 5-methoxytryptamine also show marked day–night variations in the pineal gland which are the opposite of those of melatonin, i.e. low at night. Both respond to change of duration of the photoperiod with a longer period of synthesis in short photoperiods. 5-Methoxytryptophol has also been measured in plasma of sheep, humans and hamsters (where the day–night variations correspond to those in the pineal). Entrainment by light–dark has been shown in hamsters but detailed information comparable to that of melatonin is not available. In view of the observed patterns in these two methoxyindoles it is likely that their production is determined largely by serotonin concentrations in the pineal.

REFERENCES

Arendt, J. (1985) Mammalian pineal rhythms. *Pineal Research Reviews,* *3*, 161–213.

Arendt, J. (1986) Role of the pineal gland and melatonin in seasonal reproductive function in mammals. *Oxford Review of Reproduction Biology*, **8**, 266–320.

Arendt, J. and Aldhous, M. (1988) Synchronization of a disturbed sleep–wake cycle in a blind man by melatonin treatment. *Lancet* **i**, 772–3.

Binkley, S. (1988) *The Pineal: Endocrine and Nonendocrine Function.* Prentice Hall, Englewood Cliffs, New Jersey, USA.

Bojkowski, C., Aldhous, M., English, J. *et al.* (1987) Suppression of nocturnal plasma melatonin and 6-sulphatoxymelatonin by bright and dim light in man. *Hormone and Metabolic Research*, **19**, 437–40.

Brainard, G.C., Lewy, A.J., Menaker, M. *et al.* (1985) Effect of light wavelength on the suppression of nocturnal plasma melatonin in normal volunteers. *Annals of the New York Academy of Sciences*, **453**, 376–8.

Broadway, J., Arendt, J. and Folkard, S. (1987) Bright light phase shifts the human melatonin rhythm during the Antarctic winter. *Neuroscience Letters*, **79**, 185–9.

Illnerova, H. (1988) Entrainment of mammalian circadian rhythms in melatonin production by light. *Pineal Research Reviews*, **6**, 173–217.

Illnerova, H. and Vanecek, J. (1979) Response of rat pineal serotonin *N*-acetyltransferase to one minute light pulse at different night times. *Brain Research*, **167**, 431–4.

Illnerova, H. and Vanecek, J. (1982) Complex control of the circadian rhythm in *N*-acetyltransferase activity in the rat pineal gland, in *Vertebrate Circadian Systems* (eds J. Aschoff, S. Daan and G. Groos), Springer-Verlag, Heidelberg, Berlin, pp. 285–96.

Kennaway, D.J. and Van Dorp, C.F. (1991) Free running rhythms of melatonin, cortisol, electrolytes and sleep in humans in Antarctica. *American Journal of Physiology*, **260**, R1137–1144.

Klein, D.C. and Moore, R.Y. (1979) Pineal *N*-acetyltransferase and hydroxyindole-*O*-methyl transferase: control by the retino-hypothalamic tract and the suprachiasmatic nucleus. *Brain Research*, **174**, 245–62.

Lewy, A.J. and Newsome, D.A. (1983) Different types of melatonin circadian secretory rhythms in some blind subjects. *Journal of Clinical Endocrinology and Metabolism*, **56**, 1103–7.

Lewy, A.J., Sack, R.L. and Singer, C.M. (1984) Assessment and treatment of chronobiologic disorders using plasma melatonin levels and bright light exposure, the clock-gate model and the phase response curve. *Psychopharmacology Bulletin*, **20**, 561–5.

Lewy, A.J., Wehr, T.A., Goodwin, F.K. *et al.* (1980) Light suppresses melatonin secretion in humans. *Science*, **210**, 1267–9.

Lynch, H.J., Jimerson, D.C., Ozaki, Y. *et al.* (1978) Entrainment of rhythmic melatonin secretion in man to a 12-hour phase shift in the light–dark cycle. *Life Sciences*, **23**, 1557–64.

Lynch, H.J., Rivest, R.,Ronsheim, P.M. *et al.* (1981) Light intensity and the control of melatonin secretion in rats. *Neuroendocrinology*, **33**, 181–5.

Menaker, M. and Wisner, S. (1983) Temperature compensated circadian clock in the pineal of Anolis. *Proceedings of the National Academy of Sciences, USA*, **80**, 6119–21.

Moore, R.Y. and Klein, D.C. (1974) Visual pathways and the central neural control of a circadian rhythm in pineal serotonin *N*-acetyltransferase. *Brain Research*, **71**, 17–33.

Moore-Ede, M.C., Czeisler, C.A. and Richardson, G.S. (1983) Circadian timekeeping in health and disease. Part 1. Basic properties of circadian pacemakers. *New England Journal of Medicine*, **309**, 469–76.

Poggeler, B., Balzer, I., Hardeland, R. *et al.* (1991) Pineal hormone melatonin oscillates also in the dinoflagellate *Gonyaulax polyedra. Naturwissenschaften*, **78**, 268–9.

Redman, J. (1988) The effect of exogenous melatonin on rat circadian rhythms. PhD thesis, Department of Psychology, School of Behavioral Sciences, La Trobe University, Australia.

Reiter, R.J. (1985) Action spectra, dose-response relationships and temporal aspects of light's effects on the pineal gland, in *The Medical and Biological Effects of Light* (eds R.J. Wurtman, M.J. Baum and J.R. Potts, Jr.), *Annals of the New York Academy of Sciences*, **453**, 215–30.

Reiter, R.J. (1992) Alterations of the circadian melatonin rhythm be the electromagnetic spectrum: a study in environmental toxicology. *Regulatory Toxicology and Pharmacology*, **15**, 226–44.

Rollag, M.D. Panke, E.S. and Reiter, R.J. (1980) Pineal melatonin content in male hamsters throughout the seasonal reproductive cycle. *Proceedings of the Society of Experimental Biological Medicine*, **165**, 330–40.

Shanahan, T.L. and Czeisler, C.A. (1991) Light exposure induces equivalent phase shifts of the endogenous circadian rhythms of circulating plasma melatonin and core body temperature in man. *Journal of Clinical Endocrinology and Metabolism*, **73**, 227–35.

Takahashi, J., Hamm, H.E. and Menaker, M. (1980) Circadian rhythms of melatonin release from individual superfused chicken pineal glands *in vitro*. *Proceedings of the National Academy of Sciences, USA*, **77**, 2319–22.

Underwood, H. and Goldman, B.D. (1987) Vertebrate circadian and photoperiodic systems: role of the pineal gland and melatonin. *Journal of Biological Rhythms*, **2**, 279–315.

Wehr, T.A. (1991) The durations of human melatonin secretion and

sleep respond to changes in daylength (photoperiod). *Journal of Clinical Endocrinology and Metabolism*, **73**, 1276–80.
Wever, R.A. (1979) The circadian system of man: results of experiments under temporal isolation. Springer Verlag, New York.
Wever, R.A. (1986) Characteristics of circadian rhythms in human functions, in *Melatonin in Humans* (eds R.J. Wurtman and F. Waldhauser), *Journal of Neural Transmission*, Suppl. 21, 323–74.
Wever, R.A., Polasek, J. and Wildgruber, C.M. (1983) Bright light affects human circadian rhythms. *Pflugers Archives*, **396**, 85–7.

FURTHER READING

Arendt, J., Aldhous, M., English, J. *et al.* (1987) Some effects of jet lag and their alleviation by melatonin. *Ergonomics,* **30**, 1379–93.
Arendt, J., Bojkowski, C., Folkard, S. *et al.* (1985) Some effects of melatonin and the control of its secretion in man, in *Photoperiodism, Melatonin and the Pineal* (eds D. Evered and S. Clark), Ciba Foundation Symposium 117. Pitman, London, pp. 266–83.
Arendt, J. and Ravault, J.P. (1988) Suppression of melatonin secretion in Ile de France rams by different light intensities. *Journal of Pineal Research*, **5**, 245–50.
Beck - Friis, J., von Rosen, D., Kjellman, B.F. *et al.* (1984) Melatonin in relation to body measures, sex, age, season and the use of drugs in patients with major affective disorders and healthy subjects. *Psychoneuroendocrinology*, **9**, 261–77.
Brainard, G.C., Lewy, A.J., Menaker, M. *et al.* (1988) Dose-response relationship between light irradiance and the suppression of plasma melatonin in human volunteers. *Brain Research*, **454**, 212–8.
Cassone, V. (1990) Effects of melatonin on vertebrate circadian systems. *Trends in Neuroscience*, **13**, 457–63.
Deveson, S.L., Arendt, J. and Forsyth, I.A. (1990) Sensitivity of goats to a light pulse during the night as assessed by suppression of melatonin concentrations in the plasma. *Journal of Pineal Research*, **8**, 169–77.
Elliot, J.A. (1981) Circadian rhythms, entrainment and photoperiodism in the Syrian hamster, in *Biological Clocks in Seasonal*

Reproductive Cycles (eds B.K. Follett and D.E. Follett), Wright, Bristol, pp. 203–18.

English, J., Arendt, J., Symons, A.M. *et al.* (1988) Pineal and ovarian response to 22 hour and 24 hour days in the ewe. *Biology of Reproduction*, **4**, 351–8.

Hoffman, K. (1981) The role of the pineal gland in the photoperiodic control of seasonal cycles in hamsters, in *Biological Clocks in Seasonal Reproductive Cycles* (eds B.K. Follett and D.E. Follett), Pitman Press, Bath, pp. 237–50.

Illnerova, H., Vanecek, J. and Hoffman, K. (1987) Adjustment of the rat pineal β-acetyltransferase rhythm to eight-hour shifts of the light–dark cycle: advance of the cycle disturbs the rhythm more than delay. *Brain Research*, **417**, 167–171.

Illnerova, H., Zvolsky, P. and Vanecek, J. (1985) The circadian rhythm in plasma melatonin concentration of the urbanised man, the effect of summer and winter time. *Brain Research*, **328**, 186–9.

Jimerson, D.C., Lynch, H.J., Post, R.M. *et al.* (1977) Urinary melatonin during sleep deprivation in depressed patients and normals. *Life Sciences*, **20**, 1501.

Klein, D.C., Moore, R.Y. and Reppert, S.M. (eds) (1992) *Suprachiasmatic Nucleus – The Minds Clock*, Oxford Science Publications, Oxford University Press, Oxford, UK.

Klein, D.C., Smoot, R., Weller, J.L. *et al.* (1983) Lesions of the paraventricular nucleus area of the hypothalamus disrupt the suprachiasmatic spinal cord circuit in the melatonin rhythm generating system. *Brain Research Bulletin*, **10**, 647–52.

Lewy, A.J., Sack, R.L., Miller, L.S. *et al.* (1987) Anti-depressant and circadian phase-shifting effects of light. *Science*, **235**, 352–4.

Lewy, A.J., Wehr, T.A., Goodwin, F.K. *et al.* (1981) Manic depressive patients may be supersensitive to light. *Lancet* **i,** 383–4.

Lincoln, G.A., Ebling, F.J.P. and Almeida, O.F.X. (1985) Generation of melatonin rhythms, in *Photoperiodism, Melatonin and the Pineal* (eds D. Evered and S. Clark), CIBA Foundation Symposium 117, Pitman, London, pp. 129–41.

McIntyre, I.M., Norman, T.R., Burrows, G.D. *et al.* (1989) Human melatonin response to light at different times of the night. *Psychoneuroendocrinology*, **14**, 187–193.

Menaker, M., Hudson, D.J. and Takahashi, J.S. (1981) Neural and endocrine components of circadian clocks in birds, in *Biological Clocks in Seasonal Reproductive Cycles* (eds B.K. Follett and D.E. Follett), Wright, Bristol, pp. 171–83.

Minors, D.S., Waterhouse, J.M. and Wirz-Justice, A. (1991) A human phase response curve to light. *Neuroscience Letters*, **133**, 36–40.

Okudaira, N., Kripke, D.F. and Webster, J.B. (1983) Naturalistic studies of human light exposure. *American Journal of Physiology*, **245**, R613–615.

Owen, J. and Arendt, J. (1992) Melatonin suppression in human subjects by bright and dim light in Antarctica: time and season dependent effects. *Neuroscience Letters*, **137**, 181–4.

Quay, W.B. (1963) Circadian rhythm in rat pineal serotonin and its modifications by estrous cycle and photoperiod. *General and Comparative Endocrinology*, **3**, 473–9.

Ralph, M.R., Foster, R.G., Davis, F.C. *et al.* (1990) Transplanted suprachiasmatic nucleus determines circadian period. *Science*, **247**, 975–8.

Rosenthal, N.E., Sack, D.A., Gillin, J.C. *et al.* (1984) Seasonal affective disorder. A description of the syndrome and preliminary findings with light therapy. *Archives of General Psychiatry*, **41**, 72–9.

Rusak B. and Zucker I. (1979) Neural regulation of circadian rhythms. *Physiology Reviews*, **59**, 449–526.

Thompson, C., Stinson, D. and Smith, A. (1990) Seasonal affective disorder and season-dependent abnormalities of melatonin suppression by light. *Lancet,* **ii**, 703–6.

Underwood, H. (1977) Circadian organisation in lizards: the role of the pineal organ. *Science*, **195**, 587–9.

Vollrath, L., Seidel, A., Huesgen, A. *et al.* (1989) One millisecond of light suffices to suppress night-time melatonin synthesis in rats. *Neuroscience Letters*, **98**, 297–8.

Wurtman, R.J., Axelrod, J. and Phillips, L.S. (1963) Melatonin synthesis in the pineal gland: control by light. *Science*, **142**, 1071–3.

Zimmerman, N.H. and Menaker, M. (1979) *Proceedings of the National Academy of Sciences, USA*, **76**, 999–1003.

5 Physiology of the pineal: role in photoperiodic seasonal functions

Contents

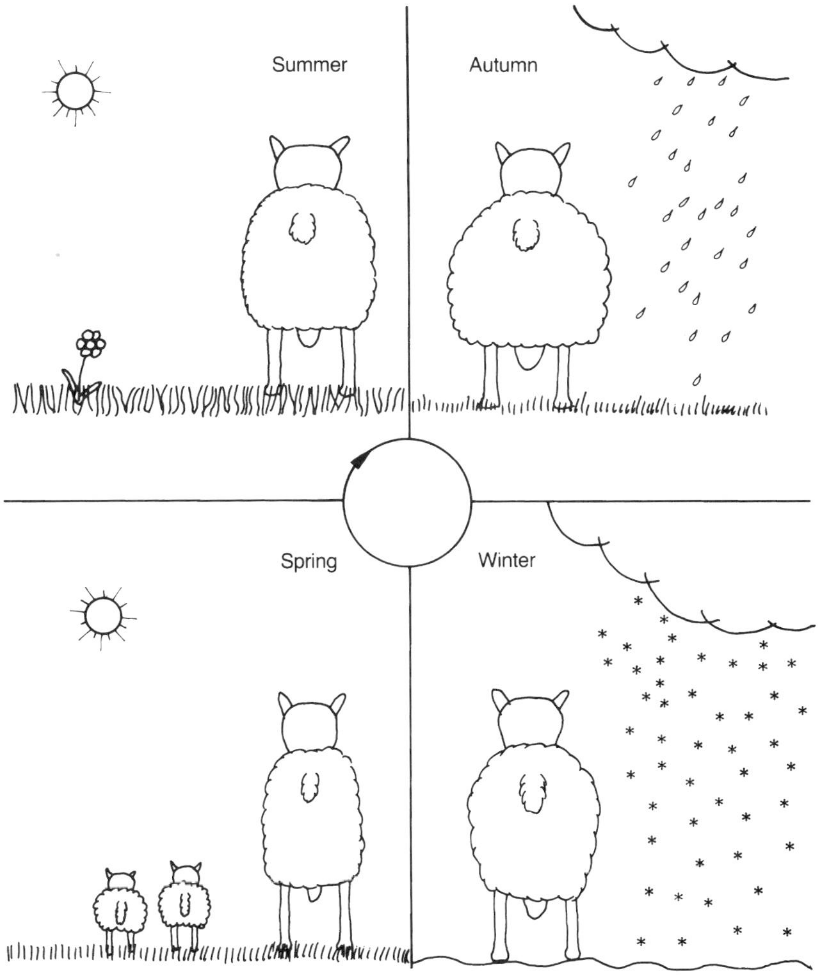

Fig. 5.1 Cartoon illustrating the seasonal cycle in sheep. Summer: reproductive quiescence (anestrous), good food availability, summer coat; autumn: growth of gonads, onset of estrous cycles, mating, pregnancy, growth of winter coat; winter: scarcity of food, seasonal inappetance (in some species), continuing gestation; spring: birth of young, lactation, loss of winter coat, gonadal involution and anestrus. (Kindly provided by Dr Fiona Gebbie, Scottish Agricultural College.)

5.1 SEASONALITY AND DAYLENGTH

Most species show seasonal variations in their physiology and behavior, even humans. The reproductive cycle, for example, is timed so that environmental conditions are propitious for the growth of the young, and variations in behavior, pelage (coat growth and colour), appetite, body weight and fat are such that survival in ambient temperature conditions is optimized and camouflage protects against predators (Figure 5.1).

When seasonal functions are primarily timed by daylength species are referred to as photoperiodic. According to gestation time, mating takes place in spring or summer (for example, hamsters, voles, ferrets, horses) or in autumn or winter (for example sheep, goats, deer). Some species, for example mink and badgers, time the birth of young by delayed implantation and others, for example wallabies, by embryonic diapause. These mechanisms are also influenced by photoperiod. In all photoperiodic mammals studied to date there is evidence that the pineal gland, via melatonin secretion, transmits daylength information for the timing of seasonal functions. Hamsters (long day breeders) and sheep (short day breeders) have been the most intensively investigated. Observations in these species have laid the foundations of our present understanding of photoperiodic time measurement (PTM). The conclusions derived are broadly applicable, with of course some variation in detail both between and within species. Most information is available concerning the reproductive system but other systems will be mentioned briefly. These phenomena clearly involve specific timing mechanisms and anticipation of future events.

5.2 PHOTOPERIODIC TIME MEASUREMENT

The most reliable environmental time cue in non-equatorial regions is the annual change in daylength or photoperiod (Figure 5.2). It is possible to drive the entire sequence of seasonal physiology in photoperiodic species using alternating periods of artificial short days and long days without controlling other variables such as temperature. It is even possible to compress the annual cycle (within limits) by using artificial photoperiod.

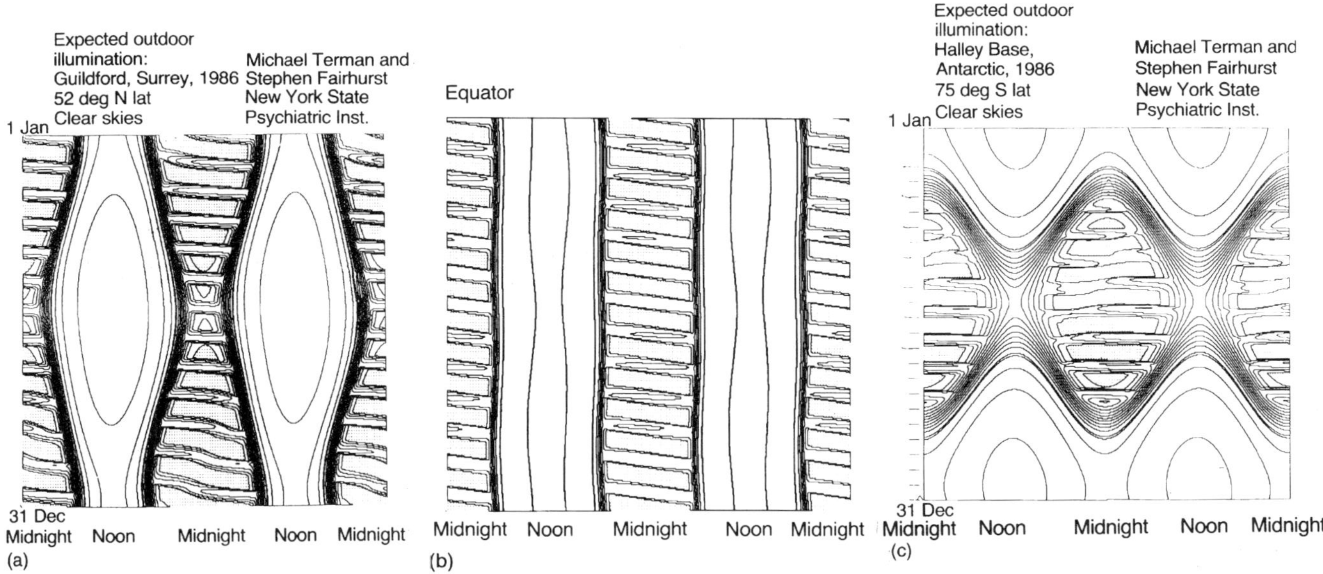

Fig. 5.2 Precision of daylength: calculated light intensity at the equator and for comparison 52 °N and 75 °S, double plotted, i.e. two days are shown across the horizontal axis. The vertical axis is a calendar beginning at the top with 1 January and ending on 31 December. The stippled area is the starlight minimum and is traversed by moonlight shown as irregular horizontal bars of very low light between sunset and sunrise. Equal step log increments in light intensity are shown. The shaded period indicates the sun's presence over the horizon. Kindly provided by Dr Michael Terman and Stephen Fairhurst, Copyright 1986, Research Foundation for Mental Hygiene Inc., New York State Psychiatric Institute. A computer-driven illumination system, outputting this algorithm, with controls over latitude, date and date progression, sunlight/moonlight options, attenuation factors, time base and daily segment(s) illuminated is available from Medic-Light Inc. (Lake Hopatcong, NJ 07849, USA). Also available, a simplified model for bedroom use, with self-contained microprocessor.

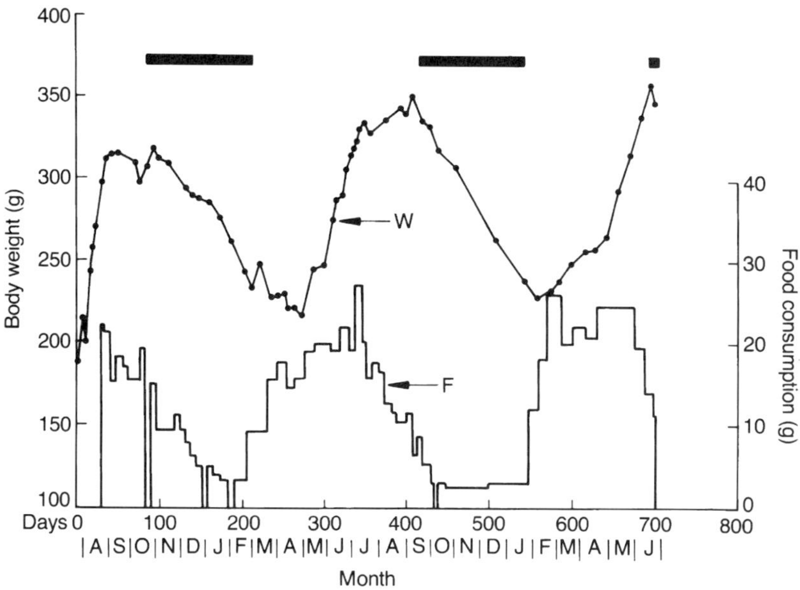

Fig. 5.3 Endogenous annual cycles: body weight (W), mean daily food consumption (F), hibernating periods (solid horizontal bars), active periods (clear spaces) of an individual *Citellus lateralis* observed daily for two years in constant conditions of 22 °C and 12L:12D, *ad libitum* food and water. (Reproduced with permission from Pengelley, E. and Asmundson, S., published by Academic Press, Inc, 1974.)

Photoperiod may be the immediate cause of a particular seasonal response, however in many long-lived species of mammals, and indeed lower vertebrates and some invertebrates, there is good evidence for endogenous circannual rhythms of seasonal functions (Figure 5.3), which are synchronized with the year by seasonal zeitgebers such as photoperiod and temperature. In general photoperiod is the major, and in some cases the only essential factor for timing of seasonal functions.

In a natural environment interactions between photoperiod and other seasonal factors such as temperature and food availability occur such as to reinforce timing effects. The extreme precision of the Syrian hamster's testicular response to photoperiod is illustrated

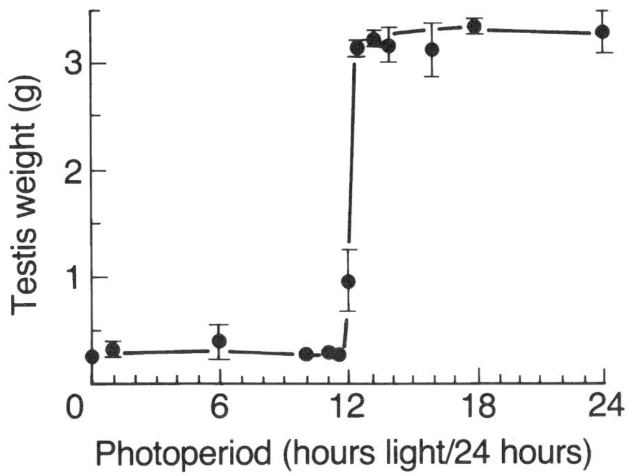

Fig. 5.4 Precision of photoperiodic time measurement in Syrian hamsters illustrated in a photoperiodic response curve. Each point represents the mean (±SEM) paired testis weight of a group of hamsters exposed to the indicated photoperiod for the previous 12–14 weeks. The minimum photoperiod required for maintenance of gonadal activity is 12.5 h. (Reproduced with permission from Elliot, J.A.; published by Butterworth Heinemann, 1981.)

in Figure 5.4 and the modification of this response in the Djungarian (Siberian) hamster by ambient temperature is shown in Figure 5.5. It is likely that, in addition to species variations, the importance of different factors is relative. In the absence or diminution of one factor (for example daylength through blinding) other factors acquire greater importance. Interactions with members of a social group play a role. It has been convincingly shown, for example, that a sighted ram can set the seasonal phase of a group of ewes deprived of photoperiod perception.

With the exceptions of the summer and winter solstices all daylengths occur twice a year. Thus a distinction must be made between increasing and decreasing daylength for the appropriate seasonal response to occur. There is now evidence for the ability to measure the direction of photoperiod change.

Photorefractoriness, or loss of an appropriate photoperiodic

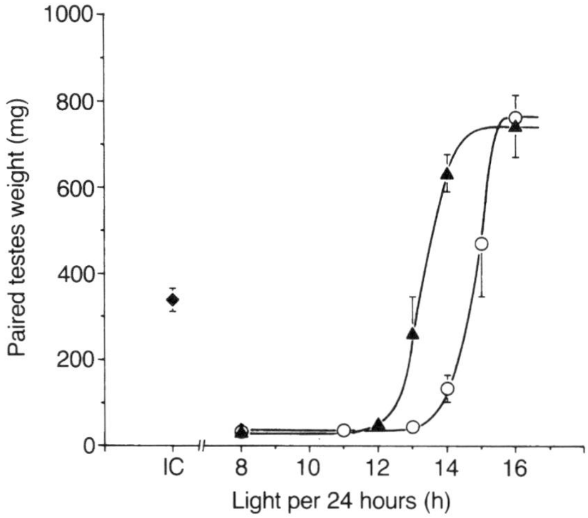

Fig. 5.5 Temperature dependency of the critical photoperiod for gonadal development in the Djungarian hamster *Phodopus sungorus*. Weights of testes from animals housed at either 22 °C (closed symbols) or 5 °C (open symbols) for 50 days. The animals were raised to 25 days of age in 16L 8D and subsequently transferred to the different photoperiod as indicated on the abscissa. Note that all animals kept under 16 h light develop active gonads irrespective of the ambient temperature. (Reproduced with permission from Steinlechner, S. and Niklowitz, P.; published by Elsevier, 1992.)

response is a feature of photoperiodicity in many species. At its simplest this involves, for example, the loss of a long day response after sufficient duration of exposure coupled with an absolute requirement for a period of shorter days before a long day response can again be induced. Photorefractoriness can avoid the problem of perceiving direction of change of daylength. Days are either long or short and the duration of exposure determines the response. However this raises the question as to how the duration of exposure is measured.

In addition to timing seasonal functions, photoperiod is often critical for the timing of pubertal development. In general puberty is only reached during the adult mating season. To this end photoperiod perception is possible *in utero* by the fetus, and can establish the time course of postnatal development.

5.3 THEORIES OF PHOTOPERIODIC TIME MEASUREMENT

At present two general theories are proposed to account for photoperiodic time measurement. Either daylength information may be accumulated over each period of 24 hours and functions as an interval timer (hourglass) or circadian rhythms are involved in the interpretation of daylength, as originally proposed by Bunning (1967). In the latter case two models – 'external coincidence' and 'internal coincidence'– have been developed.

External coincidence proposes the existence of a daily rhythm in photosensitivity, entrained to the light–dark cycle and that in an appropriate daylength the photosensitive part of the cycle would be illuminated, thus inducing the particular response (Figure 5.6). The model has been extremely productive but is complicated by the possibility of a shift in phase position of the sensitive part of the cycle in different daylengths, particularly as both dawn and dusk have been invoked as reference points.

Internal coincidence proposes that internal circadian rhythms change their phase relationship with one another in the course of changing daylength and that specific relationships are photoinductive and others are not (Figure 5.6).

The models are not necessarily mutually exclusive. There is evidence in hamsters, voles and rams for the external coincidence model of photoperiodic time measurement. Elliot (1981) showed that in the hamster 1 hour of light per circadian cycle will induce a short or a long day response depending on whether light coincides with a particular phase of the hypothetical sensitivity rhythm. Skeleton photoperiods (brief light pulses in otherwise continuous darkness or a short daylength with a brief pulse during the ensuing dark period) are also effective in inducing photoperiodic responses, reinforcing the idea that light has to fall within a particular window of sensitivity to be effective (Figures 5.6, 5.7).

Daylength can be read, physiologically speaking, as short or long depending on previous exposure. For example hamsters raised in 16L 8D and transferred to 12L 12D will respond to 12L 12D as a short photoperiod, however if they are raised in 8L 16D and moved to 12L 12D they will read this as a long photoperiod. This latter is

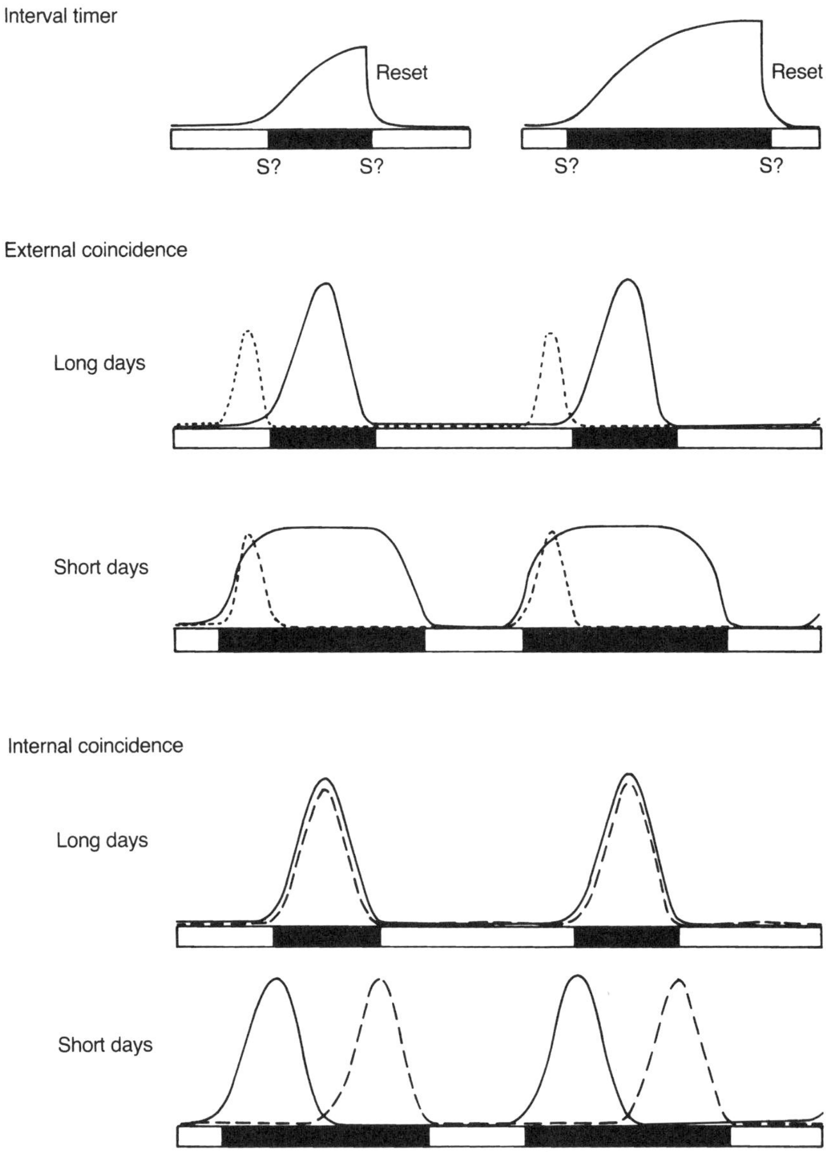
Interval timer
Reset
Reset
S?
S?
S?
S?
External coincidence
Long days
Short days
Internal coincidence
Long days
Short days

clearly a very artificial situation for an animal which is normally born in long days (usually around 14L 10D). It does not invalidate the definition of 'critical photoperiod' as the daylength above or below which a specific seasonal response will occur. However, it does illustrate a mechanism whereby the direction of daylength change can be appreciated and underlines the importance of taking prior environmental history into consideration: information must be in context. This observation is not confined to hamsters and indeed is probably a universal phenomenon.

In recent years it has become clear that the pineal gland, through the secretion of melatonin, is essential for daylength measurement. The circadian system is clearly involved in daylength perception through generation of a particular melatonin signal. The control of melatonin secretion by brief light pulses has already been discussed. The evidence to date suggests that single light pulses or skeleton photoperiods generate melatonin profiles which are then read as a particular daylength. In Siberian (Djungarian) hamsters there is also evidence for a feedback loop whereby melatonin exerts seasonal physiological effects through the SCN.

Fig. 5.6 Models of possible ways of measuring daylength or photoperiodic time. The solid bar indicates darkness. Interval timer: in this system 'product', e.g. melatonin, is generated during the night and the length of the night is measured by the duration of product generation. Light terminates and resets the mechanism. Note that a daily (circadian) signal(s) may be required for the generation of product: such signals may be lights on (dawn) or lights off (dusk), or both, indicated by S. Alternatively, product generation may be only a direct response to darkness. External coincidence: a photosensitive phase (- - -) entrained to the daily light–dark cycle coincides with the dark phase only in sufficiently short photoperiod. Interaction of dark phase 'product' at photosensitive targets induces the appropriate response. Internal coincidence: two or more circadian rhythms change their phase relationships in long and short photoperiod. The state of the phase relationships determines the photoperiodic response. The circadian system is required to generate the melatonin rhythm; however in mammals there is more evidence against a circadian rhythm of melatonin sensitivity (as in the external coincidence model) than there is for it. (Reproduced from Arendt, J. Role of the pineal gland and melatonin in seasonal reproductive function in mammals. *Oxford Reviews of Reproductive Biology*, **8**, 266–320, by permission of Oxford University Press, 1986.)

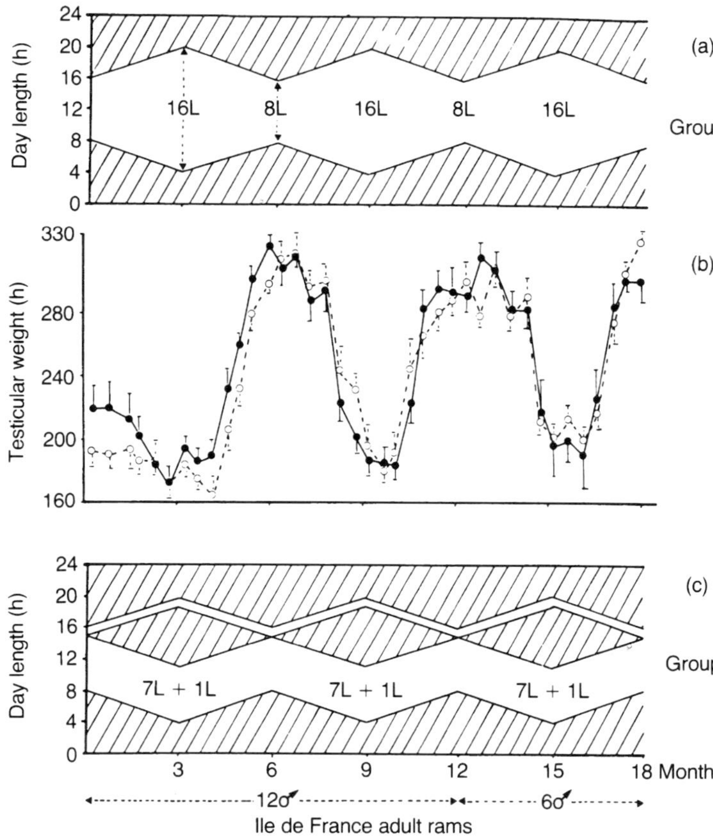

Fig. 5.7 'Skeleton' photoperiods are read as full photoperiods. Testicular weight variations (b) in control (Group 1, ●——●), and experimental (Group II, ○——○) rams according to a full photoperiod (a) and a 'skeleton' photoperiod (c) consisting of a main light period and a 1 h light pulse during the night. Testicular changes were identical in both groups. (Reproduced with permission from Pelletier, J. and Thimonier, J.; published by the *Journal of Reproduction and Fertility*, 1987.)

5.4 PINEAL-DEPENDENCE OF PHOTOPERIOD PERCEPTION

All research into seasonal breeding and other seasonal functions is exceptionally time consuming. Even though it is possible to compress the seasonal cycle by administering alternating periods of long and short artificial days to, for example, sheep and deer, there are limits beyond which physiological responses cannot be entrained. Thus to demonstrate a seasonal effect in a mammal usually requires a full cycle of observation and often several cycles.

The effects of the pineal gland in photoperiodic seasonal breeders were slow to become evident, precisely because of these constraints. Much early work, e.g. Reiter (1980), illustrated the essential role of the pineal in hamster seasonality (Figure 5.8). These observations

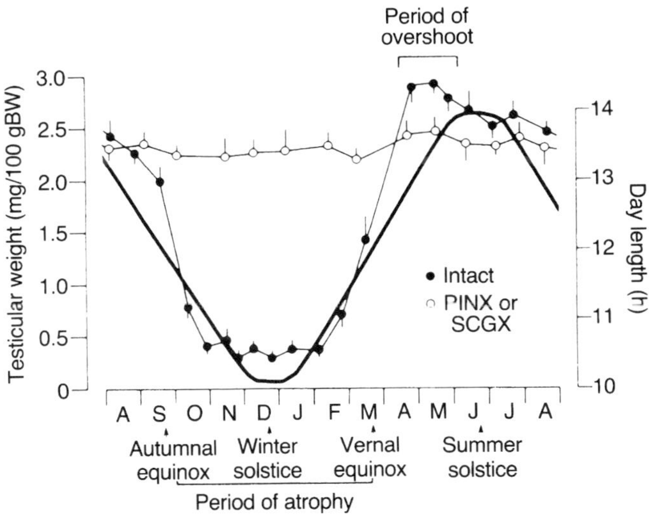

Fig. 5.8 Testicular weights in intact and pinealectomized (PINX) hamsters throughout the year when animals are kept under natural photoperiodic and temperature conditions. Either pinealectomy or superior cervical ganglionectomy (SCGX) completely prevents gonadal atrophy normally induced by short days. (Reproduced with permission from Reiter, R.J. The pineal and its hormones in the control of reproduction in mammals. *Endocr. Rev.* **1**, 109–31, 1980 © The Endocrine Society.)

suggested that the pineal had antigonadotropic effects in hamsters, because pinealectomy prevented the gonadal regression induced by short photoperiod or by blinding. Similar work in ferrets and voles supported the concept. Subsequently the term counter-anti-gonadotropic was coined to explain otherwise uninterpretable data indicating that melatonin implants also prevented gonadal regression in short days.

5.5 HAMSTERS AND FERRETS

The observations of Hoffman (1981) and Herbert (1981) and their coworkers laid more foundations for the present day understanding of pineal function in photoperiodic seasonal breeders. These two groups independently came to the conclusion, using the long day breeding Djungarian hamster and the ferret as models, that the pineal was neither specifically antigonadotropic nor pro-gonadotropic, but served to transmit the effects of daylength, whether short or long, to body physiology. This concept clarified what had appeared to be a very confusing field up to that point.

Pinealectomy prevents the reproductive response (gonadal atrophy and the concomitant endocrine changes) to short (less than 12.5 hours of light daily) days. If hamsters are maintained for a sufficiently long time on short days, testicular recrudescence occurs spontaneously. In natural light recrudescence precedes increasing spring photoperiod and hence is not daylength dependent (Figure 5.8). This is an example of photorefractoriness of physiological importance. However pinealectomy induces premature recrudescence in at least two varieties of hamster. The pineal also mediates the stimulatory effects of long days. Procedures equivalent to pinealectomy in the sense that they abolish rhythmic melatonin synthesis (exposure to continuous light, denervation of the gland and destruction of the SCN) have similar effects on reproductive function.

In ferrets pinealectomy prevents the premature onset of estrus induced by long photoperiod applied in winter and early estrus offset induced by short photoperiod in summer. Most particularly, pinealectomized ferrets kept in natural daylight for several years lose

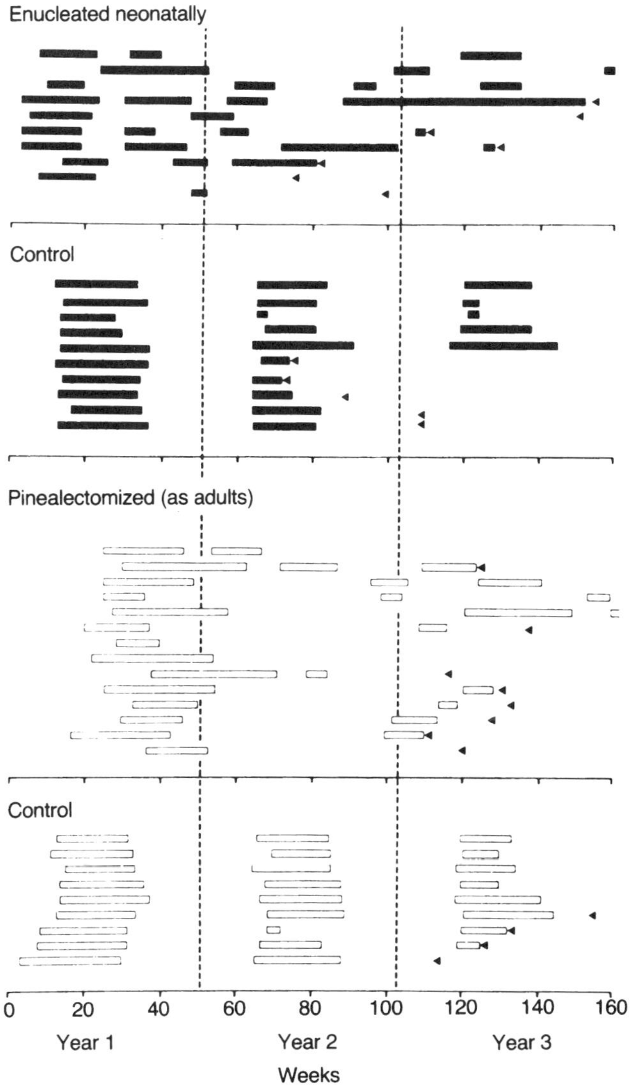

Fig. 5.9 Pinealectomy or blinding leads to desynchronization of reproductive cycles from the environmental photoperiodic cycle: Recurrent periods of estrus occurring in ferrets (photoperiodic long day breeders) kept in daylight for more than 3 years after either being pinealectomized as adults or having been enucleated neonatally. Their respective controls are also shown. Each bar represents a period of estrus for an individual animal. Year 1 starts on 1 January of the year after operation. ◀ indicates that the animal died. (Reproduced with permission from Herbert, J.; published by Wright, 1981.)

synchrony with the annual cycle and come into and out of estrus at random intervals (Figure 5.9). This was the first compelling evidence that the pineal synchronized an annual rhythm in reproductive competence in long-lived species.

5.6 SHEEP AND GOATS

The general conclusions drawn from these experiments that the pineal acts as a neuroendocrine transducer of daylength information for the organization of seasonal physiology, can now be applied to a very large number of mammals. Of the short day breeders, sheep and goats have provided much information.

In sheep pinealectomy in ewes, or superior cervical ganglionectomy in Soay rams, abolishes perception of changes in artificial daylength as assessed by reproductive function (Figure 5.10). The long-term response to either of these manipulations is not so clear.

Ganglionectomy in Soay rams reportedly led to persistent reproductive competence in alternating 16-week periods of long and short days – noted as the 'Ram for all Seasons' by Lincoln (1979). Persistence of seasonality has, however, been observed in the long term in pinealectomized Soay rams and in pinealectomized ewes kept in a natural environment with in some cases a phase advance relative to the annual cycle. The latter observation is reminiscent of phase changes in circadian rhythms when zeitgeber input is weakened. In these circumstances it is possible that other seasonal zeitgebers such as temperature and food availability acquire a greater importance. In general the mechanisms of perception of non-photic zeitgebers are unknown at present.

The ovariectomized, estradiol-implanted Suffolk ewe has proved to be an excellent model for research into the control of seasonal breeding. This animal shows marked changes in luteinizing hormone (LH) secretion with high values throughout the period of estrus and low to undetectable concentrations during anestrus in the presence of constant levels of circulating estradiol (Figure 5.11). This phenomenon amongst many others is strong evidence for a change in feedback sensitivity to steroids as a primary event in the seasonal reproductive response. The changes in LH can equally be seen in

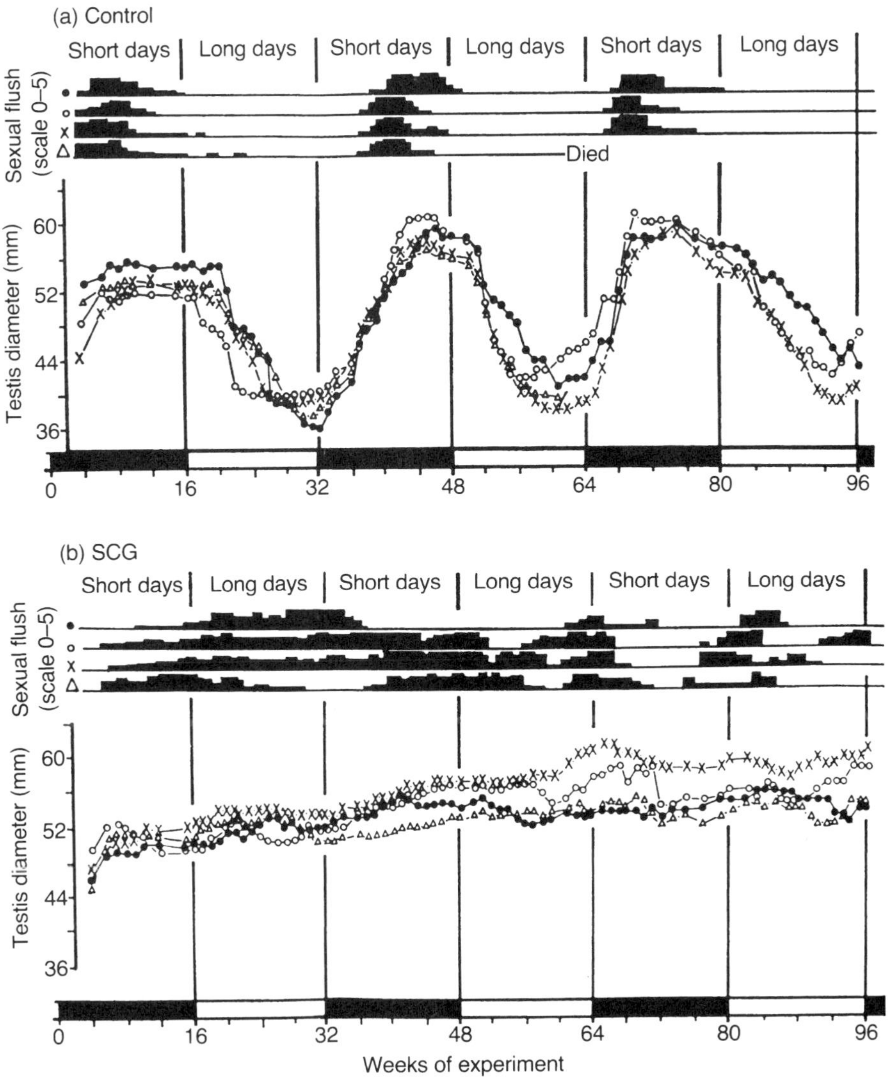

Fig. 5.10 Response to artificial photoperiods: Soay rams were kept in alternating 16-week periods of short (8L 16D) and long (16L 8D) days. Reproductive and behavioral changes are driven by the photoperiod change, illustrated here by testis diameter and sexual flush (a). Note that the complete 'annual' cycle is compressed into 32 weeks. Further compression to 8 weeks can lead to a persistent state of sexual activity. (b) If the pineal is removed or denervated by removal of the superior cervical ganglia the animals can no longer respond to artificial photoperiod change. (From Lincoln, G., 1979, reproduced by permission of the Journal of Endocrinology Ltd.)

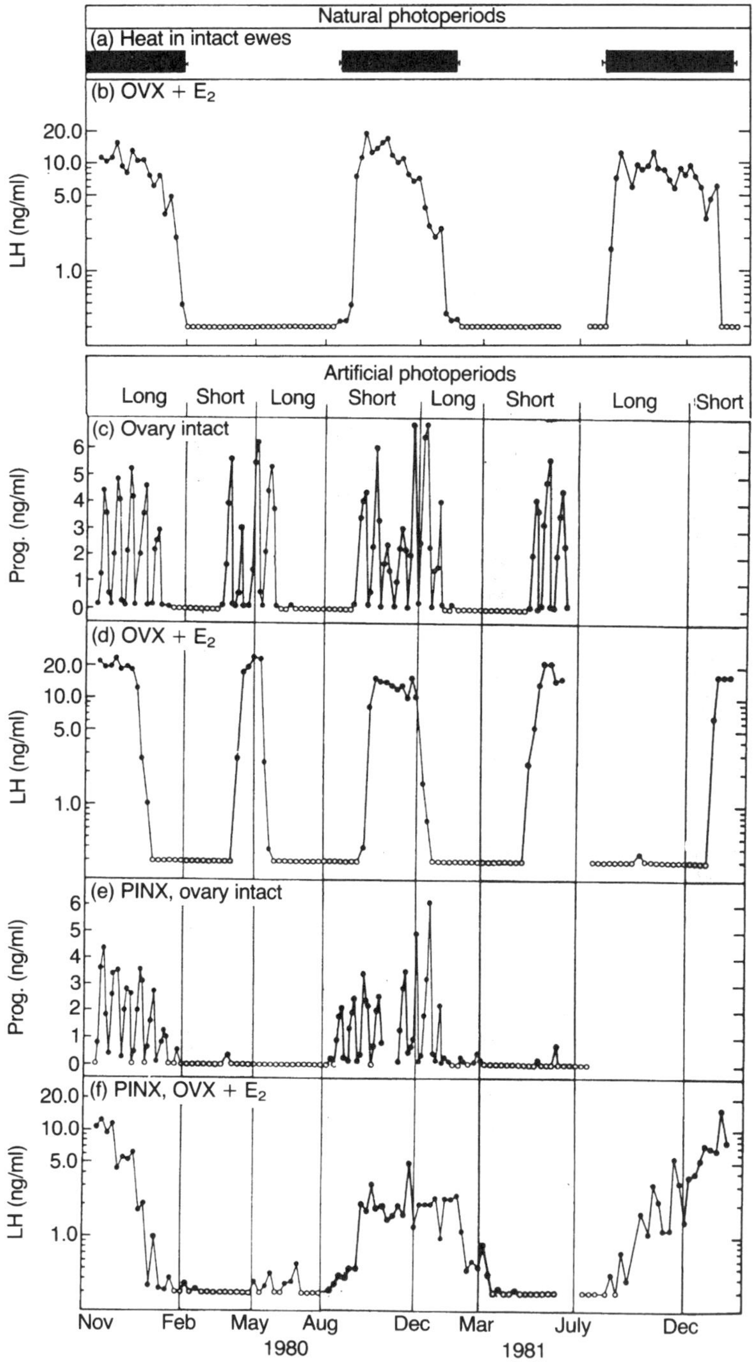

Natural photoperiods
(a) Heat in intact ewes
(b) OVX + E2
LH (ng/ml)
20.0
10.0
5.0
1.0
Artificial photoperiods
Long
Short
Long
Short
Long
Short
Long
Short
(c) Ovary intact
Prog. (ng/ml)
(d) OVX + E2
LH (ng/ml)
(e) PINX, ovary intact
Prog. (ng/ml)
(f) PINX, OVX + E2
LH (ng/ml)
Nov
Feb
May
Aug
Dec
Mar
July
Dec
1980
1981

alternating artificial short and long days and provide an endpoint in the female which is not all or nothing as in the case of estrus onset and offset.

Pinealectomy abolishes the LH response to changing artificial daylength (Figure 5.11). If such animals are kept for very long periods (years) in artificial short days they will undergo a period of 'estrous cycles' intermittently which is not related to the annual cycle, i.e. it is desynchronized. This clearly represents an endogenous capacity to cycle which is quite independent of the photoperiodic environment. Work in ground squirrels also nicely illustrates the existence of endogenous circannual rhythms. Here pinealectomy effectively changed the period of the free-running cycle.

When the endogenous annual cycle is programmed is a question of some considerable interest and remains to be answered. When photorefractoriness is encountered in long-lived breeders it may represent a non-inductive period of the endogenous cycle analogous to the 'dead zone' of a circadian phase response curve.

Partly because the ability to control reproduction is of applied interest in commercially important domestic species, this aspect of photoperiodism has received more attention than others. The winter coat of mink, arctic foxes, cashmere goats, etc. also has commercial significance and can be photoperiodically manipulated. It is beyond the scope of this book to consider photoperiodism in more detail and the reader is referred to the references and further reading list for reviews. Chapter 7 will cover photoperiodic functions in humans.

5.7 TEMPERATURE AS A ZEITGEBER

The importance of temperature as a seasonal zeitgeber has not been fully evaluated in mammals. It is important to note, however, that it

Fig. 5.11 Pinealectomy of Suffolk ewes abolishes the ability to respond to changes in artificial photoperiod. Periods of heat (a) and estrous cycles as shown by progesterone (c) in intact ewes and changes in LH in ovariectomized estradiol implanted (OVX + E_2) animals are compared with estrous cycles and LH changes in pinealectomized animals kept in alternating 16-week periods of short (8L 16D) and long (16L 8D) days. Pinealectomy abolishes the response to artificial photoperiod change leaving an endogenous annual cycle. (Reproduced with permission from Bittman, E.L.; published by Pitman, 1985.)

acts in concert with photoperiod to reinforce a particular response. Elegant studies in hamsters indicate, for example, that the gonadal response to short days is induced more rapidly in cold conditions and that ambient temperature can change the critical photoperiod (Figure 5.5).

This observation clearly indicates the flexibility of the physiological systems involved in seasonal timing with some retention of opportunistic behavior. In, for example, an unusually cold spring, reproduction could be delayed until more propitious weather conditions prevail. There are indications that in intact animals the pineal may be involved in temperature modulation of photoperiodic responses.

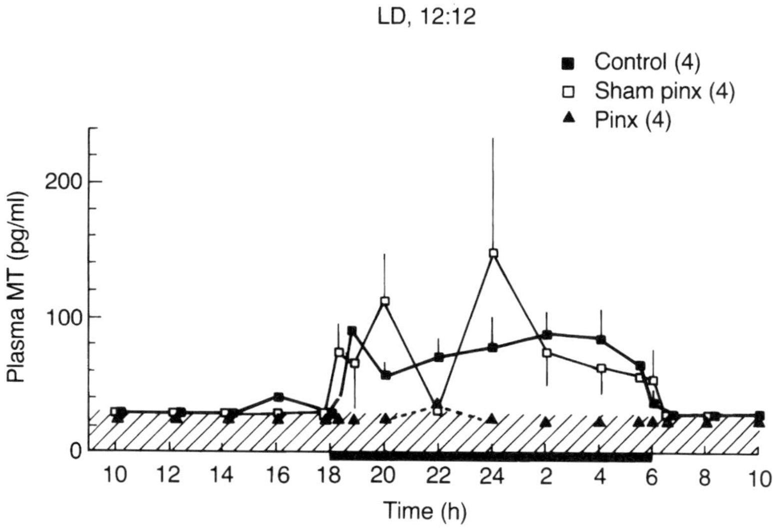

Fig. 5.12 Pinealectomy abolishes the rhythm and greatly reduces the basal levels (usually to undetectable concentrations) of plasma melatonin in mammals. Illustrated here are the effects of pinealectomy and sham pinealectomy on plasma melatonin in castrated rams. (From data in Arendt *et al.*, 1980, and Arendt, 1985, by permission.)

5.8 ROLE OF MELATONIN

Pinealectomy removes the vast majority of circulating melatonin in rodents, primates and ungulates (Figure 5.12). Although it is synthesized elsewhere, for example in the retina, other sources do not contribute significantly to circulating concentrations in mammals (although they do in birds and some lower vertebrates). It was therefore the first pineal hormone to be investigated as a photoneuroendocrine transducer. All the evidence to date indicates that it is by far the most important pineal product.

Other methoxyindoles synthesized in the pineal, 5-methoxytryptophol and 5-methoxytryptamine in particular, have some interesting physiological effects. Nevertheless the effects of pinealectomy can usually be completely reversed by appropriate melatonin administration. A major disadvantage in the investigation of melatonin is that no specific inhibitor of its synthesis exists and most deductions have been made from its administration.

The first attempts to investigate the importance of melatonin in the control of seasonal breeding employed administration protocols which gave it in pharmacological doses either by injection or by implantation. The first evidence for a role in the reproduction of hamsters and ferrets was derived from the use of implants. Subsequent work revealed that timed administration was of critical importance and results from timed administration were, and are, far simpler to interpret than the effects of implants. For an excellent review of early work see Reiter (1980).

5.9 ARTIFICIAL CREATION OF PHYSIOLOGICAL MELATONIN PROFILES

Once the pattern of daily melatonin secretion was known it became possible to devise experimental protocols that mimicked the natural secretion profile. The critical observation was that the duration of melatonin secretion was related to daylength (Figure 5.13 and see Chapter 4).

Late afternoon daily injections of melatonin to intact hamsters kept in long photoperiod induce reproductive responses that are

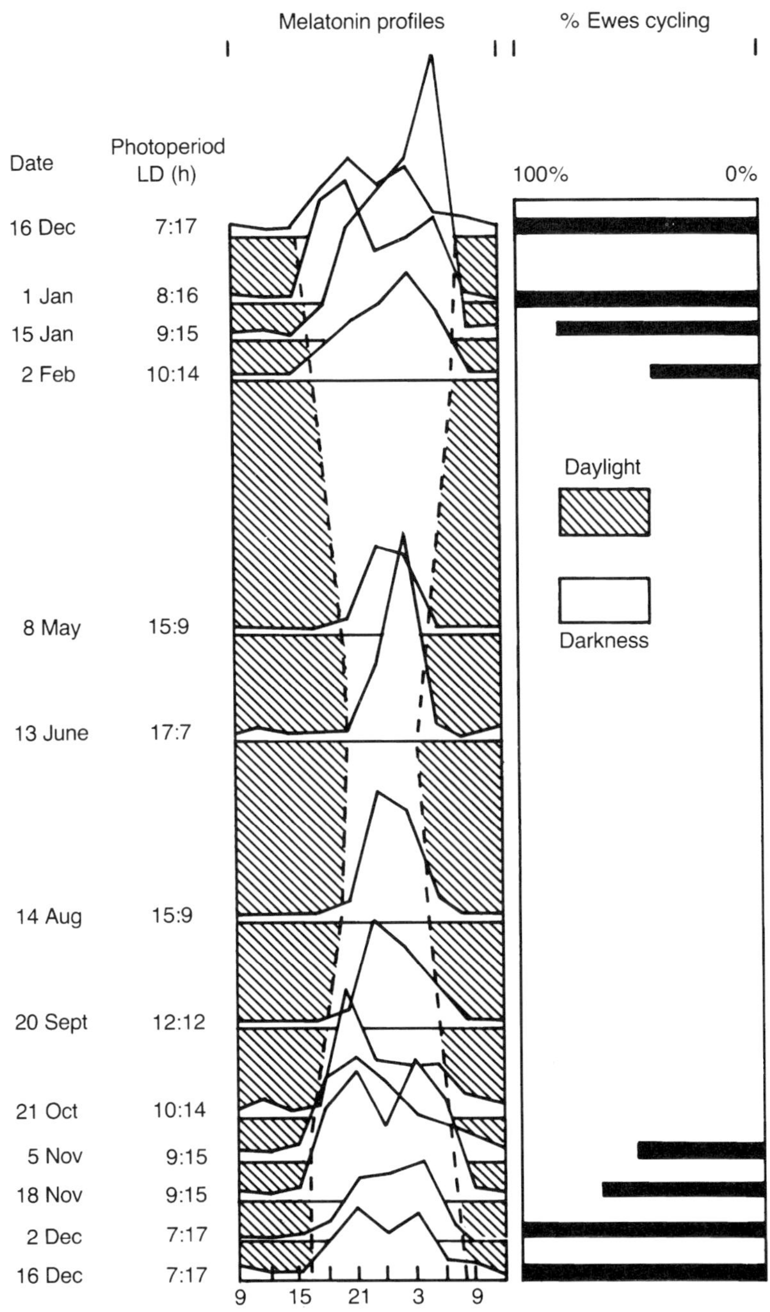
Melatonin profiles
% Ewes cycling
Date
Photoperiod LD (h)
100%
0%
16 Dec 7:17
1 Jan 8:16
15 Jan 9:15
2 Feb 10:14
Daylight
Darkness
8 May 15:9
13 June 17:7
14 Aug 15:9
20 Sept 12:12
21 Oct 10:14
5 Nov 9:15
18 Nov 9:15
2 Dec 7:17
16 Dec 7:17
9
15
21
3
9

characteristic of short photoperiod: the mechanism is essentially through extension of the duration of high night-time concentrations. Likewise, feeding melatonin to Suffolk ewes in the late afternoon from midsummer will advance estrus onset from the normal time in late September–early October by approximately six weeks: a response identical to that seen if the animals are kept in short photoperiod from midsummer (Figure 5.14). Melatonin is effectively equipotent with photoperiod.

Subsequently, it became possible to infuse melatonin to pinealectomized sheep or hamsters, and more recently other species including humans, such as to generate at will circulating profiles characteristic of particular photoperiods in the intact animal. In this way and especially from the work of Goldman, Karsch and their collaborators (Carter and Goldman, 1983; Karsch *et al.*, 1984, 1988; Bittman, 1985) we now know that melatonin duration is the necessary and sufficient condition for the induction of a given seasonal response. The melatonin signal is the same whether we are dealing with long day or short day breeders, nocturnal or diurnal species. Long duration melatonin is equivalent to short days and short duration melatonin is equivalent to long days (Figure 5.15). The interpretation of the signal, as with daylength depends on the species in question.

Like light–dark cycles the perception of the melatonin signal is relative and dependent on previous photoperiodic/melatonin duration exposure. If Syrian hamsters are transferred from 16L 8D to 12L 12D the latter is interpreted as a short day in terms of LH secretion: but transfer from 8L 16D to 12L 12D leads to a long day response in LH. The response of prolactin is not, however, so clear cut (Figure 5.16). Animals become refractory to a specific duration of melatonin as they do to a particular photoperiod. For example, a period of long days (or a long day melatonin signal) is required

Fig. 5.13 Plasma melatonin at 3-hour intervals in 4–6 ewes sampled at 2-week to 2-month intervals for one year. Mean values are shown for clarity without statistics. Solid bars indicate number of ewes showing estrous cycles. The hatched area shows the light phase. Analysis of variance indicates significant effects of time of day, time of year, number of hours after sunset and number of hours after sunrise (p<0.01–0.0001). (Redrawn from Arendt, J.; published by Alan R. Liss, 1985.)

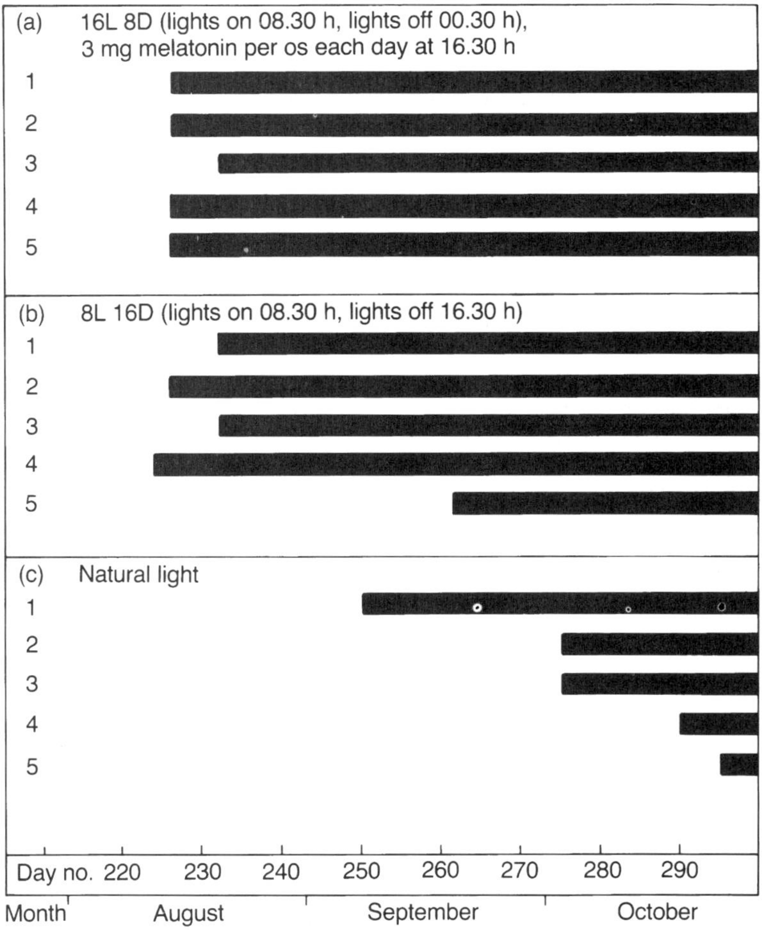

Fig. 5.14 Summer melatonin treatment advances estrus in ewes. Presence of estrous cycles (black bars) in three groups of five Suffolk cross ewes, treated with (a) 16L 8D, lights on 0830 h, 3 mg melatonin daily at 1600 h adsorbed on to a food pellet; (b) 8L 16D, lights on 0830 h; (c) natural light. Treatments commenced in mid-June. Onset of the breeding season was considered to be day 1 of the first estrous cycle with progesterone levels over 1 ng per ml. Both melatonin feeding and long dark treatments advanced estrus by approximately 2 months. (From Arendt, J. *et al.*, 1983, reproduced by permission of the Journal of Endocrinology Ltd.)

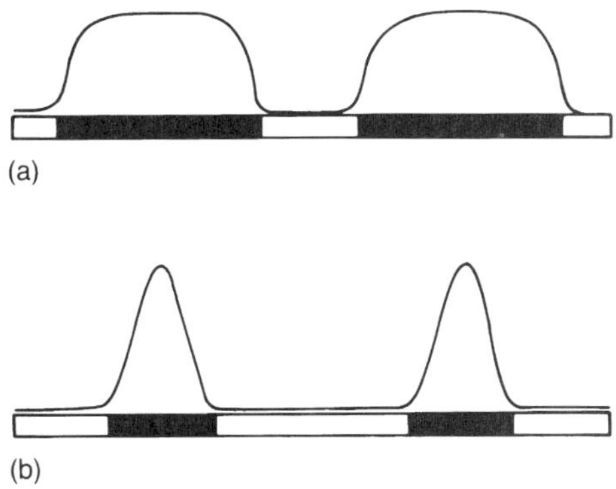

Fig. 5.15 Melatonin: transduction of photoperiodic information. Duration of melatonin secretion indicates the length of the night. Artificial creation of short or long melatonin profiles by infusion to pinealectomized animals induces the seasonal response appropriate to the equivalent short (a) or long (b) days, independently of the ambient photoperiod.

before a short day melatonin signal will advance the reproductive cycle in sheep and goats. Similarly a period of short days is necessary for a response to a long day signal. The number of long or short days required varies from species to species and no general conclusions will be drawn here.

The early interpretations of refractoriness to a specific daylength considered that either melatonin secretion was modified after lengthy exposure to the same light–dark cycle or that the signal was interpreted differently. At present experiments in sheep and hamsters indicate that the melatonin profile remains the same and processing of the information is changed after lengthy exposure to a given duration. This change may reside in receptor sensitivity or in other biochemical events and is as yet unknown.

However preliminary results in sheep show that, in continuous short days in the very long term, the melatonin profile can spontaneously revert to a long day pattern or the secretion may become undetectable for short periods. It is conceivable that this represents a rephasing of the circadian rhythm generating system through unknown mechanisms. It provides an explanation for the intermittent recurrence of estrus in sheep in constant conditions.

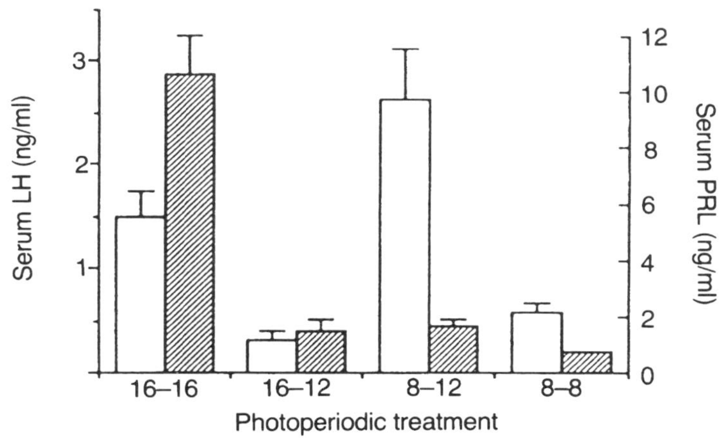

Fig. 5.16 Relative perception of photoperiod by hamsters: animals were either kept in 16L 8D (16–16), 8L 16D (8–8), transferred from 16L 8D to 12L 12D (16–12) or from 8L 16D to 12L 12D (8–12). Serum LH assayed after 7 weeks (open columns) responds as if 12L 12D was a short day in the 16–12 group and a long day in the 8–12 group. The prolactin response (hatched columns) does not show this effect of previous photoperiod exposure in that the 8–12 group do not exhibit a clear long day effect. After 7 weeks in 12L 12D the melatonin secretion profiles are the same no matter what the previous exposure and hence the differential LH response is due to interpretation of the melatonin signal. (Reproduced with permission from Hastings, M.H. *et al.*; published by the Guildford Press, 1989.)

5.10 CRITICAL ROLE OF SHORT DURATION MELATONIN IN UNGULATES

Experiments in sheep indicate that it is the long days of spring that time reproductive recrudescence (or pubertal development) in the autumn. If animals are maintained, from the summer solstice, on artificial long days they will come into estrus at the same time as animals in natural, decreasing daylength. Likewise animals kept in short days from the winter solstice will enter anestrus at the usual time.

An explanation for the estrus advancing effects of long duration melatonin from the summer solstice is that, whilst the timing of the cycle is set by long spring days, the longer days of summer exert an inhibitory effect which is abolished by long duration melatonin administration (Figure 5.17). In winter the inductive effects of short day melatonin can be counteracted by long days and the season shortened.

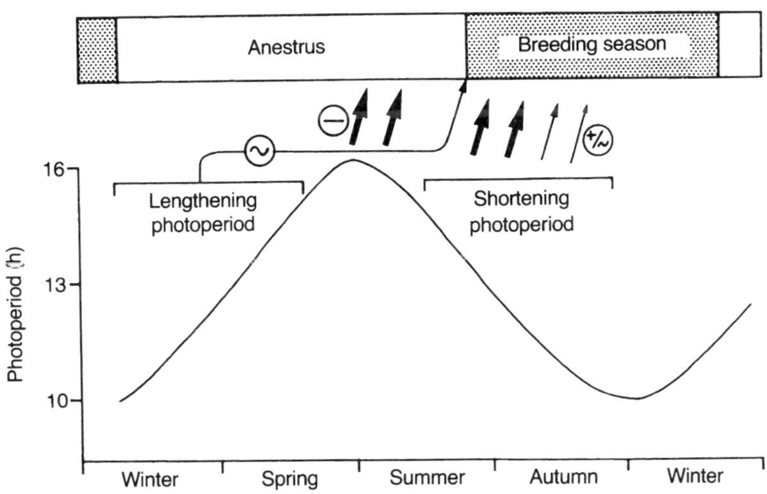

Fig. 5.17 Model for the temporal regulation of breeding seasons in ewes. Action of photoperiod to synchronize an endogenous annual process (rhythm ~); action of photoperiod to inhibit reproductive activity (—); action of photoperiod to sustain reproductive activity through synchronizing or stimulating effects (+/~). (From Malpaux, B. *et al.*, 1989, reproduced by permission of the Journal of Endocrinology Ltd.)

In pinealectomized sheep kept in constant photoperiod the whole seasonal cycle can be timed by two 70-day blocks of daily melatonin infusions, one of 16 hours' duration the other of 8 hours' duration. If these blocks are timed such that their midpoint is 365 days apart the periodicity induced is precisely annual. Moreover suitable timing of the infusions can create an annual cycle 180 ° out of phase with the natural cycle.

Karsch and coworkers (Woodfill *et al.*, 1991; Karsch *et al.*, 1988; Bittman, 1985) have tested the hypothesis that long days can set the whole seasonal cycle using this paradigm and administering a single 70-day block of daily 8-hour melatonin infusions, once a year to pinealectomized ewes. This manipulation was sufficient to create an annual cycle in otherwise desynchronized animals. The most parsimonious interpretation is that long day melatonin acts as a seasonal zeitgeber to reset an endogenous annual rhythm.

5.11 PUBERTY

Previously it has been noted that photoperiod determines the timing of puberty in some species, provided that a sufficient degree of physical maturity has been reached. Foster and colleagues (Foster, Ebling and Claypool, 1988; Foster *et al.*, 1988) have demonstrated that exposure to long days around 18–23 weeks of age is essential for puberty to occur at the natural age of 30 weeks in spring-born lambs. Neonatal exposure to short days was not necessary for subsequent recognition of long day photoperiods although the prenatal photoperiod undoubtedly plays a role. As is the case for the timing of the seasonal cycle in ewes, long days, or short duration melatonin, timed puberty which occurred in the absence of additional photoperiodic information (Figure 5.18).

In juvenile Siberian hamsters long photoperiods promote rapid maturation of testes while short photoperiods do not. However in the juvenile hamster a postnatal photoperiod of 14L 10D is seen as either stimulatory long days or inhibitory short days depending on the photoperiod (16L 8D or 14L 10D) to which the mother was exposed during gestation, another example of relative interpretation of photoperiod (Figure 5.19). Thus photoperiod perception is pres-

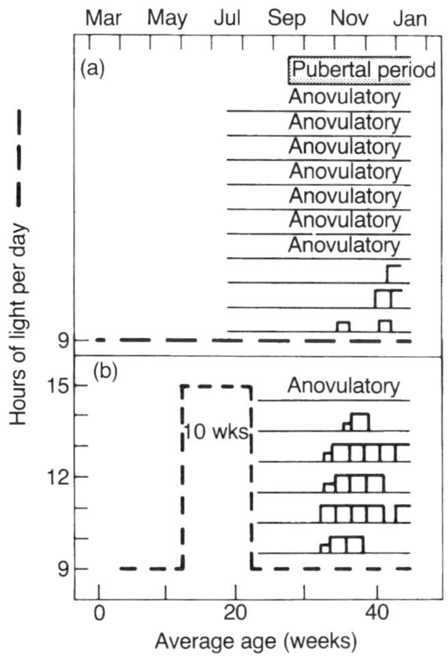

Fig. 5.18 The timing of puberty in lambs depends critically on exposure to long days (or short duration melatonin) in the first year of life. Initiation of reproductive cycles in spring-born Suffolk lambs raised in artificial photoperiod. (a) Short days only; (b) short-day–long-day sequence. Each horizontal line shows data for an individual female with large and small blocks representing normal and short luteal phases, respectively. (Reproduced with permission from Foster, D.L., Ebling, F.J.P. and Claypool, L.E.; published by INRA Publications, 1988.)

ent before birth and its effects persist postnatally. The photoperiodic signal is derived from the mother and involves both the maternal pineal via melatonin secretion, and the maternal circadian system; however the precise details of signal transduction reported by different authors are not yet consistent.

Melatonin certainly crosses the placenta in a number of species. Melatonin injections to the mother can dictate postnatal reproductive development but it is likely that maternal melatonin alone is not sufficient to program the fetus.

Ruminants differ in the flexibility of the timing of puberty. For

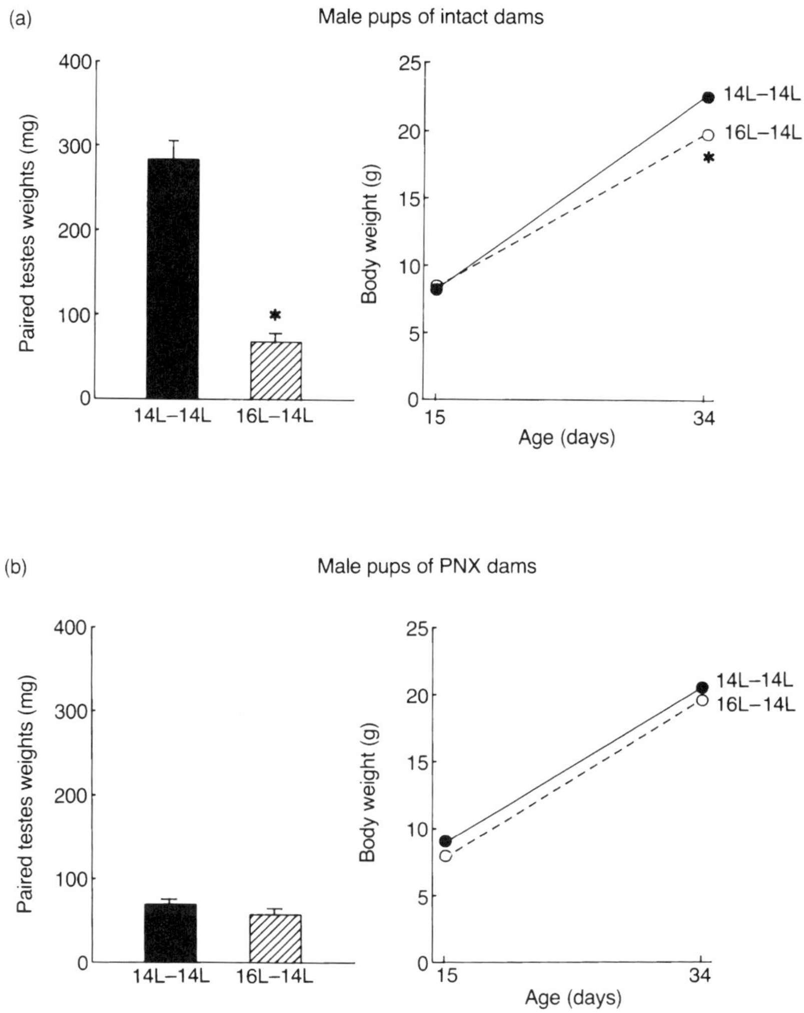

Fig. 5.19 Maternal pineal communicates daylength to the fetus in Djungarian hamsters. Mean testicular weights (left) and body weights (right) of male Djungarian hamsters from intact dams (a) and pinealectomized dams (PNX, b). Litters from dams kept in 16L during pregnancy were shifted to 14L on the day of birth (16L–14L group); another set of litters was kept in 14L during both pre- and postnatal periods (14L–14L group). In the left hand panels the vertical lines indicate SEMs. In the right hand panels SEMs are contained within the symbols for the means. n=17–35. * indicates p<0.0001. Infusion of melatonin to pinealectomized dams markedly influences postnatal testicular development (not shown). (Reproduced from Weaver, D.R. and Reppert, S.M. Maternal melatonin communicates daylength to the fetus in Djungarian hamsters. *Endocrinology*, **119**, 2861–6, 1986 © The Endocrine Society.)

example, lambs born out of season in autumn in natural light will delay puberty until the onset of the following breeding season. In short day breeding goats, kept in natural light in temperate zones, it is possible to completely reverse the natural breeding season by an initial treatment of 2 months of artificially extended natural days (20L 4D) in February and March followed by short day melatonin treatment (implant or feeding for 3 months). Animals will then mate in July/August (instead of September/October) and give birth in November/December. If a second identical long day and melatonin treatment is initiated 2 months before the expected kidding date, animals will subsequently mate in May/June and give birth in September/October. These autumn-born kids reach puberty at 12/13 weeks in natural short days in contrast to the 25/30 weeks required for spring-born goats. There would appear to be precocious development of the neuroendocrine reproductive axis in relation to body weight compared to sheep.

Goats, like a number of other species, can also perceive photoperiod prenatally. Exposure of autumn-born kids to long days (20L 4D) for 2 months prepartum significantly delays the onset of puberty in both males and females. The mechanism is presumably dependent on a maternal melatonin signal. If observations in sheep can be extrapolated partly to goats then it may be that exposure to the natural long days of summer during gestation sets the timing of puberty, and artificial long days prior to birth delay this by a mechanism which may be related to phase delay by light in circadian cycles.

5.12 PUBERTY IN RATS

The laboratory rat is only marginally photoperiodic. Nevertheless as an adult it can be rendered sensitive to photoperiod by means such as olfactory bulbectomy, androgen sterilization or food restriction and its pubertal development is subject to manipulation by both light and melatonin. These comments justify its inclusion in this chapter.

Young rats kept in either 16L 8D or 8L 16D from birth show small but significant differences in testis weight with an increase in long photoperiod. Moreover if pregnant rats are kept for the week before

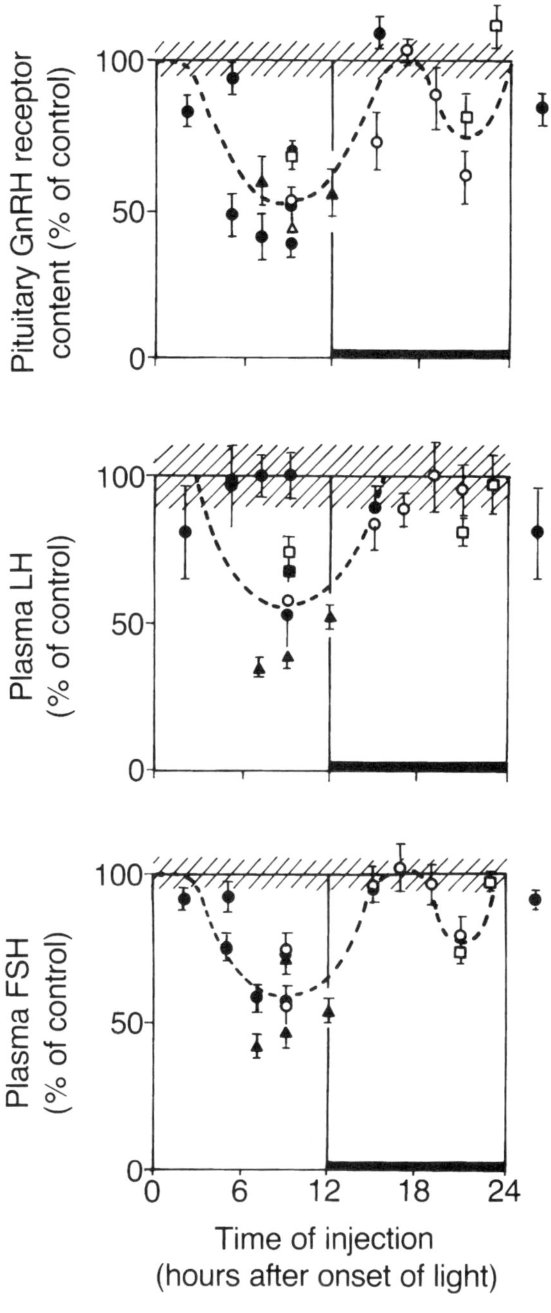

Pituitary GnRH receptor content (% of control)
Plasma LH (% of control)
Plasma FSH (% of control)
100
50
0
0
6
12
18
24
Time of injection (hours after onset of light)

parturition and thereafter in 16L 8D or 8L 16D, at 8 days of age differences in the duration of melatonin synthesis were present and on day 15 testes and brown fat were heavier in the long day group. If rats reared on 12L 12D are given late afternoon injections of melatonin specifically during the period of pubertal development then reproductive maturity is delayed in all relevant aspects – gonadal growth, hormone levels, pituitary GnRH receptor content, etc. in both males and females (Figure 5.20). Full sexual maturity is eventually achieved; thus the system is not permanently compromised. The minimum amount of melatonin needed is close to the range producing physiological night-time concentrations after injection.

In addition to late afternoon injections, there is a small window of sensitivity to melatonin in the late dark phase (Figure 5.20). The interpretation of these results is clearly that a long profile of melatonin is generated by artificially extending the endogenous night-time secretion such that it corresponds to a short day. 6-Hydroxy-melatonin and 5-methoxytryptamine were also able to reproduce this effect but a ten-fold increase in dose was required compared to melatonin. *In vitro* melatonin will inhibit LHRH-induced LH release by cultured rat pituitary glands from prepubertal animals, suggesting that its ability to delay puberty may involve a direct action on the pituitary.

Not all strains of rats respond similarly to melatonin administration during puberty. Nevertheless there is good evidence in general for the presence of photoperiodic machinery which, possibly due to selection processes, is not always manifest. In species without adult photoperiodic reactions it is possible that photoperiodism is expressed during fetal and neonatal life, and early development. Circulating

---

Fig. 5.20 Melatonin delays puberty in rats as illustrated by the decline in pituitary GnRH receptor content, plasma LH and plasma FSH on day 41 as a function of the time of daily injections of melatonin (100 μg) to male rats from days 20–40 of life. The dotted lines represent suggested patterns of sensitivity to melatonin. Note that there is one principal window in the late afternoon and a second minor window in the late dark phase. Similar results were obtained for plasma testosterone, testes and seminal vesicle weights. Similar delays are found in females and although the dose of melatonin here is pharmacological, near physiological amounts are also effective. (Reproduced with permission from Lang, V. *et al.* Influence of exogenous melatonin on melatonin secretion and the neuroendocrine reproductive axis of intact male rats during sexual maturation. *Endocrinology*, **112**, 1578–84, 1983 © The Endocrine Society.)

melatonin concentrations decline in the prepubertal period and high concentrations may exert a restraining action on sexual maturation via a residual photoperiodic effect. The role of the pineal and melatonin during human puberty will be considered in Chapter 7.

5.13 PROLACTIN

Prolactin is of particular interest in terms of photoperiodic response in animals. In all seasonal mammalian species, regardless of whether they are short or long day breeders, long days increase prolactin secretion and short days suppress it. The effect is very marked. In sheep and mink for example, summer plasma concentrations can be 10 to 20-fold higher than in winter (Figure 5.21).

Prolactin is secreted with a daily rhythm which is partly endogenous. Most is secreted during the night. Sleep is a major determinant of the secretion profile, at least in humans, where daytime naps will elevate prolactin and arousal at night suppresses production. Ruminants often show a double peak of secretion (Figure 5.21), although the profiles are subject to stress and metabolic variations and are rather noisy. The peaks appear closely related to dusk and dawn.

Short day melatonin treatment suppresses prolactin in summer in all observations to date. In sheep basal secretion dropped, the dusk peak was retained, with much lower amplitude and the major night-time peak was abolished (Figure 5.22). This leads to the possibility that the dusk peak is a strictly circadian manifestation of prolactin secretion, whereas basal secretion and the major peak correspond to a strictly seasonal response.

In spite of this seasonality prolactin does not appear to play an important role in the timing of seasonal breeding in sheep or goats. Different species appear to put prolactin to different uses. The concept that it is not hormones which have evolved but the uses to which they are put is essentially untrue of melatonin but it applies to prolactin. For example in mink prolactin is luteotropic and in wallabies it inhibits the corpus luteum. In some seasonal mammals it may be important in functions such as changes in coat constitution and growth rate (see later).

The prolactin response to photoperiod or melatonin is rapid and

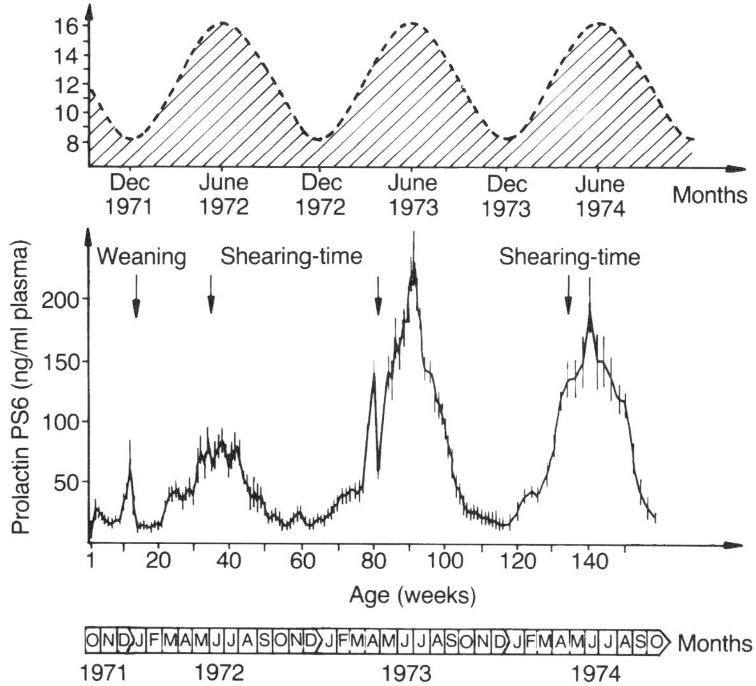

Fig. 5.21 Prolactin shows marked photoperiod and temperature dependent seasonal variations in many species. Seasonal variations of plasma prolactin in Ile de France rams (mean ± SEM, n=9).(Reproduced with permission from Ravault, J.; published by Rhodos International, 1976.)

visible with a week of initiating treatment in ruminants. Hence in contrast to most photoperiodic effects it provides a convenient end point for the assessment of different manipulations. It can be differentiated from the reproductive response by appropriate timing of treatment. Very early spring (May) treatment of Suffolk cross ewes with melatonin initially suppresses prolactin which then, possibly through photorefractoriness, subsequently increases after 3 months of treatment. However the reproductive system does not respond by estrus advance to initial melatonin treatment in May without prior exposure to artificial long photoperiods in Suffolk cross sheep or Saanen goats.

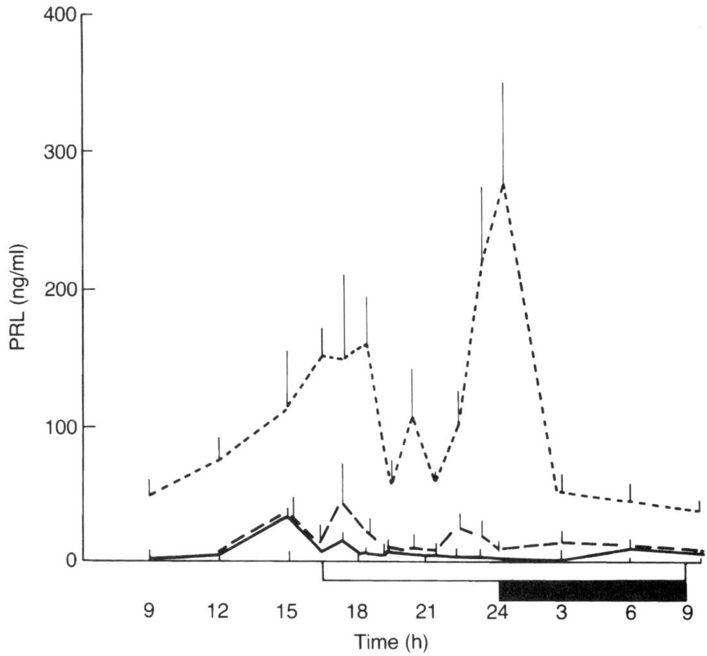

Fig. 5.22 Modification of prolactin level and circadian rhythm by melatonin or extension of darkness. Plasma prolactin levels in three groups of five ewes: (- - -) = 3 mg melatonin by mouth at 1630 h, 16L 8D, lights on 0830 h; (———) = 8L 16D, lights on 0830 h; (······) = natural summer photoperiod (8L 16D). Treatments commenced in mid-June, sampling was in mid-August. Melatonin induces a winter photoperiod prolactin rhythm in spite of prevailing summer photoperiod. (From Symons, A.M., Arendt, J. and Laud, C.A., 1983, reproduced by permission of the Journal of Endocrinology.)

5.14 LACTATION

A major possible disadvantageous 'side effect' of the manipulation of seasonal breeding with melatonin is that it may decrease milk yields in lactating animals. This is not only a potential problem for the farmer, but also for suckling young when mothers are receiving melatonin treatment.

5.15 COAT GROWTH

A number of photoperiodic seasonal breeders produce pelage of commercial interest, for example, the mink and the cashmere goat. It is possible to advance and delay moulting and winter coat growth with suitable artificial photoperiod treatment. A complete reversal of the normal coat growth cycle can accompany induced reversal of the breeding season (Figure 5.23). This implies a winter coat in summer and a summer coat in winter. This is clearly disadvantageous to animals kept outdoors.

The control of coat growth cycles depends on the melatonin signal as the primary information molecule; however, subsequent biochemical events are only partly clarified. In hamsters, sheep and mink, prolactin has been proposed as the hormone responsible for the regulation of coat characteristics; however, the evidence is less convincing for a role of prolactin in goats.

5.16 OTHER PHOTOPERIOD DEPENDENT FUNCTIONS

The pineal gland via melatonin secretion probably plays a role in all photoperiod dependent functions in mammals. Evidence exists to substantiate this statement with respect to behavior, body weight, coat constitution and color (for example the white winter coat of some polar species), antler growth, thyroid activity, appetite, thermoregulation, delayed implantation, embryonic diapause and hibernation. The general principles of action derived from studies in reproduction are very likely to apply to all these functions, and they will not be treated in detail. The interested reader is referred to the reference and further reading lists at the end of this chapter for further information.

5.17 OTHER ORDERS

It is beyond the scope of this book to consider non-mammalian species in any detail. In general the most consistent functions attributable to the pineal of non-mammalian vertebrates are the

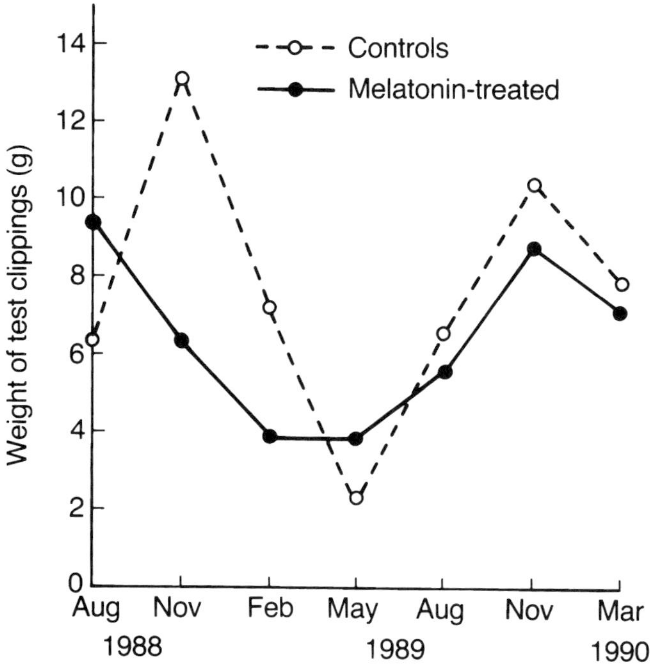

Fig. 5.23 Melatonin changes the timing of seasonal changes in coat growth. Variations in coat weight in British Saanen dairy goats (*n*=6) under natural conditions and after combined long days (20L 4D, February and March) and melatonin (subcutaneous implant, April and May) treatment to advance estrus. Note that a summer coat was present in winter (November to February). By judicious timing of the light and melatonin treatment it is possible to advance estrus but leave sufficient time for regrowth of the winter coat in winter. (Deveson, Arendt and Forsyth, 1992, and Gebbie, Forsyth and Arendt, 1993, unpublished.)

synchronization of endogenous daily and seasonal rhythms which are driven and phased by photoperiod. However a distinct role for the pineal in photoperiodic seasonal events in birds has not been defined. In contrast the circadian system in many birds is pineal dependent. Reviews of this area can be found in, for example, Binkley (1988), Cassone (1990) and Collin, Arendt and Gern (1988).

Of particular interest is a recent report that in the unicellular alga *Gonyaulax* melatonin is able to induce characteristic photoperiodic responses, thus reinforcing the concept that melatonin is a universal indicator of light–dark conditions.

5.18 CHARACTERISTICS OF THE MELATONIN SIGNAL

If melatonin is administered with a 24-hour periodicity as a single daily block of a given duration, then a specific number of daily signals are required in order to induce a photoperiodic response. This number depends on species and even within species, on photoperiodic environment, age, temperature and possibly other variables. The amplitude of the secretion or infusion profile does not appear to play a significant role provided that a threshold limit of physiological night-time levels is exceeded. Whilst episodic secretion of melatonin has been described in sheep and rabbits, this pulsatility equally cannot be important for its physiological effect as non-pulsatile infusions work as effectively in the induction of seasonal responses as endogenous secretion.

The requirement for a single block of melatonin secretion or infusion of a particular duration has been investigated in hamsters. It appears that the maximum break during a block of secretion is around 1 hour. For example in photostimulated hamsters a 10 hour melatonin infusion will induce gonadal regression. If it is split into two blocks of 5 hours separated by a 2-hour gap the two blocks are not additive, and a long day is perceived.

It is important to note that melatonin infusions given with a 24-hour periodicity work regardless of the ambient photoperiod in which the animals are kept. This is true for both hamsters and sheep and argues against the external coincidence theory of photoperiodic time measurement. It is presumed that a hypothetical sensitive phase will be entrained by, and have a consistent relationship to, the ambient light–dark cycle. If this were so then infusions would only be effective under specific lighting regimes.

Further evidence has accumulated against this theory. For

example, ewes will respond to 22-hour cycles of long artificial photoperiods in winter by early estrous offset and if melatonin is infused to photostimulated pinealectomized male hamsters with non-24-hour periodicity whilst the animals are kept in 24-hour light–dark cycles the infusions will induce gonadal regression even if the periodicity is as short as 20 hours.

However if the number of signals is kept constant but frequency is increased to 18 hours or reduced sufficiently beyond 25 hours the response is lost. Whilst this might be taken as evidence for participation of a daily photosensitive phase, Hastings and coworkers (Maywood *et al.*, 1992) interpret these results as evidence for the importance of the 'melatonin free interval' between secretion or infusion episodes. In the author's opinion none of these experiments conclusively refute the hypothesis of a sensitive phase which is itself entrainable by melatonin.

5.19 MELATONIN IMPLANTS: APPLICATIONS TO AGRICULTURE

Much emphasis has been placed in the preceding discussion on the rhythmic characteristics of melatonin secretion or infusion. Subcutaneous implants and other slow-release melatonin preparations such as oral soluble glass boluses are able to generate steady plasma melatonin levels overlaid with the endogenous night-time rise (Figure 5.24). Early work showing clear reproductive effects of melatonin in hamsters was accomplished with such subcutaneous implants. Surprisingly a short day melatonin signal can be given by continuous release, depending on the photoperiodic history of the animal.

For example, ewes receiving subcutaneous melatonin implants in midsummer respond exactly as if they were given long duration rhythmic melatonin by feeding or infusion, i.e. estrus is advanced. Rams receiving 16-week periods of melatonin implants followed by 16-week periods with the implant removed will show alternately short and long day seasonal responses respectively. If implants are left in place for longer periods refractoriness develops and gonadal regression develops in the presence of the implant.

This phenomenon is of considerable interest, not only for the

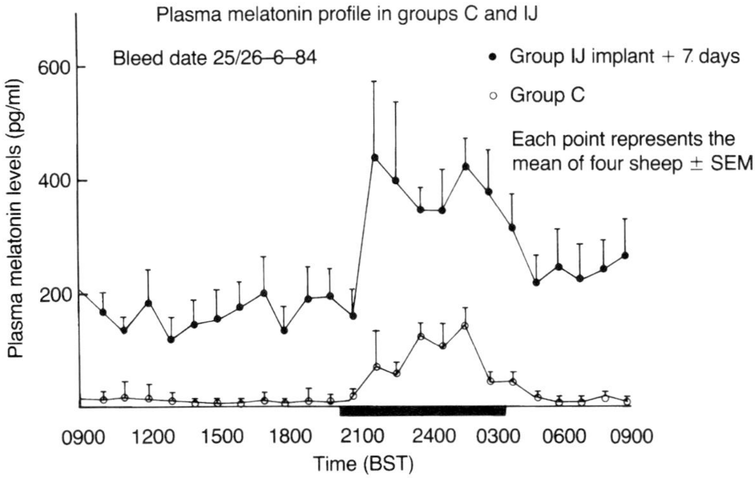

Fig. 5.24 Melatonin implants (in this case 1 g in silastic subcutaneously placed) maintain night-time plasma levels during the day with the endogenous night-time rise superimposed. ○—○, controls: no treatment, ●—● melatonin implant. (Reproduced with permission from English, J. *et al.*; published by the *Journal of Reproduction and Fertility*, 1986.)

manipulation of the breeding season in domestic ruminants but also for acceleration or delay of coat growth in mink, cashmere goats, and foxes. For practical applications implants are considerably more convenient to use than daily feeding or injection of melatonin. Early lambs command higher market prices than those born in a normal season. Lactation in goats is unequally distributed with a lack of milk in winter and reversal of the breeding season would spread milk production throughout the year. Accelerated early growth of the winter coat in mink decreases the costs of husbandry. Control of the breeding season in race horses has very substantial potential benefits in order to achieve earlier maturity of two and three-year-old colts and fillies.

As a result several commercial preparations of melatonin have been or are being developed. They include subcutaneous silicone-based implants, slow-release soluble glass boluses which remain in the rumen for extended periods, and more recently transdermal preparations. A secondary advantage may be improved fertility at least in sheep. At present their availability is limited. Further recent developments including melatonin agonists and antagonists are discussed in Chapter 9. They are destined to be of therapeutic use in humans as well as in the agricultural industry.

8.20 MECHANISM OF ACTION OF MELATONIN

The target sites and mechanisms of action of melatonin in the control of seasonal breeding are only beginning to be identified at the time of going to press (see Chapter 9). Reproductive and other responses to melatonin involve changes in multiple systems. Whilst there is very strong evidence that the primary site of action is on central neuroendocrine control mechanisms, the evidence undoubtedly exists for effects at pituitary and peripheral target organ level. In almost every endocrine system that has been investigated melatonin application has been shown to have some consequence. For reviews see Reiter (1980), Relkin (1976).

Intrahypothalamic administration of melatonin can induce photoperiodic effects in rodents and sheep whereas in other parts of the brain it is ineffective. This clearly implicates the hypothalamus as the central target area in these species. A consensus of opinion would undoubtedly state that melatonin's primary action in the control of reproductive function is to modulate LH secretion. It is well established that changes in LH levels and pulse frequency characterize the breeding and non-breeding season in sheep. In ovariectomized ewes bearing estradiol implants, and in rams, high frequency LH pulses are found in short days during estrus and lower frequency higher amplitude pulses are characteristic of long days and anestrus. LH pulsatility reflects pulsatile release of LHRH and it is reasonable to assume that melatonin acts directly or indirectly on the so-called LHRH pulse generator.

The major central rhythm generator situated in the SCN might therefore be a target site for seasonal effects of melatonin. In the Siberian hamster SCN lesions block the ability of short day type melatonin infusions to induce regression of the gonads, decrease body and fat pad weights and inhibit prolactin secretion. However the effects of long day melatonin have not so far been shown to be abolished in lesioned Siberian hamsters. In the Golden hamster by contrast, long but not short duration daily infusions will elicit gonadal regression in animals lesioned either in the SCN or the PVN. In this species lesions of the anterior hypothalamic area will block short day effects of melatonin.

High-affinity, high-specificity binding sites for ^{125}I-melatonin, which may be receptors of physiological importance have been found within the SCN region of (*inter alia*) rodents, primates, and according to some authors in ruminants, thus reinforcing the potential of the SCN as a target area. However binding is also found localized in a number of other regions (see Chapter 9) and is particularly intense in all mammalian species in the pars tuberalis of the pituitary stalk. The biochemical machinery necessary for a response through inhibition of cAMP has been demonstrated in both the SCN and the pars tuberalis.

It has so far proved impossible to specifically lesion the pars tuberalis without irremediably damaging the portal blood supply to the pituitary. Observations as to its importance in seasonal breeding are therefore circumstantial. Its function(s) is unknown; however gonadotrophes are present in detectable numbers within the structure. A recent report (Nakazawa, Marubayashi and McCann, 1991) has shown that *in vitro* melatonin will inhibit cultured median eminence-pars tuberalis release of LH whilst stimulating LHRH release from the median eminence. The authors hypothesize that melatonin inhibits LH release from pars tuberalis, which in turn, by short loop feedback, removes the inhibitory effect of LH on the LHRH neurons of the median eminence.

These observations are complicated by the fact that in other studies in whole animals (rats, sheep and humans) exogenous melatonin in large doses either does not modify or inhibits LH release. Moreover it also inhibits LHRH and Ca^{2+} induced LH release from dispersed pituitary cell cultures in prepubertal and pubertal rats (Figures 5.25, 5.26). Nevertheless it is clear that biochemical mechanisms exist whereby melatonin can influence LH release.

It has been shown that the content of melatonin binding sites in the pars tuberalis varies with season in hamsters and with time of

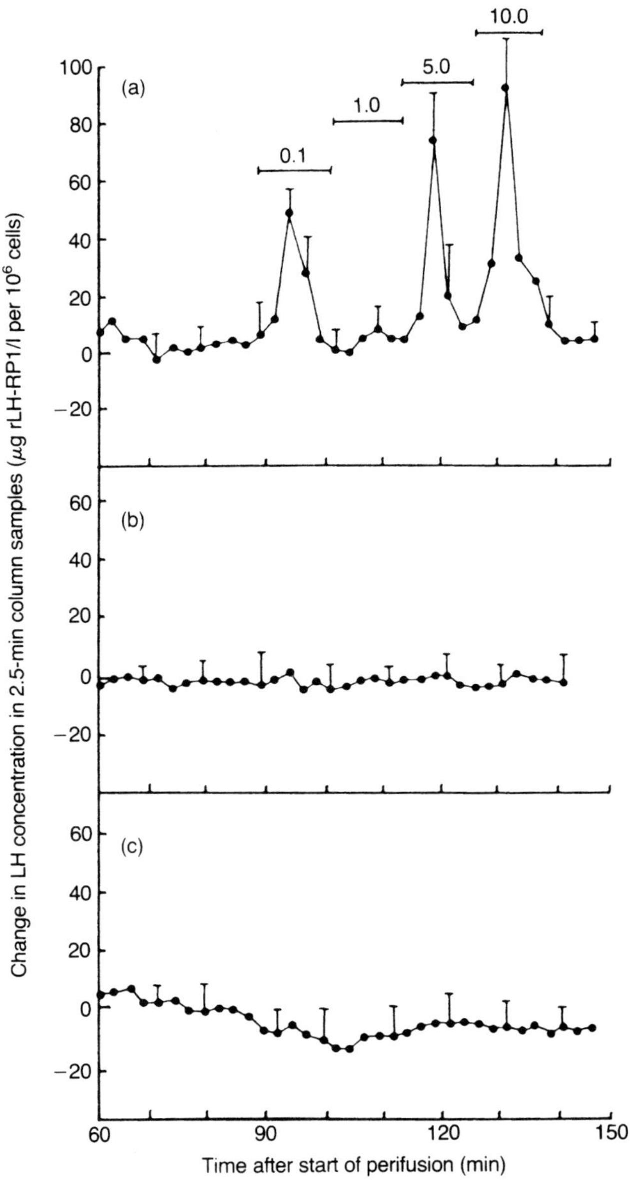
10.0
5.0
1.0
0.1
(a)
(b)
(c)
100
80
60
40
20
0
−20
Change in LH concentration in 2.5-min column samples (μg rLH-RP1/l per 10⁶ cells)
60
90
120
150
Time after start of perifusion (min)

day in hamsters and rats. In some reports, castration of female rats induces a decline in binding site numbers which are restored by estrogen administration. This implies that steroid feedback sensitivity could be modulated by melatonin acting through the pars tuberalis.

Nakazawa and coworkers (1991) have proposed that all anterior pituitary hormones may be produced by the pars tuberalis and this short loop feedback could be a common mechanism of control, susceptible to melatonin. During ontogenesis pituitary hormones first appear in the pars tuberalis: the pineal may influence the developing reproductive system in less photoperiodic species more than in the adult and once again conceptually this short loop pathway may be one primary route of its effects. Melatonin implants placed within the pars tuberalis have now been shown to influence photoperiodic responses in sheep (Lincoln, personal communication).

At present it seems likely that melatonin acts on several brain sites, affecting the central rhythm generating systems, and also downstream on the hormonal expression of electrical signals. There is equally evidence for an effect on neuropeptide pathways within the brain, in particular those of opioid peptides (in seasonal breeders and during pubertal development) and neuropeptide Y (during pubertal development).

One difficulty in relating acute *in vivo* or *in vitro* actions of melatonin to systems physiology is the (usually) slow time course of seasonal changes in target organs. The time course of changes in receptor sensitivity and indeed of the development of tolerance to opioids fits well with the slower manifestation of seasonal responses. In the photorefractory state hamsters and sheep are unresponsive to a particular melatonin duration. This suggests that target tissue receptor down regulation is a consistent factor in seasonal timing mechanisms.

Fig. 5.25 Melatonin in physiological concentrations inhibits LHRH (12.5-min consecutive pulses of 0.1, 1, 5, and 10 nmol LHRH, horizontal lines) induced LH release in a dispersed cell perfusion column preparation of anterior pituitary cells: (a) no melatonin, (b) 0.25 nmol melatonin, (c) 1 nmol melatonin. Values are changes in LH concentration from baseline, mean ± SEM for every 5th sample, n=4. (From Symons, A.M., Arendt, J. and Pryde, S.J., 1985, reproduced by permission of the Journal of Endocrinology Ltd and see Martin and Klein, 1975.)

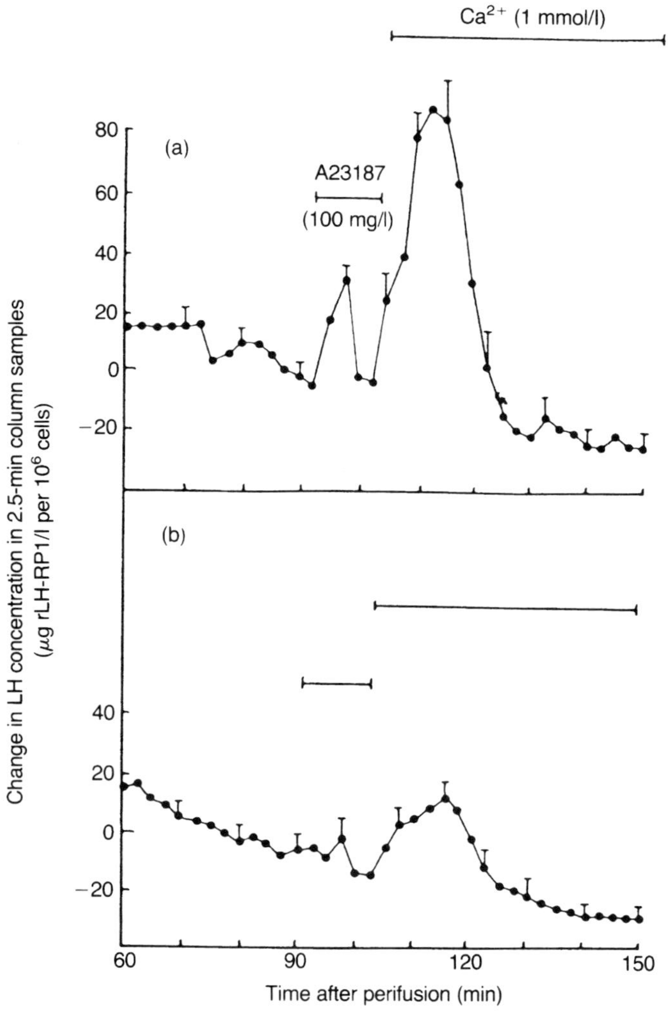

Fig. 5.26 Melatonin in physiological concentrations inhibits Ca^{2+} evoked (12.5-min pulse of the calcium ionophore A23187 followed by Ca^{2+}) LH release in a dispersed cell perfusion column preparation of anterior pituitary cells: (a) no melatonin, (b) 0.25 nmol melatonin. Values are changes in LH concentration from baseline, mean ± SEM for every 5th sample, *n*=4. (From Symons, A.M., Arendt, J. and Pryde, S.J., 1985, reproduced by permission of the Journal of Endocrinology Ltd.)

Lincoln and coworkers (1994) have convincingly shown in sheep that melatonin-dependent photoperiodic variations in prolactin are independent of hypothalamic neural input to the pituitary and probably are a result of a direct action of melatonin on the par tuberalis.

A more detailed account of melatonin receptors is given in Chapter 9.

REFERENCES

Arendt, J. (1985) Mammalian pineal rhythms. *Pineal Research Review*, **3**, 161–213.

Arendt, J. (1986) Role of the pineal gland and melatonin in seasonal reproductive function in mammals. *Oxford Reviews of Reproductive Biology*, **8**, 266–320.

Arendt, J., Brown, W.B., Forbes, J.M. *et al.* (1980) Effect of pinealectomy on immunoassayable melatonin in sheep. *Journal of Endocrinology*, **85**, 1–2.

Arendt, J., Symons, A.M., Laud, C.A. *et al.* (1983) Melatonin can induce early onset of the breeding season in ewes. *Journal of Endocrinology*, **97**, 395–400.

Binkley, S. (1988) *The Pineal: Endocrine and Nonendocrine Function*, Prentice Hall, Englewood Cliffs, New Jersey, USA.

Bittman, E.L. (1985) The role of rhythms in the response to melatonin, in *Photoperiodism, Melatonin and the Pineal* (eds D. Evered and S. Clark), Ciba Foundation Symposium 117, Pitman, London, pp. 149–69.

Carter, D.S. and Goldman, B.D. (1983) Antigonadal effects of timed melatonin infusion in pinealectomized male Djungarian hamsters (*Phodopus sungorus*); Duration is the critical parameter. *Endocrinology*, **113**, 1261–7.

Cassone, V.M. (1990) Melatonin: time in a bottle. *Oxford Reviews of Reproductive Physiology*, **12**, 319–67.

Collin, J.P., Arendt, J. and Gern, W.A. (1988) Le "Troisieme Oeil". *La Recherche*, **203**, 1154–65.

Deveson, S., Arendt, J. and Forsyth, I. (1992) Role of the pineal and melatonin rhythms in reproductive performance in female domestic ungulates. *Reviews of Animal Reproduction Science*, **30**, 113–34.

Elliot, J.A. (1981) Circadian rhythms, entrainment and photoperiodism in the Syrian hamster, in *Biological Clocks in Seasonal Reproductive Cycles* (eds B.K. Follett and D.E. Follett), Scientechnica, Bristol, pp. 203–18.

English, J., Poulton, A.L., Arendt, J. *et al.* (1986) A comparison of the efficiency of melatonin treatments in advancing estrus in ewes. *Journal of Reproduction and Fertility*, **77**, 321–7.

Foster, D.L., Ebling, F.J.P. and Claypool, L.E. (1988) Timing of puberty by photoperiod. *Reproduction, Nutrition, Development*, **28(2B),** 349–64.

Foster, D.L., Ebling, F.J.P., Claypool, L.E. *et al.* (1988) Cessation of long day melatonin rhythms time puberty in a short day breeder. *Endocrinology*, **123**, 1636–41.

Hastings, M.H., Walker, A.P., Powers, J.B. *et al.* (1989). Differential effects of photoperiodic history on the responses of gonadotrophins and prolactin to intermediate daylengths in the male Syrian hamster. *Journal of Biological Rhythms*, **4**, 335–50.

Herbert, J. (1981) The pineal gland and photoperiodic control of the ferret's reproductive cycle, in *Biological Clocks in Seasonal Reproductive Cycles* (eds B.K. Follett and D.E. Follett), Wright, Bristol, pp. 261–76.

Hoffman, K. (1981) Photoperiodism in vertebrates, in *Handbook of Behavioural Neurobiology (Suppl 4)*, (ed J. Aschoff J), pp. 449–73.

Karsch, F.J., Bittman, E.L., Foster, D.L. *et al.* (1984) Neuroendocrine basis of seasonal reproduction. *Recent Progress in Hormone Research*, **40**, 185–232.

Karsch, F.J., Malpaux, B., Wayne, N.L. *et al.* (1988) Characteristics of the melatonin signal that provide the photoperiodic code for timing seasonal reproduction in the ewe. *Reproduction Nutrition Development*, **28**, 459–72.

Lang, U., Aubert, M.L., Conne, B.S. *et al.* (1983) Influence of exogenous melatonin on melatonin secretion and the neuroendocrine reproductive axis of intact male rats during sexual maturation. *Endocrinology*, **112**, 1578–84.

Lincoln, G. (1979) Photoperiodic control of seasonal breeding in the ram: participation of the cranial sympathetic nervous system. *Journal of Endocrinology*, **82**, 135–47.

Lincoln, G.A. (1994) Duration-variable melatonin signal acts directly in the pituitary gland to mediate effects of photoperiod in the ram. Abstract 136, Society for Research in Biological Rhythms, Florida, May 1994.

Maywood, E.S., Grosse, J., Lindsay, J.O., Karp, J.D. *et al.* (1992) The effect of signal frequency on the gonadal response of male Syrian hamsters to programmed melatonin infusions. *Journal of Neuroendocrinology*, **4**, 3–6.

Malpaux, B., Robinson, J., Wayne, N. *et al.* (1989) Regulation of the onset of the breeding season of the ewe: importance of long days and of an endogenous reproductive rhythm. *Journal of Endocrinology*, **122**, 269-N78.

Martin, J.E. and Klein, D.C. (1975) Melatonin inhibition of neonatal pituitary response to luteinising hormone-releasing factor. *Science*, **191**, 301–2.

Nakazawa, K., Marubayashi, U. and McCann, S.M. (1991) Mediation of the short-loop feedback of luteinising hormone (LH) on LH-releasing hormone release by melatonin-induced inhibition of LH release from the pars tuberalis. *Proceedings of the National Academy of Sciences, USA*, **88**, 7576–9.

Pengelley, E. and Asmundson, S. (1974) in *Circannual Clocks, Annual Biological Rhythms* (ed. E. Pengelley), Academic Press, New York, p. 105.

Pelletier, J. and Thimonier, J. (1987) The measurement of daylength in the Ile-de-France ram. *Journal of Reproduction and Fertility*, **81**, 181–6.

Ravault, J. (1976) Prolactin in the ram: seasonal variations in the concentration of blood plasma from birth until three years old. *Acta Endocrinologica*, **83**, 720–5.

Reiter, R.J. (1980) The pineal and its hormones in the control of reproduction in mammals. *Endocrinology Reviews*, **1**, 109–31.

Relkin, R. (1976) *The Pineal.* Eden Press, Montreal and Lancaster.

Rivest, R.W., Lang, U., Aubert, M.L. and Sizonenko, P.C. (1985) Daily administration of melatonin delays rat vaginal opening and disrupts the first estrus cycles: evidence that these effects are synchronised by the onset of light. *Endocrinology*, **116**, 779–87.

Steinlechner, S. and Niklowitz, P. (1992) Impact of photoperiod and melatonin on reproduction in small mammals. *Animal Reproduction Science*, **30**, 1–28.

Symons, A.M., Arendt, J. and Laud, C.A. (1983) Melatonin feeding decreases prolactin levels in the ewe. *Journal of Endocrinology*, **99**, 41–6.

Symons, A.M., Arendt, J. and Pryde, S.J. (1985) Differential effects of melatonin on the stimulated release of LH from dispersed pituitary cells of the prepubertal female rat. *Journal of Endocrinology*, **107**, 107–12.

Weaver, D.R. and Reppert, S.M. (1986) Maternal melatonin communicates daylength to the fetus in Djungarian hamsters. *Endocrinology*, **119**, 2861–3.

Woodfill, C.J.I., Robinson, J.E., Malpaux, B.M. and Karsch, F.J. (1991) Synchronisation of the circannual reproductive rhythm of the ewe by discrete photoperiodic signals, *Biology of Reproduction*, **45**, 110–21.

FURTHER READING

Badia, P., Myers, B. and Murphy, P. (1992) Melatonin and thermoregulation, in *Melatonin: Biosynthesis, Physiological Effects, and Clinical Applications* (eds R.J. Reiter and H.S. Yu), CRC Press, Boca Raton, Florida.

Bittman, E.L. (1993) The sites and consequences of melatonin binding in mammals. *American Zoologist*, **33**, 200–1.

Bunning, E. (1967) *The Physiological Clock*, Longmans-Springer-Verlag, New York.

Evered, D. and Clark, S. (eds) (1985) *Photoperiodism, Melatonin and the Pineal*, Ciba Foundation Symposium 117, Pitman, London.

Deveson, S., Forsyth, I.A. and Arendt, J. (1992) Retardation of pubertal development by prenatal long days in goat kids born in autumn. *Journal of Reproduction and Fertility*, **95**, 629–37.

Duncan, M.J. and Goldman, B.D. (1984) Hormonal regulation of the annual pelage color cycle in the Djungarian hamster. *Phodopus sungorus*. II. Role of prolactin. *Journal of Experimental Zoology*, **230**, 97–103.

Ebling, F.J.P. and Foster, D.L. (1989) Pineal melatonin rhythms and the timing of puberty in mammals. *Experientia*, **45**, 946–55.

Fiske, V.M. (1941) Effect of light on sexual maturation, estrous cycles, and anterior pituitary of the rat. *Endocrinology*, **29**, 187–96.

Glass, J.D. (1988) Neuroendocrine regulation of seasonal reproduction by the pineal and melatonin. *Pineal Research Reviews*, **6**, 219–59.

Hastings, M., Herbert, J., Martenz, N.D. *et al.* (1985) Melatonin and the brain in photoperiodic mammals, in *Photoperiodism, Melatonin and the Pineal*, Ciba Symposium Foundation 117, Pittman, London, pp. 57–70.

Kennaway, D.J., Dunstan, E.A. and Staples, L.D. (1987) Photoperiodic control of the onset of breeding activity and fecundity in ewes. *Journal of Reproduction and Fertility*, **34**, 187–99.

Kennaway, D.J., Gilmore, T.A. and Seamark, R.F. (1982) Effect of melatonin feeding on serum prolactin and gonadotrophin levels and the onset of seasonal oestrous cyclicity in sheep. *Endocrinology*, **110**, 1766–1772.

Lincoln, G.A. and Forbes, J.M. (1984) Seasonal physiology of pinealectomised rams. *Society for the Study of Fertility*, Abstract No. 33.

Lincoln, G.A. and Short, R.V. (1980) Seasonal breeding: Nature's contraceptive. *Recent Progress in Hormone Research*, **36**, 1–52.

Loudon, A.S.I., Curlewis, J. and English, J. (1985) The effect of melatonin on the seasonal embryonic diapause in the Bennett's wallaby. *Journal Zoology London (A)*, **206**, 35–9.

Martinet, L. and Allain, D. (1985) Role of the pineal gland in the photoperiodic control of reproductive and non-reproductive functions in mink (*Mustela vison*), in *Photoperiodism, Melatonin and the Pineal*, (eds D. Evered and S. Clark), Ciba Foundation Symposium 177, Pitman, London, pp. 170–87.

Morgan, P.J., Williams, L.M., Davidson, G. *et al.* (1989) Melatonin receptors on ovine pars tuberalis: characterisation and autoradiographical localisation. *Journal of Neuroendocrinology*, **1**, 1–4.

Ortavant, R., Pelletier, J., Ravault, J.-P. *et al.* (1985) Photoperiod: main proximal and distal factor of the circannual cycle of reproduction in farm mammals. *Oxford Reviews of Reproductive Biology*, **7**, 305–45.

Parkes, A.S. (1976) *Patterns of Sexuality and Reproduction*, Oxford University Press, Oxford.

Poulton, A.L., Symons, A.M., Kelly, M.I. *et al.* (1987) Intra-ruminal soluble glass boluses containing melatonin can induce early onset of ovarian activity in ewes. *Journal of Reproduction and Fertility*, **80**, 235–9.

Reiter, R.J. (1989) Melatonin: its sources, its message, and the interpretation of the message, in *Advances in Pineal Research*, Vol. 3 (eds R.J. Reiter and S.-P. Pang), John Libbey, London.

Rivest, R.W., Aubert, M.L. and Lang U. (1986) Puberty in the rat: Modulation by melatonin and light, in *Melatonin in Humans* (eds R.J. Wurtman and F. Waldhauser), *Journal of Neural Transmission* (Suppl 21).

Robinson, J.E. and Karsch, F.J. (1984) Refractoriness to inductive day lengths terminates the breeding season of the Suffolk ewe. *Biological Reproduction*, **31**, 656–63.

Sizonenko, P.C., Lang, U., Rivest, R.W. *et al.* (1985) The pineal and pubertal development, in *Photoperiodism, Melatonin and the Pineal* (eds D. Evered and S. Clark), Ciba Foundation Symposium 117, Pitman, London, pp. 208–230.

Tamarkin, K., Baird, C.J. and Almeida, O.F.X. (1985) Melatonin: a coordinating signal for mammalian reproduction. *Science*, **227**, 714–20.

Vriend, J. (1983) Evidence for pineal gland modulation of the neuroendocrine-thyroid axis. *Neuroendocrinology*, **36**, 68–78.

Williams, L.M., Martinoli, M.G., Titchener, L.T. *et al.* (1991) The ontogeny of central melatonin binding sites in the rat. *Endocrinology*, **128**, 2083–90.

Zisapel, N., Nir, I. and Laudon, M. (1988) Circadian variations in melatonin-binding sites in discrete areas of the male rat brain. *FEBS Letters*, **232**, 172.

Zisapel, N., Oaknin, S., Anis, Y. *et al.* (1991) Melatonin receptors in discrete areas of the rat and Syrian hamster brain: modulation by melatonin, pinealectomy, testosterone and the photoperiod, in *Advanced Pineal Research*, **5**, 175–81 (eds J. Arendt and P. Pevet).

Role of the pineal gland and melatonin in circadian rhythms

6

Contents

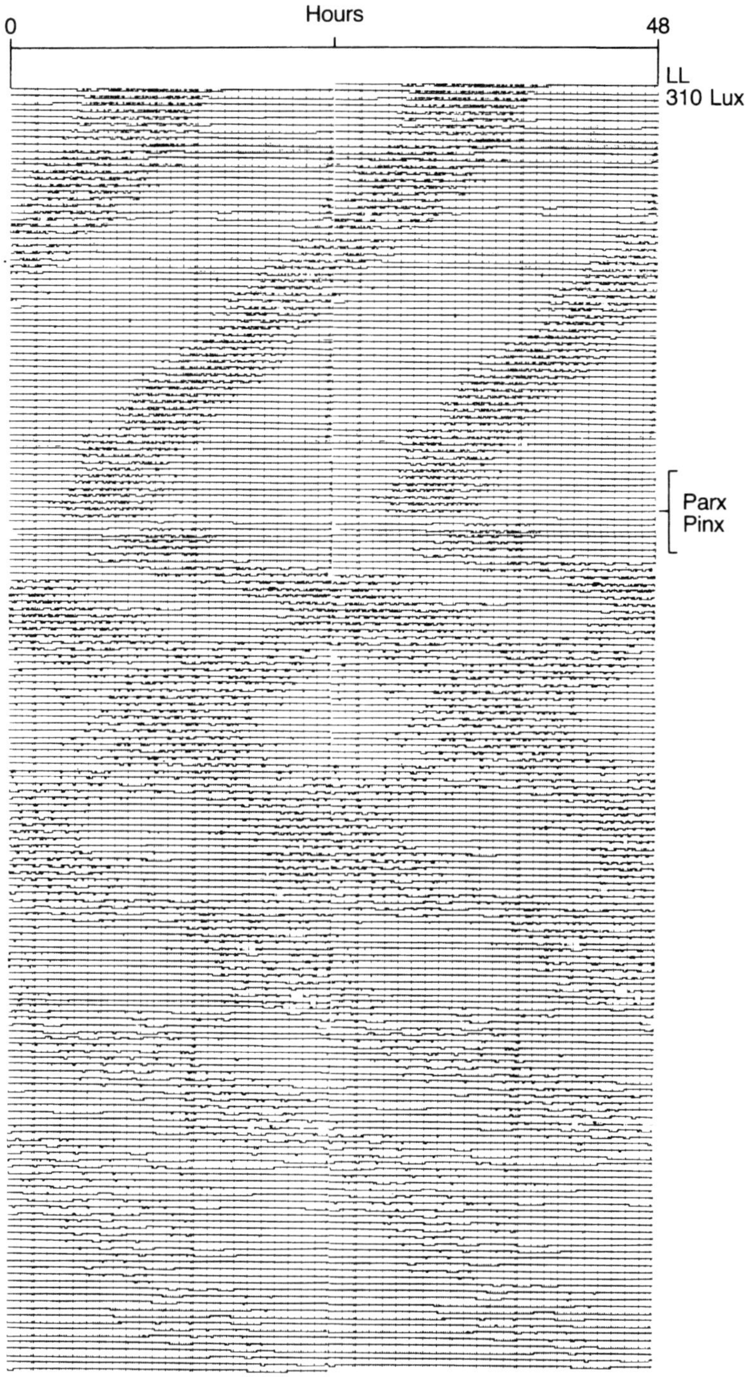

Hours
0
48
LL
310 Lux
Parx
Pinx

6.1 THE PINEAL AND CIRCADIAN RHYTHMS IN NON-MAMMALIAN VERTEBRATES

There is extensive literature on the importance of the pineal (and also the retina) to the control of circadian rhythms in lower vertebrates. This is beyond the scope of this book to consider in any detail and good reviews exist (Cassone, 1990). A brief account is necessary however, as it is through the importance of the pineal in lower vertebrates that we have come to understand much of its circadian importance in mammals.

In several sauropsid (reptiles and birds) species kept in constant conditions pinealectomy induces arrhythmic behavior (but not necessarily all behaviors: in the European starling for example, feeding rhythmicity persists after pinealectomy) or changes in tau. In others where pinealectomy alone does not suffice, blinding and pinealectomy abolish rhythmicity. In the lizard *S. olivaceous* pinealectomy leads to the manifestation of two or more endogenous rhythms of activity which are not coupled and run with different periodicities (Figure 6.1).

Thus the pineal may act as a central rhythm generator, a coupling agent or a combination of both. Phase is communicated humorally at least in sparrows: section of pineal innervation does not affect the endogenous rhythmicity in this species and the circadian phase of a donor passerine bird can be transferred through implantation of its pineal to the anterior chamber of the eye of a recipient bird (Figure 6.2).

Melatonin is produced rhythmically by both the pineal and the retina in many lower vertebrates and probably serves as the common humoral signal for circadian organization. There is a great variety of attribution of function with one organ assuming part or all of the circadian functions of the other from species to species. Pineals from some birds and lizards can be cultured over several days and continue to produce melatonin rhythmically with no external temporal input

Fig. 6.1 The pineal complex in lizards can function as a coupling device for the circadian system. Double plotted free-running rest–activity cycle of *S. olivaceous* in constant light showing splitting of the activity rhythm into two circadian components with different periods after removal of the pineal system (Parx, Pinx). (Reproduced with permission from Underwood, H.; published by Wright, 1981).

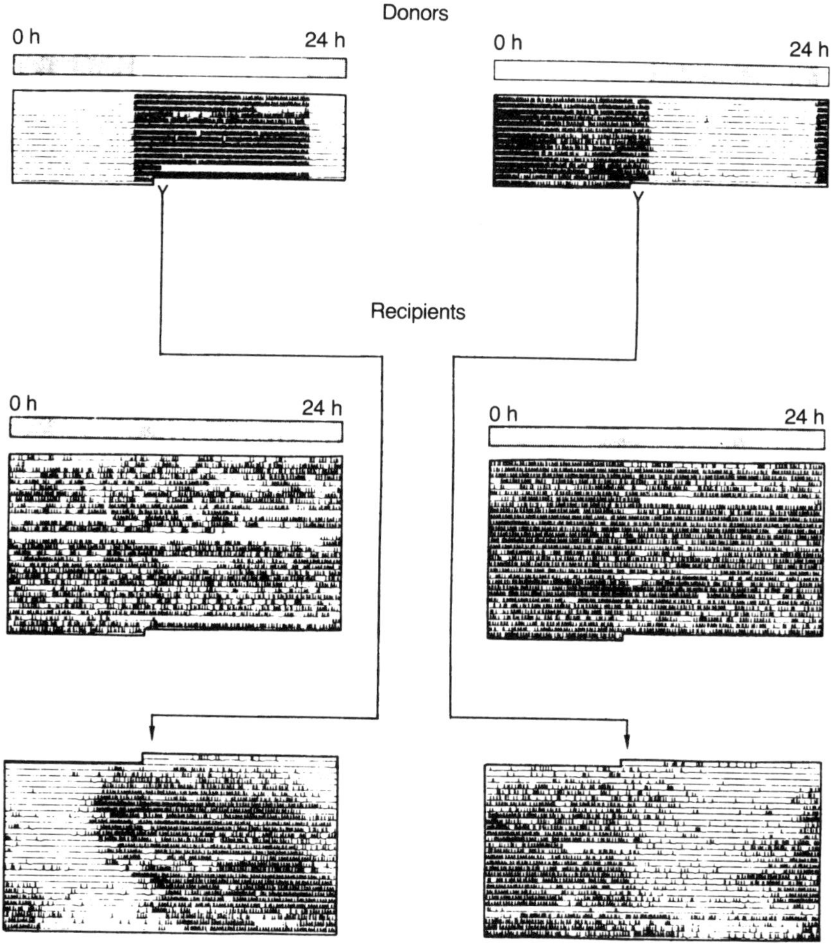

Fig. 6.2 The pineal in some birds can act as a master clock or circadian rhythm generator. Transplantation transfers the phase of the donor to pinealectomized sparrows, *Passer domesticus*, whose circadian rhythms of activity (and deep body temperature) are abolished by pinealectomy. Experimental design for transplantation of pineals from donors on different light schedules. The donors' light cycles and activity records are shown in the top panel. The pinealectomized hosts were kept in constant darkness. Their activity records before transplantation are shown in the middle panel and after transplantation in the bottom panel. (Reproduced with permission from Zimmerman, N.H. and Menaker, M.; published by the National Academy of Sciences, 1979).

(see Chapter 4). However they will entrain *in vitro* to LD cycles. They clearly contain the machinery of a pacemaker or biological clock together with the capacity for photoreception.

Whilst the pineal does possess the capacity to act as a central pacemaker it is likely that in the intact animal the SCN are also involved.

Melatonin administration has profound effects on activity rhythms in sauropsids. Subcutaneous or intraperitoneal implantation of melatonin shortens circadian period and in higher doses can abolish rhythmicity altogether. Timed daily administration of melatonin will entrain rest activity cycles in some species and the phase relationship assumed is related to the phase relationship of rest–activity with endogenous melatonin secretion. It is of particular interest that much higher concentrations of melatonin binding sites are present in the brains of those birds, reptiles and fish which have been investigated in comparison to mammals. This of course probably reflects the greater importance of melatonin to non-mammalian vertebrate physiology.

There may appear to be consistency in these observations; however, it is not possible to generalize to all lower vertebrates or even to all sauropsids. Some fish and lizards may not possess an endogenous circadian rhythm of melatonin whether it be generated in the pineal the SCN or elsewhere. Discrete pineal glands may even be absent, for example in crocodiles. Nevertheless it is evident that melatonin is concerned in one way or another with circadian organization and in evolutionary terms its functions probably predate the biological clock as we know it today.

6.2 ROLE OF THE PINEAL IN MAMMALIAN CIRCADIAN RHYTHMS

Until quite recently received opinion was that the pineal did not have a role in the mammalian circadian system. This conclusion was essentially based on one approach to function: the effects of pinealectomy on free-running rest–activity cycles. According to many careful experiments of this nature pinealectomy has no effect on tau in rats and hamsters. However as long ago as 1970 Quay reported that when rats are subjected to forced phase shifts of the light–dark cycle,

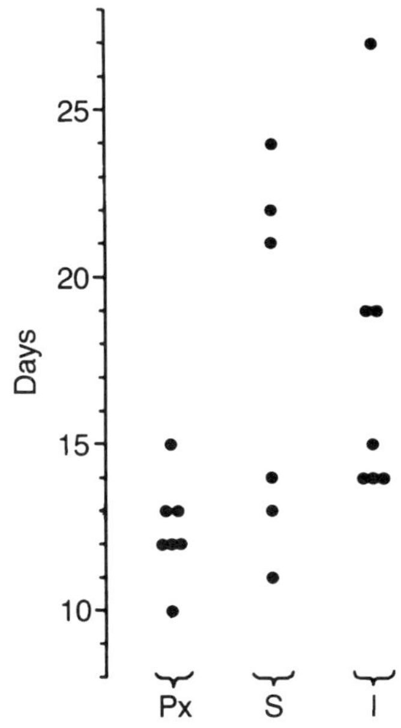

Fig. 6.3 Pinealectomy enhances the rate of re-entrainment of the rest–activity cycle in rats. Number of days taken to re-entrain activity rhythms after a 5 h advance of the light–dark cycle in pinealectomized (Px), sham-operated (S) and intact (I) rats that adapted by phase advance. (Reproduced with permission from Armstrong, S. and Redman, J.; published by Pitman, 1985.)

pinealectomy increased the rate of re-entrainment. He even suggested that the rate of adaptation to phase shift could be a diagnostic approach to the detection of destructive lesions of the pineal in humans. Armstrong and Redman (1985) have confirmed this report (Figure 6.3).

As these observations could not be reconciled with the results of free-running experiments they were largely ignored for a decade. More recently Cassone (1992) has shown that pinealectomy of hamsters in constant light leads to a major disruption of the circadian system (Figure 6.4).

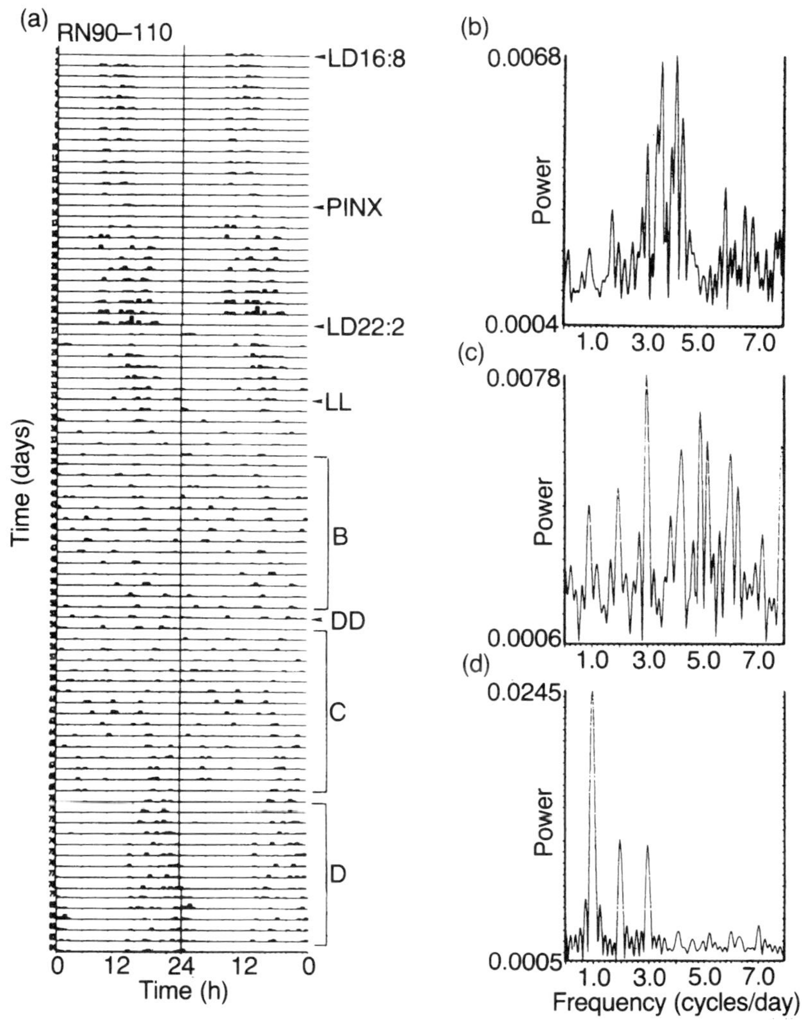

Fig. 6.4 Pinealectomy disrupts circadian patterns of wheel-running activity in rats maintained in LL. A continuous, double plotted actogram of a typical rat that had been pinealectomized (PINX). Although the length of activity increased after surgery in this particular rat in 8L 16D no consistent effects of surgery were found in general. In LL this rat, like most, expressed high-frequency, ultradian patterns of activity which merged into a single circadian bout in DD. (Reprinted from Cassone, V., 1992, by permission of the Guildford Press.)

In the late 1970s accumulated experience suggested to several investigators that administration of melatonin to mammals might provide more rewarding information concerning its role in circadian rhythms than pinealectomy. Several threads of information led to these initiatives.

It was becoming clear that whilst the pineal had an important role in the transmission of daylength information, it required either very long-term experiments as, for example, in ferrets (Chapter 5), or a direct challenge in the form of changing artificial daylength (analogous to the forced circadian phase shift employed by Quay) for its effects to be manifested. Melatonin administration on the other hand had marked and rapid effects on reproductive physiology. Secondly it was evident that changes in circadian physiology accompany photoperiodic responses, for example, splitting of activity–rest cycles in suitable photoperiods and major changes in the waveform of prolactin secretion (see Chapter 5) and body temperature (Figure 6.5). Thus it was reasonable to suppose that melatonin could bring about changes in circadian physiology.

Thirdly adaptation to change in daylength clearly involves changes in phase relationship of circadian rhythms with respect to the 24-hour day. In particular the onset and offset of melatonin are respectively advanced and delayed during lengthening nights.
One interpretation of the latter observation is that shortening daylength removes suppression of melatonin secretion at the beginning and/or the end of the night and that the released melatonin secretion might then act to phase advance and/or phase delay the circadian rhythm generating system. If two oscillators (dawn and dusk) are concerned this would imply a phase advance of the dusk and a phase delay of the dawn oscillator with a lengthening night.

There is no doubt that, at least in photoperiodic responses the pineal via melatonin indicates relative changes in daylength. If it is concerned with change in phase of the circadian cycle then it is not surprising that experiments in constant conditions were in general so unrewarding.

One further important point must be made. Implants of melatonin were initially used in many mammalian circadian experiments. The availability of melatonin is then continuous: a situation

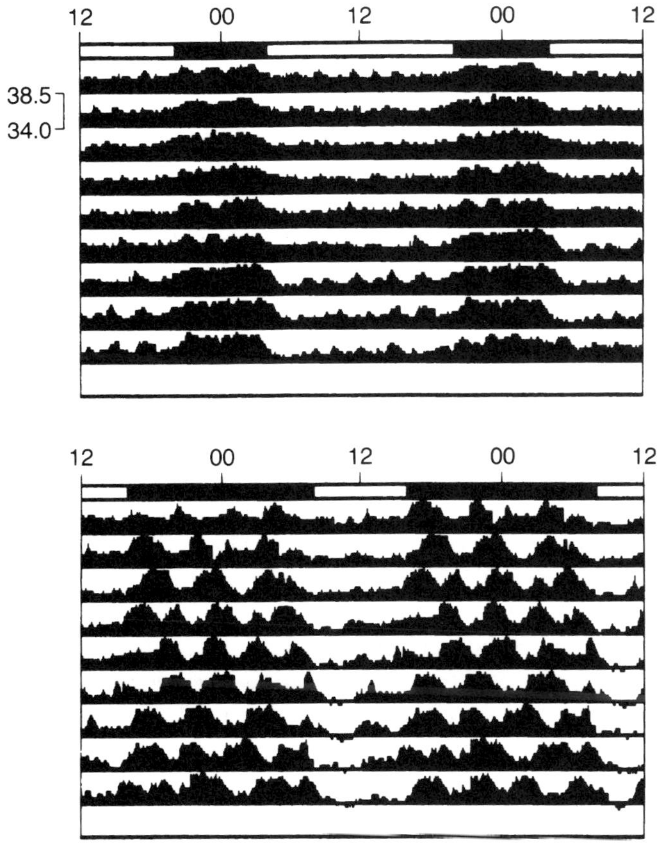

Fig. 6.5 Changes in circadian waveform with photoperiod. Double plots of body temperature in a hamster kept in long photoperiod (upper panel) and short photoperiod (lower panel). Abscissa = time of day (h). Ordinate = body temperature (°C). Each line represents two consecutive days. Differences between body temperature and the 34 °C line are filled in to aid visualization. As an example the temperature scale for days 2 and 3 is shown. (Reproduced with permission from Ruf, T., Steinlechner, S. and Heldmaier, G.; published by Peter Lang, 1987.)

never found in intact animals and the interpretation of results is very difficult. Nevertheless Murakami and colleagues showed in 1983 that melatonin implants near the SCN in rats hastened adaptation of cortisol rhythms to forced phase shift. This effect may well be akin to the effects of pinealectomy as described by Quay (see earlier) by abolishing the perceived melatonin rhythm.

6.3 TIMED ADMINISTRATION OF MELATONIN

The observations of two groups working simultaneously and independently with rats and humans, and noting the circadian effects of timed daily melatonin administration, showed for the first time that melatonin had entraining and phase shifting effects respectively in mammals.

Armstrong and coworkers (Redman, Armstrong and Ng, 1983; Armstrong, 1989) injected melatonin at 24-hour intervals into free-running rats and obtained entrainment of the rest–activity cycle when saline control injections were usually ineffective. Arendt and coworkers (1984; 1985) gave daily melatonin orally in the late afternoon to human volunteers and obtained a phase advance of the endogenous melatonin rhythm, of evening fatigue and of prolactin morning offset. The work in humans will be considered in more detail in Chapter 8; however, it is of interest to note in the light of the earlier discussion on morning and evening oscillators that the onset of secretion was advanced more than the offset.

6.4 EFFECTS OF MELATONIN ON RODENT ACTIVITY RHYTHMS

The entrainment of rats occurred, when melatonin injections corresponded with activity onset, by phase advance of the onset, with the offset requiring several cycles to stabilize and entrain thereafter (Figure 6.6a). This is further evidence that offset and onset are

Fig. 6.6(a) Melatonin entrains free-running activity–rest cycles in rats when the injection time coincides with activity onset. Double-plotted running-wheel activity records of rats housed under DD and free-running with periods greater than 24 h. A–D, Stage 1, pre-injection, Stage 2, daily melatonin injection (1 mg/kg) as indicated by the arrow, Stage 3 postinjection. E–H controls injected with vehicle in Stage 2. One control rat entrained. (Reproduced with permission from Armstrong, S. and Redman, J.; published by Pitman, 1985.)

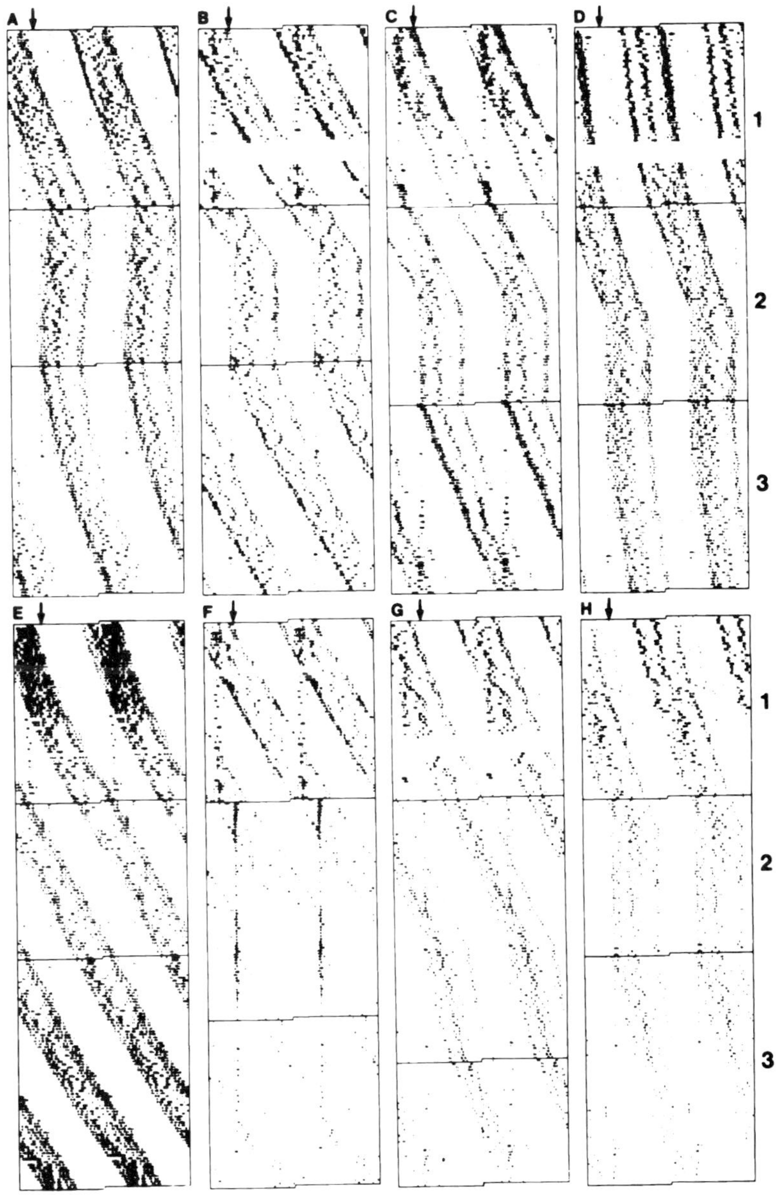
A
B
C
D
E
F
G
H
1
2
3
1
2
3

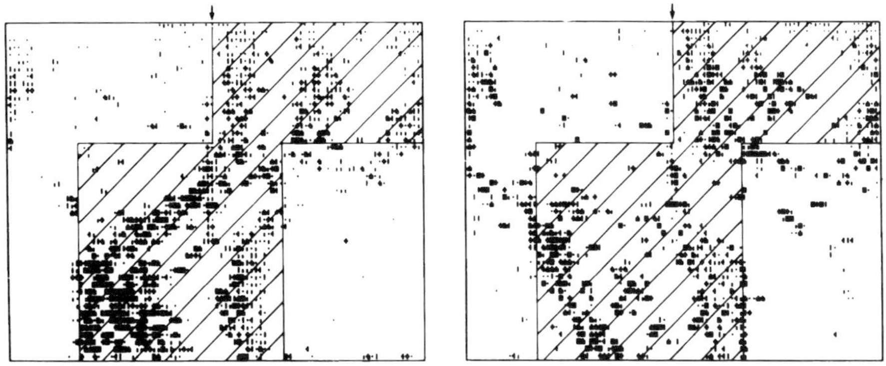

Fig. 6.7 Melatonin can change the direction of re-entrainment. Activity plot for a melatonin injected (1 mg/kg, daily indicated by arrows) rat that phase advanced (a) and a control (vehicle injected) rat that phase delayed (b) after an 8-h phase advance of the light–dark cycle. 10/10 melatonin rats advanced and 10/10 control rats delayed. (Reproduced with permission from Armstorong, S. and Redman, J.; published by Pitman, 1985.)

There are also reports of similar results in Golden hamsters by single injections, and entrainment in free-running Siberian and European hamsters by daily infusion of melatonin. Although entrainment by injection requires what appear to be large amounts of melatonin, entrainment by infusion follows administration of near-physiological night-time levels. It is still not known what the physiological concentration of melatonin is at its target sites and it is possible, given the very short half-life of injected material and the large first pass effect, that supraphysiological plasma levels do not necessarily exceed physiological target site concentrations.

Moreover melatonin in pineal efflux of the confluens sinus is an order of magnitude more concentrated than in the peripheral plasma and may reach the brain partly by diffusion from this source.

Fetal hamsters can be entrained by maternal injections of melatonin

Fig. 6.8 Phase response curve (PRC) for the effect of melatonin (50 μg/kg s.c.) on activity–rest cycles in rats. Up to one hour advances are obtained between CT 9–11 and no phase delays. This is compared with the PRC for iguanid lizards where both advances and delays are present but primarily delays. (Reproduced with permission from Armstrong, S.; published by John Libbey, 1991.)

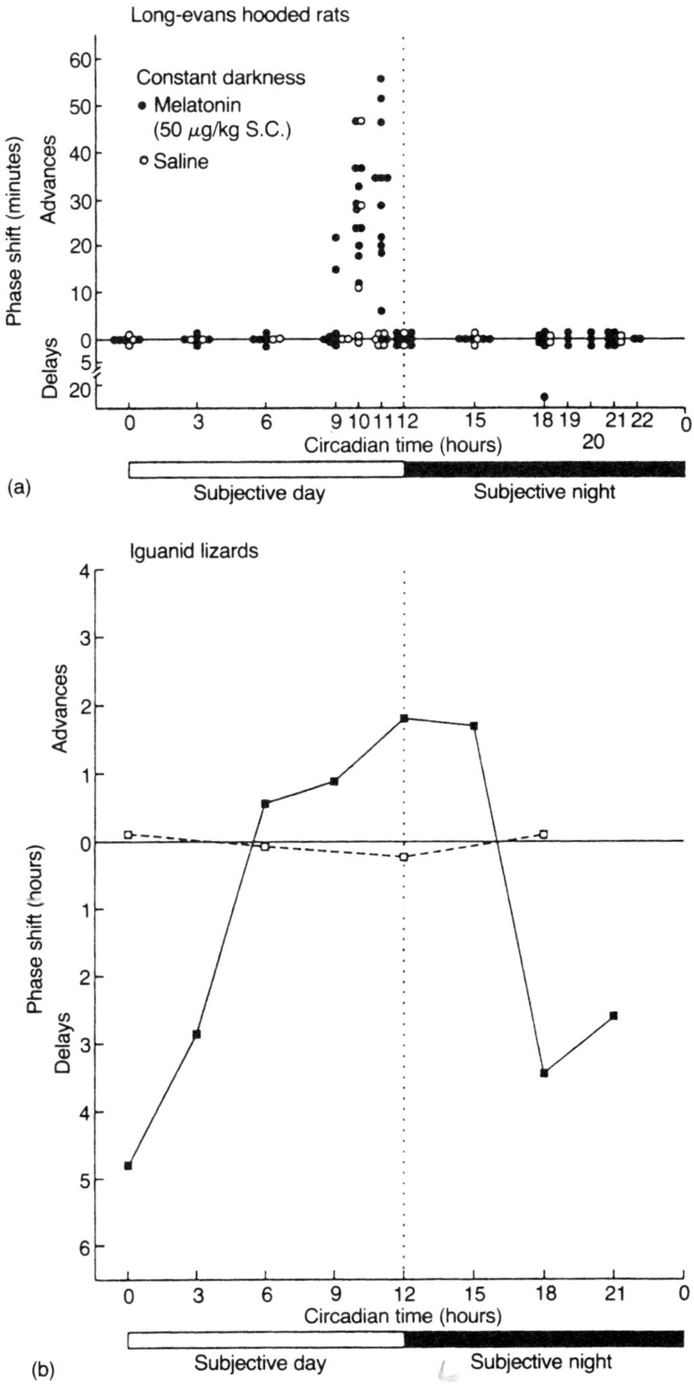
Long-evans hooded rats
Constant darkness
• Melatonin
(50 μg/kg S.C.)
○ Saline
Phase shift (minutes)
Advances
Delays
60
50
40
30
20
10
0
5
20
0
3
6
9 10 11 12
15
18 19
21 22
20
0
Circadian time (hours)
Subjective day
Subjective night
(a)
Iguanid lizards
Phase shift (hours)
Advances
Delays
4
3
2
1
0
1
2
3
4
5
6
0
3
6
9
12
15
18
21
0
Circadian time (hours)
Subjective day
Subjective night
(b)

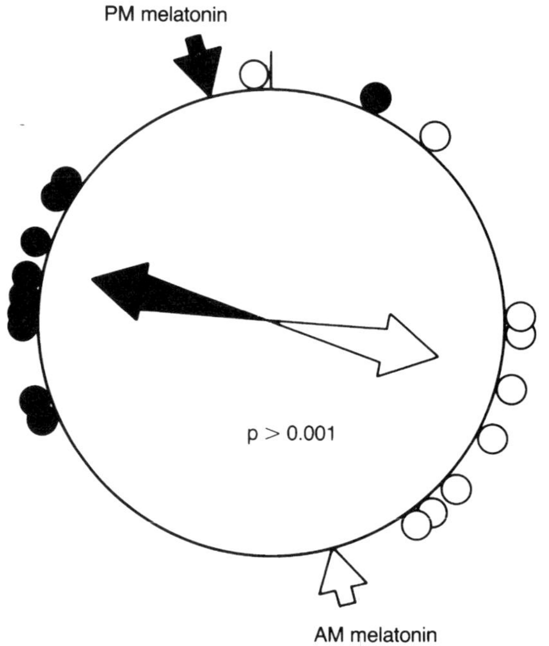

Fig. 6.9 Melatonin administration during gestation can set the circadian phase of hamster pups. Average phases of all litters born to hamsters that were injected with melatonin (10 or 50 μg daily) during gestation plotted on a circle representing day of weaning. ●=Phases of litters from hamsters injected with melatonin at 2300 h (pm); ○=phases of litters from hamsters injected with melatonin at 1100 h (am). Arrows outside circle=times of melatonin injections during gestation, arrows inside circle=average phase and scatter of groups. (Reproduced with permission from Davis, F.C. and Mannion, J.; published by the American Physiological Society, 1988.)

at 24-hour intervals at specific circadian phases (Figures 6.9, 6.10). The phase position of the fetus is assessed postnatally. One interesting observation is that whilst SCN lesioned Golden hamsters did not entrain to melatonin injections, animals whose rhythmicity was restored by fetal hypothalamic grafts did entrain to melatonin.

In species such as these with free running tau usually greater than 24 hours, the small phase advances induced by melatonin are theoretically sufficient to synchronize the rest–activity cycle to 24 hours. However the intact rat or hamster will never in principle see a rise in endogenous melatonin at CT 9, 10 and 11 as endogenous melatonin is secreted with a substantial delay relative to dark onset.

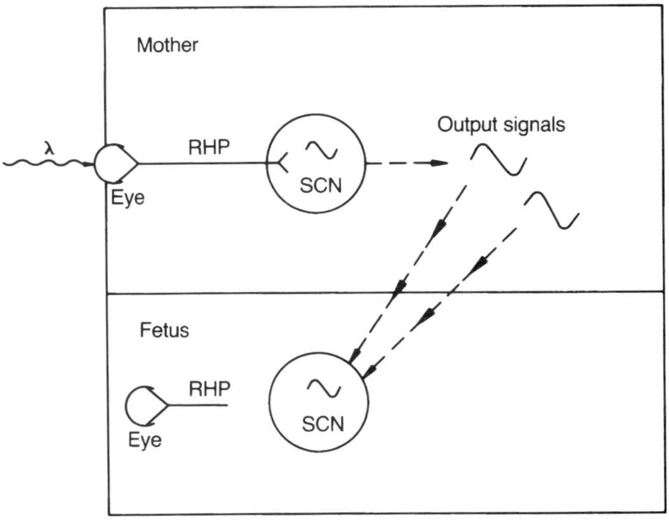

Fig. 6.10 Conceptual model of maternal fetal communication of circadian phase. Light-induced neural signals are conveyed to the maternal SCN by her retinohypothalamic pathway (RHP), entraining her circadian rhythms. Maternal output signals then entrain the fetal clock at a time when the innervation of the fetal SCN by the RHP is incomplete. (Reproduced with permission from Reppert, S.; published by John Libbey, 1991.)

Thus the physiological significance remains ambiguous.

In contrast to these nocturnal species, entrainment in the diurnal sauropsids occurs with the rest period following melatonin injection rather than activity (Figure 6.11). A phase response curve to melatonin in an iguanid lizard is rather different to the rat, being spread out such that large phase delays are found in the late subjective dark and early light phase and small phase advances are found in the late light and early dark phase. At the time of writing the only other complete phase response curves to melatonin have been performed in humans and will be treated in Chapter 8.

Melatonin can induce sleep in diurnal species and activity in nocturnal species. There has been discussion as to whether its entraining effects depend on acute induction of activity, as forced activity itself can phase shift the clock. Redman (personal communication) has refuted this objection by showing that in rats entrainment is possible with restricted activity. However, the situation with regard to other nocturnal and diurnal species (i.e. entrainment secondary to sleep or activity induction) has not been fully clarified.

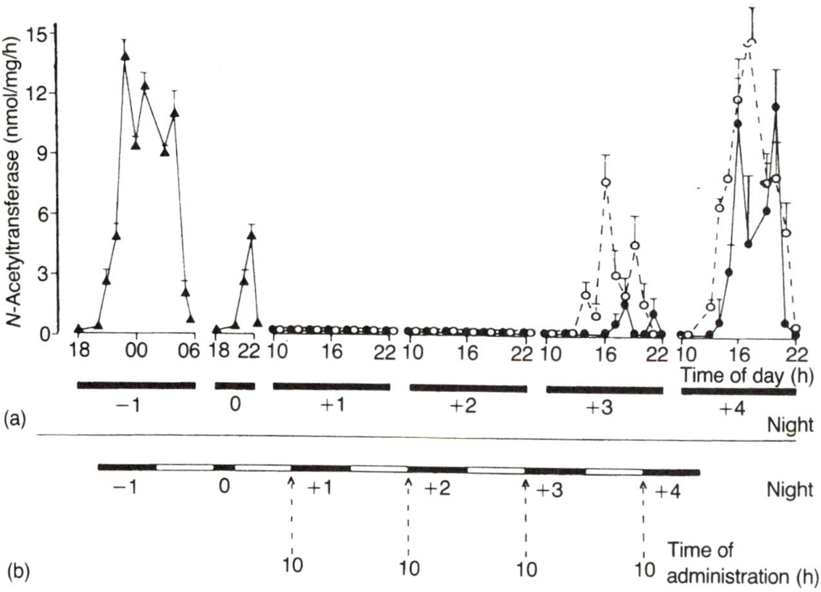

Fig. 6.11 Entrainment of free-running rest–activity cycles to melatonin (injection time indicated by the stars) in diurnal and nocturnal vertebrates. Note that the phase of entrainment by exogenous melatonin coincides to the phase of the rest–activity cycle and not to the endogenous melatonin rhythm. (Reproduced with permission from Armstrong, S.; published by John Libbey, 1991.)

6.5 EFFECTS OF MELATONIN ON OTHER CIRCADIAN RHYTHMS: TEMPERATURE

If melatonin acts on the central biological clock, and indeed this is the most parsimonious interpretation of a wealth of data, then all endogenous rhythms should be affected similarly to activity–rest in conditions where masking effects are eliminated. To date, whilst there are numerous reports of its effects on physiological systems, very few have been assessed from the point of view of circadian rhythms. An extensive list of such melatonin-induced effects can be found in Relkin (1976). This discussion will be restricted to the most

Erratum

The figure appearing on page 178 as Fig. 6.11 is incorrect. The correct figure is shown below.

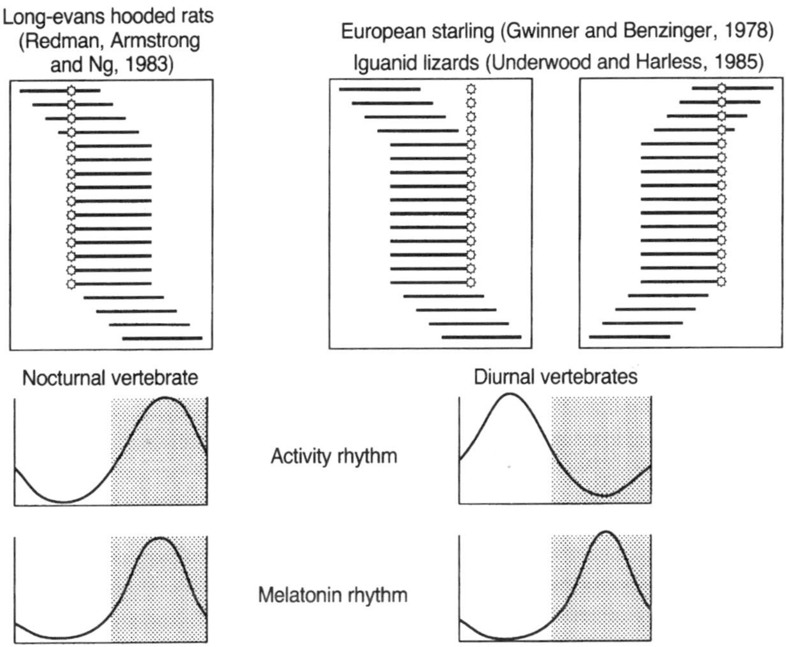

Fig. 6.11 Entrainment of free-running rest–activity cycles to melatonin (injection time indicated by the stars) in diurnal and nocturnal vertebrates. Note that the phase of entrainment by exogenous melatonin coincides to the phase of the rest–activity cycle and not to the endogenous melatonin rhythm. (Reproduced with permission from Armstrong, S.; published by John Libbey, 1991.)

Melatonin and the Mammalian Pineal Gland, Josephine Arendt. Published in 1995 by Chapman & Hall. ISBN 0 412 53600 5.

interesting observations which are interpretable on the grounds of rhythm theory.

Two circadian rhythms are of particular interest in animals (human data will be discussed in Chapter 8). These are the daily rhythm in core body temperature and the circadian rhythm of melatonin itself.

In the case of temperature there is a substantial body of work implicating melatonin in thermoregulation (see Ralph *et al.*, 1979 and Badia, Myers and Murphy, 1992 for reviews), with seasonal changes in thermoregulation being the most obvious photoperiod dependent (via melatonin) functions.

Changes in the circadian waveform of the temperature rhythm in the course of changing daylength in hamsters are certainly melatonin dependent (Figure 6.5). Moreover, daily torpor and hibernation may be controlled by photoperiod in some species – again an effect of the pineal. Non-shivering thermogenesis by brown adipose tissue is photoperiodically controlled in some rodents and melatonin implants can increase brown fat pad weight in photostimulated animals, although contradictory reports exist and the effects of pinealectomy are still confusing. Similar effects have been found in hamsters, ground squirrels and white footed mice.

Melatonin increases cold resistance in rats and hamsters and may be involved in the seasonal lowering of body temperature in some birds. Interestingly, it can modify behavioral thermoregulation in some ectotherms by changing the preferred environmental temperature. Many thermoregulatory effects of melatonin may involve the thyroid gland. Some thyroid responses to photoperiod change can be mimicked by appropriate melatonin administration.

Pinealectomy, like continuous light, abolishes the rhythm of deep body temperature in sparrows and pinealectomy, together with enucleation, likewise abolishes temperature rhythms in pigeons. Melatonin injections induce roosting and lowered body temperature in sparrows. Synchronization of the rest–activity cycle in rats by melatonin is accompanied by synchronization of the temperature rhythm.

The most consistent short-term results in mammals and birds indicate that melatonin acutely lowers body temperature, pinealectomy can increase it, and melatonin can reverse the effects of pinealectomy.

It is possible that melatonin has two effects on temperature – an immediate action to lower body temperature and a secondary entrainment of the rhythm. Whether the two are separable is a question of some importance. The lowering of body temperature by melatonin may in fact be an integral part of its entrainment properties. Curiously, some of the most interesting circadian effects of melatonin on body temperature have been seen in humans (Chapter 8).

6.6 EFFECTS OF MELATONIN ON THE ENDOGENOUS MELATONIN RHYTHM

The best endocrine circadian rhythm marker is that of melatonin itself. As previously observed, it is resistant to stress and metabolic variations, in contrast to cortisol which historically has been used as an endocrine marker rhythm. Investigation of the endogenous melatonin rhythm in small rodents is, however, very difficult. Frequent sampling is necessary to define a circadian rhythm and the best methods of measuring melatonin need volumes of plasma such that an individual small animal would be rapidly exsanguinated. One alternative is to take a suitable number of animals at each time point throughout 24 hours. This approach has been used to assess the entrainment of the melatonin synthesis rhythm (*N*-acetyltransferase activity) in rats after a forced advance phase shift of 8 hours. Timed administration increased the rate of resynchronization and, interestingly, increased the amplitude of the rhythm in parallel (Figure 6.12).

A simpler approach is to use 6-sulfatoxymelatonin (aMT6s), the urinary metabolite of melatonin, as an index of the endogenous melatonin rhythm. This requires continuous urine collection in metabolic cages and only species producing urine at frequent intervals throughout 24 hours are useful. Rats fulfill this condition. Melatonin itself and a number of analogs all hasten the resynchronization of the 6-sulfatoxymelatonin rhythm in rat urine after forced phase shift of the light–dark cycle when administration is appropriately timed. In addition to the speed of resynchronization, the amplitude of the endogenous 6-sulfatoxymelatonin rhythm is enhanced, specifically after melatonin treatment. This is not a direct feedback effect as it persists after cessation of treatment.

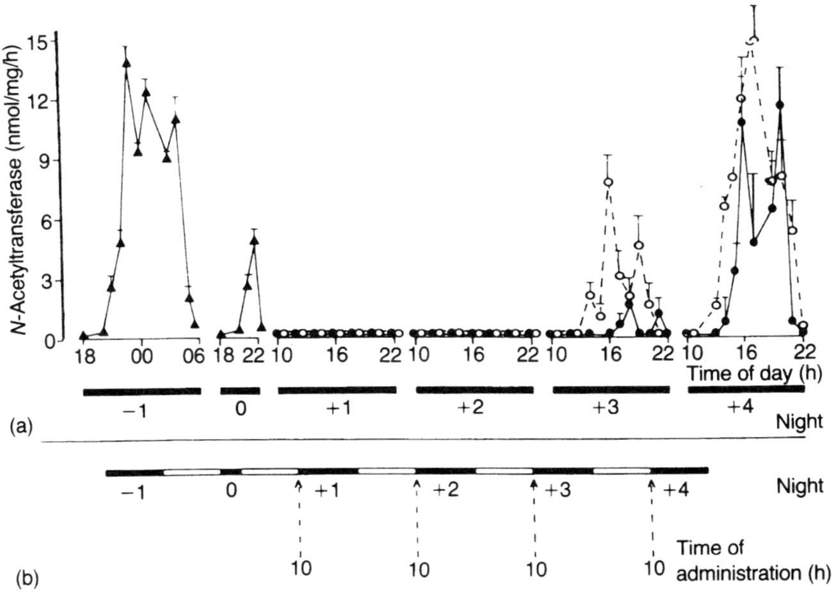

Fig. 6.12 Melatonin hastens the re-entrainment of the *N*-acetyltransferase rhythm after advance phase shift of 8 hours. (a) Effect of melatonin (1 mg/kg) i.p. applied at new dark onset on re-entrainment. O–O Rats treated with melatonin, ●–● rats treated with vehicle. Prior to photoperiodic manipulation, as indicated by (b), rats were adapted to 12L 12D. Note that in addition to the phase advance induced (nights 3 and 4) the amplitude of the rhythm is also increased by melatonin. (Reproduced from Illnerova, H. and Vanecek, J. Pineal *N*-acetyltransferase: a model to study properties of biological clocks, in *Fundamentals and Clinics in Pineal Research* (ed. G. Trentini) pp. 165–78 by permission of Raven Press, 1987.)

Circadian amplitude is an integral part of the expression of the status of the biological clock. In rats, advance phase shifts of the light–dark cycle of 8 hours are followed by a major decrease or complete flattening ('crushing') of the amplitude of melatonin secretion (see Figure 6.12) which requires several days to recover. Similar results have been reported for other circadian rhythms in humans. It is possible that this phenomenon results from transitory and totally inappropriate phase relationships within the core of the central rhythm generating system. Weak coupling in the absence of strong zeitgebers is also accompanied by lowered amplitude as well as changed phase relationships to the external zeitgeber. The enhancement of amplitude by melatonin is probably a reflection of strengthened internal coupling and relates back to the coupling function of the pineal in the lizard.

6.7 MELATONIN, THE PINEAL AND AGING

One of the most consistent observations in pineal research is a decline in the amplitude of the melatonin rhythm in old age (Figure 6.13). In circadian terms this may reflect inadequate coupling to the external zeitgebers. Clearly the defect may exist at several levels. Environ-

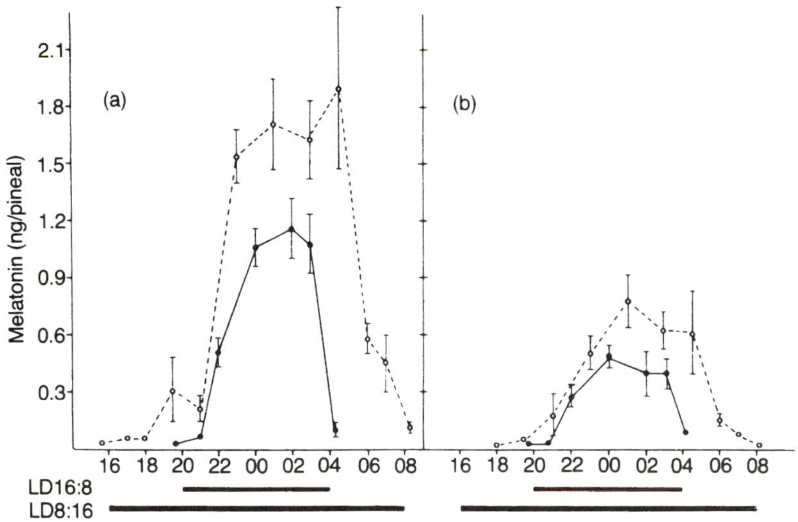

Fig. 6.13 The amplitude of the rhythm of melatonin declines with age. Rhythms of pineal melatonin concentration in young adult and old Djungarian hamsters. 3–6-month-old females, reared in 16L 8D from birth (a), and 17–22-month-old females, maintained in 16L 8D for at least a year before the beginning of the experiment (b) were either sacrificed in 16L 8D (●—●), or transferred to 8L 16D and sacrificed 2 weeks later (○—○). A lower amplitude rhythm is evident in the older animals but the duration increase on transfer to short photoperiod is present in both groups. (Reproduced with permission from Hoffmann, K., Illnerova, H. and Vanecek, J.; published by Elsevier, 1985.)

mental sensors lose acuity through the aging process and thus signal strength is depleted. The pineal itself may lose some capacity to synthesize melatonin. There is, however, some evidence that the primary defect may be at SCN level, as transplants of fetal SCN to old hamsters will restore the ability to respond appropriately to the phase shifting drug triazolam.

There exist many associations of the pineal and melatonin with the aging process which lend themselves to a circadian interpretation. There is a substantial potential for enhancement of well being in old age which arises as a consequence of these associations (see also Chapter 8).

Many years ago (1959) Swedish workers (cited in Armstrong and Redman, 1991) reported that pinealectomy shortened life span in rats. Then scattered reports indicated that pineal extracts inhibited some age dependent processes. More recently Swiss workers (Pierpaoli and Maestroni, 1987) described a prolongation of life by 20% by nightly melatonin administration to mice. They considered that their results were due to a stimulation of the immune system and an anti-stress action, both operating via the effects of melatonin on endogenous opiates. There is accumulating evidence for the possibility that the pineal and melatonin may act on the immune system and on initiation and development of cancer. The relationships of melatonin with the opioid system in puberty and seasonal breeding have been referred to earlier. Both effects are probably related to photoperiodic function.

Other effects of pinealectomy in rodents suggest an acceleration of the aging process. Hypertension and a diabetes-like state are induced. Rapid eye movement sleep is dissociated from slow wave sleep, blood cholesterol and alkaline phosphatase activity are elevated, sodium balance is upset and prostaglandin synthesis modified. Some of these can be counteracted by administration of melatonin in the late subjective day. Several cancers and biliary cirrhosis have been suggested to be related to pineal deficiency.

The period of free-running rhythms changes with age and the range of entrainment is decreased. Repeated shifting of zeitgebers and exposure to non-24-hour days will decrease the survival time of blowflies and exposure to LL (intensity dependent) has the same effect in fruitflies. In rats repeated shifting of the LD cycle suppresses the immune system. All of these threads suggest that perturbation of circadian rhythms has deleterious effects on survival.

In addition to melatonin other circadian rhythms become 'less robust' in old age (most of this information is derived from human work). Disintegration of the structure of sleep is a useful example. The question arises as to whether a decline in the amplitude of melatonin is a cause or a consequence of circadian problems. It is possible that the stability of the circadian system relates positively to its amplitude. Greater amplitude may contribute to rigidity of the system and resistance to phase change. Of course this is not necessarily advantageous as flexibility to daylength change in animals and to circadian phase shifts in humans may well confer biological benefits.

The amplitude of the melatonin rhythm probably reflects the strength of circadian coupling. As melatonin has coupling properties with regard to disrupted activity–rest cycles in rats and lizards it may well operate endogenously to enhance coupling. This suggests that properly timed exogenous melatonin will reinforce the circadian system and in old age recreate optimal phase relationships which may have been lost. An analogy with the pendulum of a clock, repeatedly tapped to maintain its angular momentum such that it does not damp out, is possibly apt.

Armstrong and Redman (1991) have recently incorporated the available information in this area into a generalized hypothesis that melatonin has anti-aging properties.

The story is not quite so simple however. Maximum life span is clearly genetically determined. The Bible quotes four score years and ten for humans and this has hardly been improved upon in 2000 years. Some species of mice have no detectable melatonin but are not reported to have notably different life spans. Any effect of melatonin on the aging SCN is likely to depend on a form of melatonin receptor and receptor populations may decline with age.

Declining melatonin and declining melatonin receptors are theoretically interdependent. Interindividual variations in melatonin amplitude are large and receptivity may adjust on an individual basis by up or down regulation in the short term. Loss of receptors may follow an overall decline in the available ligand concentration. 'Use it or lose' it as the old wives' tale relates. The author is not convinced that melatonin has anti-aging properties having taken it daily in the evening for many years during middle age and in the course of these years turning very grey-haired indeed!

It may, however, be useful in alleviating rhythm disturbance in old age.

6.8 GESTATION AND PARTURITION

Gestation length in some species, for example mink, is photoperiodically determined, via melatonin, by delayed implantation. In the rat gestation length depends on the light–dark cycle in which the animal is kept. Bosc (1987) has concluded a series of elegant experiments with pregnant rats in light–dark cycles of slightly less or slightly more than 24 hours. He was able to show a shortening of gestation length in 23.45-hour cycles and a lengthening in 24.15-hour cycles. The effect is not large and can be interpreted on the grounds that the timing of parturition is a circadian phenomenon. The effects could be replicated by administering melatonin at the appropriate periodicity, although the results of pinealectomy were not striking.

6.9 ESTROUS CYCLE

Estrous cycles in rodents clearly have a circadian basis. In gonadectomized animals a daily surge of LH appears, precisely timed with respect to the 24-hour day or, in constant conditions, to circadian time. There is also evidence in other species, especially primates, for a circadian timing mechanism involved in the ovulatory LH surge. Lesions of the SCN destroy the timing mechanisms controlling the LH surge. Phase shifting strategies using light–dark cycles or drugs shift the timing of the LH surge in parallel with the rest–activity cycle. This timing is therefore environmentally determined, but generated endogenously.

As the pineal is involved in circadian timing the presumption must be that it is concerned with the timing of the LH surge and indeed with general estrous timing. For example, there is evidence that in rats the estrous cycle can be lengthened by increasing photoperiod length – and there are few effects of photoperiod that are not pineal dependent.

Most attempts to relate the pineal and melatonin directly to the control of the estrous cycle have given rather ambiguous results although some of the very earliest observations of pineal function are pertinent. There is an extensive literature on the mediation by the pineal of the effects of light on the gonads in rats, reviewed by Reiter amongst others (1980). Moreover melatonin administered into the lateral ventricles (albeit in very large amounts) during the 'critical period' on the day of pro-estrus is able to block ovulation in rats.

Timed melatonin administration can mimic the effects of extending the light–dark cycle on the timing of the LH surge. Melatonin injections 11 hours after lights on in 12L 12D retards vaginal opening and is accompanied by an advance in the timing of the pro-estrous surge of gonadotropins and estradiol.

Some of the most interesting observations have been concerned, not with an effect of melatonin, but with the curious rhythmicity of HIOMT during the estrous cycle in rats. Wallen and Yochim (1974) proposed that two component rhythms with periodicities of 24 hours and 20 hours could combine to explain the waveform which included a drop in amplitude during estrus.

Observations of the melatonin rhythm itself show a decreased amplitude during pro-estrus in rats (Figure 6.14) and hamsters but with conflicting reports in other species. Estrous cycle effects in rats have been overshadowed by the much more dramatic effects of melatonin on the reproductive system in seasonal breeders such as hamsters. This whole area needs review in the light of current knowledge as to the phase response curve to melatonin. If light is important in the control of the human menstrual cycle and fertility as seems likely (Chapter 7) a rich reward might be reaped by further careful experimentation.

6.10 METABOLIC RELATIONSHIPS OF MELATONIN

The metabolic effects of melatonin have rarely been studied from a circadian point of view. Pinealectomy in the rat in the long term

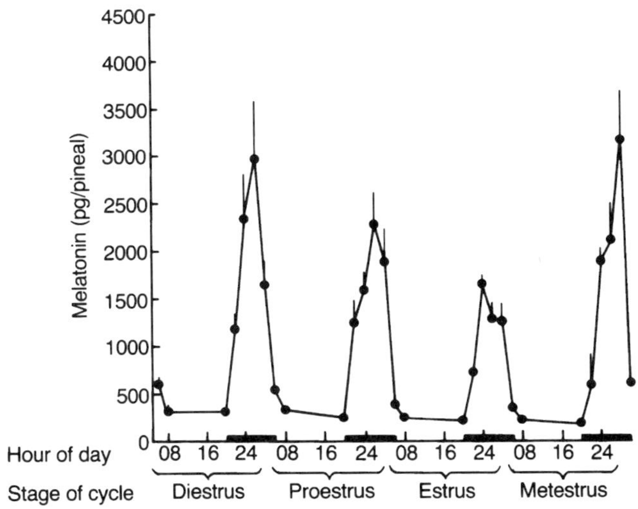

Fig. 6.14 Pineal melatonin content is lower during estrus in rats. Each point is the mean of values from 8–12 animals ± SEM. (Reproduced with permission from Johnson, L. *et al.*; published by the Society for Experimental Biology and Medicine, 1982.)

increases plasma glucose levels, plasma glucagon, food intake and body weight while decreasing insulin. All of these parameters have circadian and/or seasonal rhythms. For example, the insulin response to a meal and insulin resistance is highest during the night and early morning, food intake is largely nocturnal in the rat, and body weight varies with season. It is not unreasonable to suppose that the observed effects of pinealectomy result from disturbed circadian and seasonal organization.

By far the most interesting, circadian rhythm related, metabolic effect of melatonin is the inhibition of 2-deoxyglucose uptake into the rat SCN *in vivo*. SCN 2-deoxyglucose uptake was inhibited by melatonin (1 mg/kg) at CT10 when uptake is normally high, but there was no effect at CT14 when uptake is low (Figure 6.15).

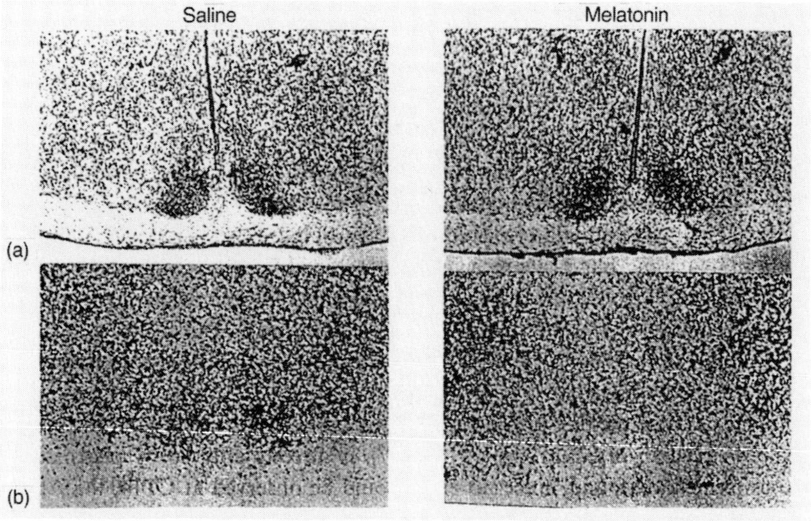

Fig. 6.15 Melatonin (1 mg/kg s.c.) inhibits 2-deoxyglucose uptake into the suprachiasmatic nuclei (SCN) when given at CT 10 2 hours before subjective night onset. Autoradiography and accompanying histological sections of SCN from rats receiving either melatonin or saline vehicle. (a) Cresyl-violet stained histological sections of SCN from intact rats sacrificed at CT 10. (b) Autoradiographs of SCN depicted in (a) from rats sacrificed at CT 10. All rats were treated 15 min after melatonin or saline with 150 μCi/kg 2-deoxyglucose i.m. (Reproduced with permission from Cassone, V.M., Roberts, M.H. and Moore, R.Y.; published by the American Physiological Society, 1988.)

6.11 *IN VITRO* PHASE SHIFTS

Whilst the evidence for entrainment and phase shift of the rest–activity cycle by melatonin is very strong, it is still not clear whether, in rodents, melatonin will entrain all other circadian variables to the same extent. In other words does it shift the pacemaker or only some of the downstream overt rhythms such as rest activity?

This question can be addressed in a number of different ways, but perhaps the most convincing is to show that it affects the activity of the SCN itself. Demaine and Kann (1979) were the first to demonstrate that melatonin modulates the electrical activity of the hypothalamus, in particular the pre-optic and anterior hypothalamic areas.

Subsequently the metabolic activity of the SCN *in vivo* and the electrical activity of various *in vitro* SCN preparations was shown to be modified by melatonin. As well as the inhibition of 2-deoxyglucose uptake into the nuclei in late subjective day it also inhibits electrical activity during late subjective day.

The most convincing evidence, however, comes from long-term (days) cultures of rat SCN in a hypothalamic slice. These preparations show marked endogenous circadian rhythmicity in electrical activity and survive for sufficient time to perform both control and test experiments on the same preparation. Gillette and coworkers (McArthur, Gillette and Prosser, 1991) have elegantly demonstrated the phase advancing effects of melatonin on the circadian rhythm of electrical activity in this system (Figure 6.16). Melatonin was by no means the

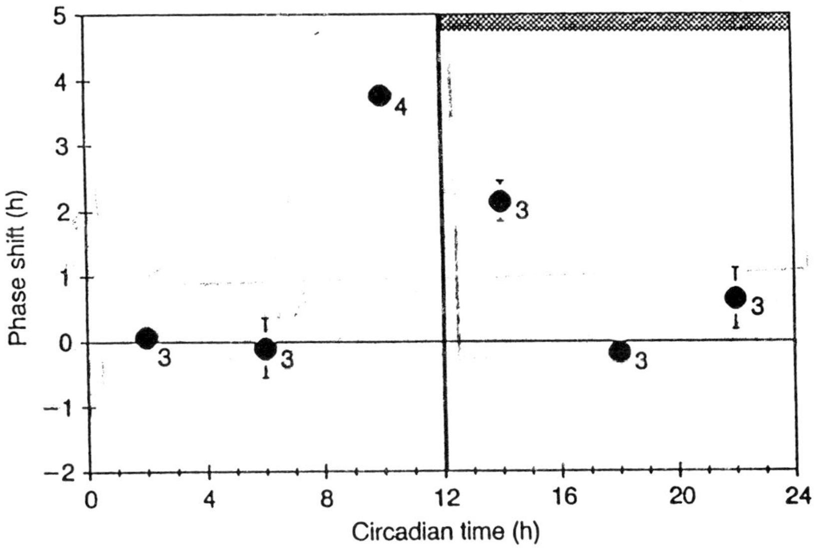

Fig. 6.16 Melatonin directly resets the rat suprachiasmatic circadian clock *in vitro*. Using the maintained circadian electrical activity rhythm of SCN slices *in vitro*, McArthur and coworkers have shown a phase response curve to melatonin application (10^{-9} M, for 1 hour) with only phase advances. The diagram shows the results of at least three separate experiments per time point. The average magnitude of the phase shift ± SEM is plotted according to circadian time (e.g. CT 2 = 2 h after lights on in the animal colony on a 12L 12D light schedule). (Reproduced with permission from McArthur, A.J., Gillette, M.U. and Prosser, R.A.; published by Elsevier, 1991.)

only phase shifting agent: serotonin for example was effective, however different compounds showed different phase response curves. The effect with melatonin was large and acute with shifts of several hours being observed.

Thus melatonin acts directly on a central biological clock to change its phase. The presence of melatonin receptors within the SCN adds considerable force to this statement (see Chapter 9).

Does this imply that all SCN-controlled variables are responsive to melatonin? Probably not. For example, the rest–activity cycle of one blind man was entrained by melatonin without synchronizing the cortisol and temperature rhythms (see Chapter 8). This may be a special case as there is good evidence that melatonin affects the temperature rhythm in a number of species. Moreover the multi-oscillator control of circadian rhythms is a matter for much debate. As the SCN itself is not homogeneous with respect to neurotransmitter and neuromodulator content, it may be differentially responsive to different agents. Daan (personal communication) has even suggested that the left and right SCN may be evening and morning oscillators. It is not yet known whether melatonin binds to subpopulations of SCN cells and if so to which. All of these areas are under intense investigation and will no doubt be clarified in the next few years.

6.12 RETINAL RHYTHMS

Melatonin is present and synthesized in the retina of many species as has been previously noted. There are very large species differences in quantitative and qualitative aspects and the rhythm does not always show the same day–night variations as are seen in the pineal. In birds and lower vertebrates retinal melatonin usually contributes significantly to circulating plasma concentrations (Figure 6.17).

However, in mammals pinealectomy reduces plasma melatonin to undetectable levels. If melatonin does have a function in the mammalian retina it must act locally and be rapidly metabolized *in situ*. It is conceivable that photodestruction is one mechanism of inactivation.

There is good evidence for rhythms in several retinal functions.

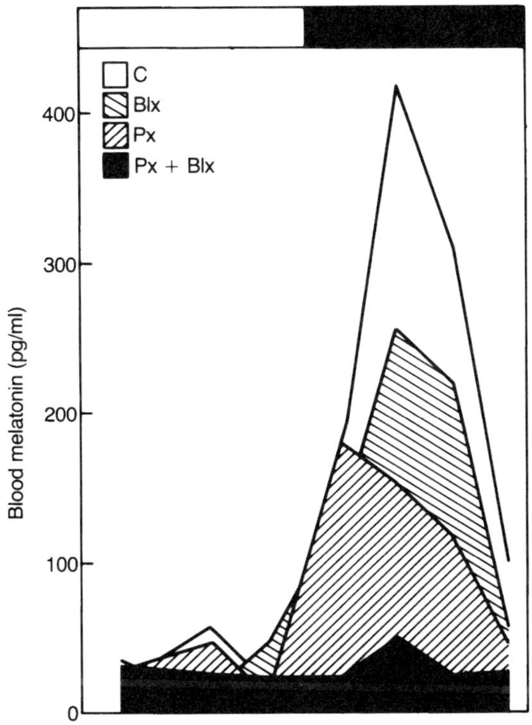

Fig. 6.17 Relative contribution of the pineal gland and the eyes to blood melatonin levels in Japanese quail held in 12L 12D. Eye removal (Blx) and pinealectomy (Px) depressed blood melatonin levels by 33 % and 54 % respectively, relative to controls (C). After both blinding and pinealectomy (Px + Blx), a blood melatonin rhythm was almost completely absent, and the small rhythm remaining is believed to be due to melatonin secretion from remnants of pineal stalk remaining after pinealectomy. (Reprinted from Underwood, H. and Goldman, B., 1987, by permission of the Guildford Press.)

Visual sensitivity varies with the time of day, increasing during the course of the night. According to Rémé, Wirz-Justice, Terman and coworkers (1991) the retina contains an endogenous clock which maintains the rhythm of visual sensitivity after SCN lesion and destruction of virtually all other circadian rhythms. Intra-ocular melatonin increases retinal sensitivity in rabbits.

Light sensitive metabolic processes are important for the renewal of photoreceptor cells. For example, at the tip of the outer segments of rod cells disk shedding takes place with subsequent degradation by pigment epithelial cells. Renewal occurs at the base. This process is circadian in nature and, at least in the frog eye cup, is activated by melatonin.

Photomechanical processes accompany light–dark transitions in the retina. These have mostly been studied in lower vertebrates. In darkness cones elongate, rods contract and in the pigment epithelium pigment granules aggregate, whereas in the light the opposite processes occur. The dark-induced reactions are induced by melatonin and the light effects by dopamine. The dispersal of pigment in light is thought to act as a screen and it is of interest to note that precisely this sort of process is seen in amphibian melanophores in the light, with aggregation in the dark. The movements of photoreceptors are thought to take place in order to optimize light perception. In the dark therefore melatonin may increase retinal sensitivity.

The second physiological melatonin receptor (the first was the amphibian melanophore) was found in the retina of rabbits and chickens by Dubocovich (1985). In this case melatonin acts to inhibit Ca^{2+} dependent dopamine release, and thus a reciprocal relationship exists between these two mediators of retinal function. High-affinity melatonin binding sites are present on retinal membranes in some species.

6.13 INTERACTIONS OF THE RETINA, THE SCN AND THE PINEAL

The perception of light, whether for vision or for circadian and seasonal organization, is influenced by melatonin apparently at all levels. Reciprocally melatonin is controlled by light–dark cycles acting through the SCN. Moreover the SCN may well influence retinal activity through the sympathetic nervous system. These observations have been integrated into a model, originally for the avian circadian system and modified by Steinlechner (1991) for mammals

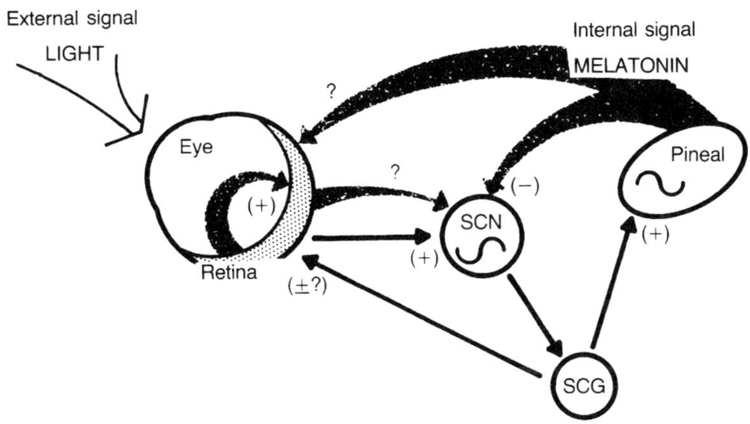

Fig. 6.18 Neuro-endocrine loop model for the mammalian circadian system. Arrows = neuronal connections; thick, shaded arrows = endocrine connections. The SCN may exert feedback control on its visual input via the sympathetic nervous system. There is evidence that melatonin has both inhibitory and phase shifting effects on SCN activity. Both pineal and retinally derived melatonin may influence light perception at retinal level. Retinal melatonin may also reach the SCN by a direct, possibly neural route. (Adapted from Steinlechner, S.; published by John Libbey, 1991.)

(Figure 6.18). This encompasses the mutual interactions of the pineal, the SCN and the retina.

Light entering the eye synchronizes the endogenous rhythm of the SCN (and possibly the retinal clock) and entrains and suppresses melatonin. In the course of the night melatonin in turn may increase retinal sensitivity thus strengthening the relative perception of light at dawn and increasing the coupling strength of light with respect to the circadian system. With a tau greater than 24 hours, morning light will act to phase advance the circadian system. Evening onset of melatonin secretion may reinforce the phase advance and help maintain 24 hour synchronization. With tau less than 24 hours we may hypothesize that afternoon light acts to phase delay the circadian system and early morning melatonin may reinforce this effect (Chapter 8).

6.14 SUMMARY

It would be reasonable to conclude that in adult mammals melatonin serves to modulate phase and strengthen coupling within the circadian system. In fetal and neonatal mammals it serves to program the circadian system and, in combination with environmental input, to dictate the timing of developmental stages especially puberty. It may have a role in determining life span in association with many other factors.

Timing mechanisms are an integral part of homeostasis, gestation, development, reproductive cycles and aging. The pineal via melatonin is clearly concerned one way or another with all of these. We await more precise definitions and in the meantime explore its therapeutic possibilities in humans.

REFERENCES

Arendt, J., Bojkowski, C., Folkard, S. *et al.* (1985) Some effects of melatonin and the control of its secretion in man, in *Photoperiodism, Melatonin and the Pineal* (eds D. Evered and S. Clark), Ciba Foundation Symposium 117, Pitman, London, pp. 266–83.

Arendt, J., Borbely, A.A., Franey, C. *et al.* (1984) The effect of chronic small doses of melatonin given in the late afternoon on fatigue in man: a preliminary study. *Neuroscience Letters*, **45**, 317–21.

Armstrong, S. (1991) Entrainment of vertebrate circadian rhythms by melatonin, in *Advances in Pineal Research* (eds J. Arendt and P. Pevet), John Libbey, London, pp. 259–66.

Armstrong, S.M. (1989) Melatonin and circadian control in mammals. *Experientia*, **45**, 932–9.

Armstrong, S. and Redman, J. (1985) Melatonin administration: effects on rodent circadian rhythms, in *Photoperiodism, Melatonin and the Pineal* (eds D. Evered and S. Clark), Ciba Foundation Symposium 117, Pitman, London.

Armstrong, S.M. and Redman, J. (1991) Melatonin: a chronobiotic with anti-ageing properties. *Medical Hypotheses*, **34**, 300–9.

Badia, P., Myers, B. and Murphy, P. (1992) Melatonin and thermoregulation, in *Melatonin: Biosynthesis, Physiological Effects, and*

Clinical Applications (eds R.J. Reiter and H.S. Yu), CRC Press, Boca Raton, Florida, pp. 349–64.

Bosc, M.J. (1987) Time of parturition in rats after melatonin administration or change of photoperiod. *Journal of Reproduction and Fertility*, **80**, 563–8.

Cassone, V. (1990) Effects of melatonin on vertebrate circadian systems. *Trends in Neurosciences*, **13**, 457–63.

Cassone, V. (1992) The pineal gland influences rat circadian activity rhythms in constant light. *Journal of Biological Rhythms*, **7**, 27–40.

Cassone, V.M., Roberts, M.H. and Moore, R.Y. (1988) Effects of melatonin on 2-deoxy-(1-14-C)-glucose uptake within rat suprachiasmatic nucleus. *American Journal of Physiology*, **255**, R332.

Chesworth, M.J., Cassone, V.M. and Armstrong, S.M. (1987) Effects of daily melatonin injections on activity rhythms of rats in constant light. *American Journal of Physiology*, **253**, R101–7.

Davis, F.C. and Mannion, J. (1988) Entrainment of hamster pup circadian rhythms by prenatal melatonin injections. *American Journal of Physiology*, **255**, R439–48.

Demaine, C. and Kann, H.C. (1979) Modification of the electrical activity of hypothalamic neurones by pineal indoles. *Progress in Brain Research*, **52**, 373–5.

Dubocovich, M.I. (1985) Characterisation of a retinal melatonin receptor. *Journal of Pharmacology and Experimental Therapeutics*, **234**, 395–401.

Hoffmann, K., Illnerova, H. and Vanecek, J. (1985) Comparison of pineal melatonin rhythms in young adult and old Djungarian hamsters under long and short photoperiods. *Neuroscience Letters*, **56**, 39–43.

Illnerova, H. (1991): in *Suprachiasmatic Nucleus: The Minds Clock* (eds D.C. Klein, R.Y. Moore and S.M. Reppert), Oxford University Press, New York, Oxford, pp. 197–219.

Johnson, L., Vaughan, M., Richardson, B. *et al.* (1982) Variation in pineal melatonin content during the estrous cycle of the rat. *Proceedings of the Society for Experimental Biology and Medicine*, **169**, 416–9.

McArthur, A.J., Gillette, M.U. and Prosser, R.A. (1991) Melatonin directly resets the rat suprachiasmatic circadian clock *in vitro. Brain Research*, **565**, 158–61.

Murakami, N., Hayafuji, C., Sasaki, Y. *et al.* (1983) Melatonin accelerates the re-entrainment of the circadian adrenocortical rhythm in inverted illumination cycle. *Neuroendocrinology*, **36**, 385–91.

Pierpaoli, W. and Maestroni, G.J.M. (1987) Melatonin: a principal neuroimmunoregulatory and anti-stress hormone: its anti ageing effects. *Immunology Letters*, **16**, 355.

Quay, W.B. (1970) Precocious entrainment and associated characteristics of activity patterns following pinealectomy and reversal of photoperiod. *Physiology and Behaviour*, **5**, 1281–90.

Ralph, C.L., Firth, B.T., Gern, W.A. *et al.* (1979) The pineal complex and thermoregulation. *Biological Reviews*, **54**, 41.

Redman, J., Armstrong, S. and Ng, K.T. (1983) Free-running activity rhythms in the rat: entrainment by melatonin. *Science*, **219**, 1089–91.

Reiter, R.J. (1980) The pineal and its hormones in the control of reproduction in mammals. *Endocrine Review*, **1**, 109–31.

Relkin, R. (1976) *The Pineal*, Eden Press, Montreal.

Rémé, C.E., Wirz-Justice, A. and Terman, M. (1991) The visual input stage of the mammalian circadian pacemaking system: 1. Is there a clock in the mammalian eye? *Journal of Biological Rhythms*, **6**, 5–29.

Reppert, S. (1991) Development of a circadian timing system: melatonin as a time giver from mother to fetus, in *Advances in Pineal Research 5* (eds J. Arendt and P. Pevet), John Libbey, London pp. 251–7.

Ruf, T., Steinlechner, S. and Heldmaier, G. (1987) Influence of photoperiod on body temperature rhythms in the Djungarian hamster, in *Chronobiology and Chronomedicine* (eds G. Hildebrandt, R. Moog and F. Raschke), pp. 79–84.

Steinlechner, S. (1991) In search of a physiological role for retinal melatonin, in *Advanced Pineal Research* (Suppl. 5), (eds J. Arendt and P. Pevet), pp. 123–128.

Underwood, H. (1981) Circadian clocks in lizards: photoreception, physiology and photoperiodic time measurement, in *Biological Clocks in Seasonal Reproductive Cycles* (eds B.K. Follett and D.E. Follet), John Wright & Sons, Bristol, pp. 137–52.

Underwood, H. and Goldman, B. (1987) Vertebrate circadian and

photoperiodic systems: role of the pineal gland and melatonin. *Journal of Biological Rhythms*, **2**, 279–315.
Wallen, E.P. and Yochim, J.M. (1974) Rhythmic function of pineal hydroxyindole-*O*-methyltransferase during the estrous cycle: an analysis. *Biology of Reproduction*, **10**, 461–6.
Zimmerman, N.H. and Menaker, M. (1979) *Proceedings of the National Academy of Sciences, USA*, **76**, 999–1003.

FURTHER READING

Arendt, J. (1988) Melatonin. *Clinical Endocrinology*, **29**, 205–29.
Bartsch., C., Bartsch., H., Fuchs, U. *et al.* (1989) Stage-dependent depression of melatonin in patients with primary breast cancer. *Cancer*, **64**, 426–33.
Binkley, S. (1988) *The Pineal: Endocrine and Nonendocrine Function*, Prentice Hall, Englewood Cliffs, New Jersey, USA.
Blask, D.E., Hill, S.M. and Pelletier, D.B. (1988) Oncostatic signalling by the pineal gland and melatonin in the control of breast cancer, in *The Pineal Gland and Cancer* (eds D. Gupta, A. Attanasio and R.J. Reiter), Müller Bass, Tubingen, Germany, pp. 195–206.
Brzezinski, A. and Wurtman, R.J. (1988) The pineal gland: its possible roles in human reproduction. *Obstetrics and Gynecology Survey*, **43**, 197–207.
Cassone, V.M., Chesworth, M.J. and Armstrong, S.M. (1986) Entrainment of rat circadian rhythms by daily injections of melatonin depend upon the hypothalamic suprachiasmatic nuclei. *Physiology and Behavior*, **36**, 1111.
Cramer, H. (1978) Melatonin and sleep. *Fourth European Congress on Sleep Research.* Tirgu-Mures. Karger, Basel, pp. 204–10.
Edwards, R.G. (1981) Test-tube babies. *Nature*, **293**, 253.
Elliot, J.A. (1981) Circadian rhythms, entrainment and photoperiodism in the Syrian hamster, in *Biological Clocks in Seasonal Reproductive Cycles* (eds B.K. Follett and D.E. Follett), Wright, Bristol, pp. 203–18.
Folkard, S., Arendt, J., Aldhous, M. *et al.* (1990) Melatonin stabilises sleep onset time in a blind man without entrainment

of cortisol or temperature rhythms. *Neuroscience Letters*, **113**, 193–8.

Fraschini, F.O., Mess, B. and Martini, L. (1968) Pineal gland, melatonin and the control of luteinizing hormone secretion. *Endocrinology*, **82**, 919–24.

Gupta, D., Attanasio, A. and Reiter, R.J. (eds) *The Pineal Gland and Cancer*, Müller & Bass, Tubingen, Germany. pp. 195–206.

Gwinner, E. (1989) Melatonin in the circadian system of birds: model of internal resonance, in *Circadian Clocks and Ecology* (eds T. Hiroshige and K. Honma), Hokkaido University Press, Sapporo, pp. 127–53.

Gwinner, E. and Benzinger, I. (1978) Synchronisation of a circadian rhythm in pinealectomised European starlings by daily injections of melatonin. *Journal of Comparative Physiology*, **127,** 209–13.

Illnerova, H. and Vanecek, J. (1987) Pineal *N*-acetyltransferase: a model to study properties of biological clocks, in *Fundamentals and Clinics in Pineal Research* (ed G. Trentini), Raven Press, New York. pp. 165–78.

Maestroni, G.J., Conti, A. and Pierpaoli, W. (1988) Role of the pineal gland in immunity. III. Melatonin antagonises the immunosuppressive effect of acute stress via an opiatergic mechanism. *Immunology*, **63**, 465–9.

Maestroni, G.J.M., Conti, A. and Pierpaoli, W. (1989) Melatonin, stress and the immune system. *Pineal Research Reviews*, **7**, 268.

Menaker, M., Hudson, D.J. and Takahashi, J.S. (1981) Neural and endocrine components of circadian clocks in birds, in *Biological Clocks in Seasonal Reproductive Cycles* (eds B.K. Follett and D.E. Follett), Wright, Bristol, pp. 171–83.

Morin, L.P. (1988) Age-related changes in hamster circadian period, entrainment, and rhythm splitting. *Journal of Biological Rhythms*, **3**, 237.

Ozaki, Y., Wurtman, R.J., Alonso, R. *et al.* (1978) Melatonin secretion decreases during the proestrous stage in the rat estrous cycle. *Proceedings of the National Academy of Sciences, USA*, **75**, 531–4.

Ralph, M.R., Foster, R.G., Davis, F.C. *et al.* (1990) Transplanted suprachiasmatic nucleus determines circadian period. *Science*, **247**, 975–8.

Rusak, B. and Zucker, I. (1979) Neural regulation of circadian rhythms. *Physiology Reviews*, **59**, 449–526.

Schneider, T., Thalau, P., Wiltschko, W. *et al.* (1991) The pineal hormone melatonin guides a sensitive period in the pied flycatcher, in *Advances in Pineal Research* (Suppl. 5), (eds J. Arendt and P. Pevet), pp. 245–7.

Symons, A.M., Arendt, J. and Laud, C.A. (1983) Melatonin feeding decreases prolactin levels in the ewe. *Journal of Endocrinology*, **99**, 41–6.

Takahashi, J.S. (1987) Circadian oscillators at cellular levels. *Proceedings from XVIII International Conference on Chronobiology*, Leiden, July, Pergamon, Oxford.

Tamarkin, L., Baird, C.J. and Almeida, O.F.X. (1985) Melatonin: a coordinating signal for mammalian reproduction. *Science*, **227**, 714–20.

Tamarkin, L., Cohen, M., Roselle, D. *et al.* (1981) Melatonin inhibition and pinealectomy enhancement of 7,12-dimethylbenz(a)anthracene-induced mammary tumours in the rat. *Cancer Research*, **41**, 4432–6.

Testart, J., Frydman, R. and Roger, M. (1982) Seasonal influence of diurnal rhythms in the onset of the plasma luteinising surge in women. *Journal of Clinical Endocrinology and Metabolism*, **55**, 374–7.

Turek, F.W. (1987) Pharmacological probes of the mammalian circadian clock: use of the phase response curve approach. *Trends in Pharmaceutical Science Reviews*, **8**, 212–7.

Underwood, H. (1977) Circadian organisation in lizards: the role of the pineal organ. *Science*, **195**, 587–9.

Vanecek, J.A., Pavlik, H. and Illnerova, H. (1987) Hypothalamic melatonin receptor sites revealed by autoradiography. *Brain Research*, **435**, 359–62.

Vivien-Roels, B. and Pevet, P. (1983) The pineal gland and the synchronisation of reproductive cycles with variations of the environmental climatic conditions, with special reference to temperature. *Pineal Research Reviews*, **1**, 92–144.

Vriend, J. (1983) Evidence for pineal gland modulation of the neuroendocrine-thyroid axis. *Neuroendocrinology*, **36**, 68–78.

Walker, R.F., McCamant, S. and Timiras, P.S. (1982) Melatonin and the influence of the pineal gland on timing of the LH surge in rats. *Neuroendocrinology*, **39**, 37–42.

Walsh, L., Vandenberg, R.J., Marai, N.E. and Rietveld, W.J. (1987) Electrophysiology of cultured neurones from the suprachiasmatic nucleus. Abstract 245, *XVIII International Conference on Chronobiology,* Leiden.

Weaver, D.R. and Reppert, S.M. (1986) Maternal melatonin communicates daylength to the fetus in Djungarian hamsters. *Endocrinology*, **119**, 2861–3.

The pineal in human physiology and pathology

7

Contents

7.1 INTRODUCTION

The human pineal occupies the depression between the superior colliculi of the mesencephalon. It is very variable in structure, being more or less lobular with different amounts and distribution of connective tissue, and with a normal weight range of around 50–200 mg (Figure 7.1). It calcifies in the majority of people in the course of development. By the end of puberty calcification is detectable by radiography in most individuals. This is almost certainly an underestimate as fine granules (brain sand) of calcification cannot be detected in this way. The presence of calcification is of considerable use to radiologists as a reference point (Figure 7.2); however it delayed an appreciation of human pineal function. Many assumed that calcification indicated that the gland was inactive. In fact we now know that in spite of calcification it remains active throughout life with some decline in metabolism in old age. Tapp and Huxley (1972) found that 76% of the gland in males and 82% in females still consisted of parenchymal cells in the 75–90-year age group. Studies designed to relate human melatonin production to degree of calcification have shown no relationship in adults under 50 years old.

Whether calcification has any functional significance remains an open question. As it is deposited in rings it presumably reflects rhythmic activity. It is conceivable that human functions which are influenced by the pineal (possibly of particular importance in early

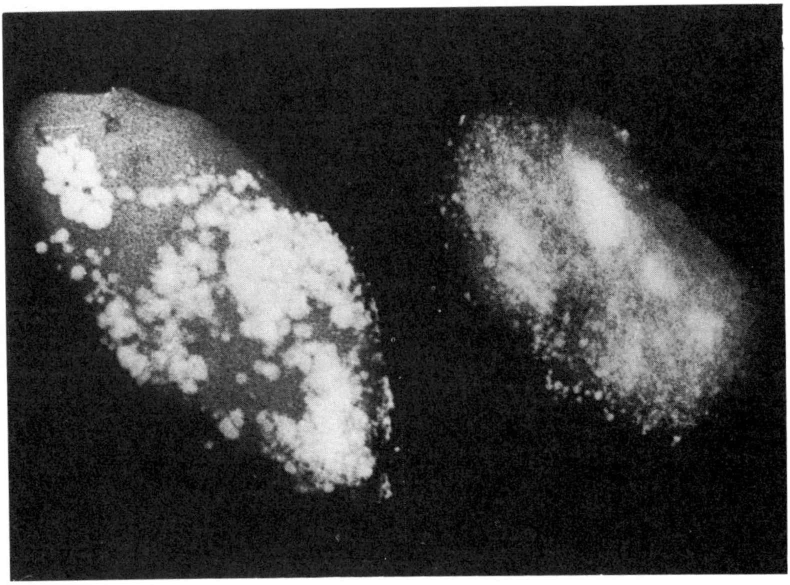

Fig. 7.1 X-ray photo of two isolated human pineal glands. The white areas represent calcifications and even when they are widespread the gland contains melatonin-producing tissue as has been proven in organ culture systems. (Reproduced with permission from Wetterberg, L., Melatonin in humans; physiological and clinical studies. *Journal of Neural Transmission* (suppl.), **13,** 389–410; published by Springer-Verlag, Wein, 1978.)

life) may in some way be affected by the degree of calcification. The hydroxy apatite which forms the chemical basis of the granules appears to combine with a number of trace metals of which zinc is the most prominent. Again the significance of this is unknown.

7.2 PINEALECTOMY

Classical endocrine approaches to determination of function, i.e. extirpation of the gland and observation of consequent effects, followed by replacement of the presumed active principle, are evidently not possible in humans. No specific inhibitor of melatonin synthesis is available and studies in humans have relied heavily on peripheral measurement of pineal metabolites to assess function.

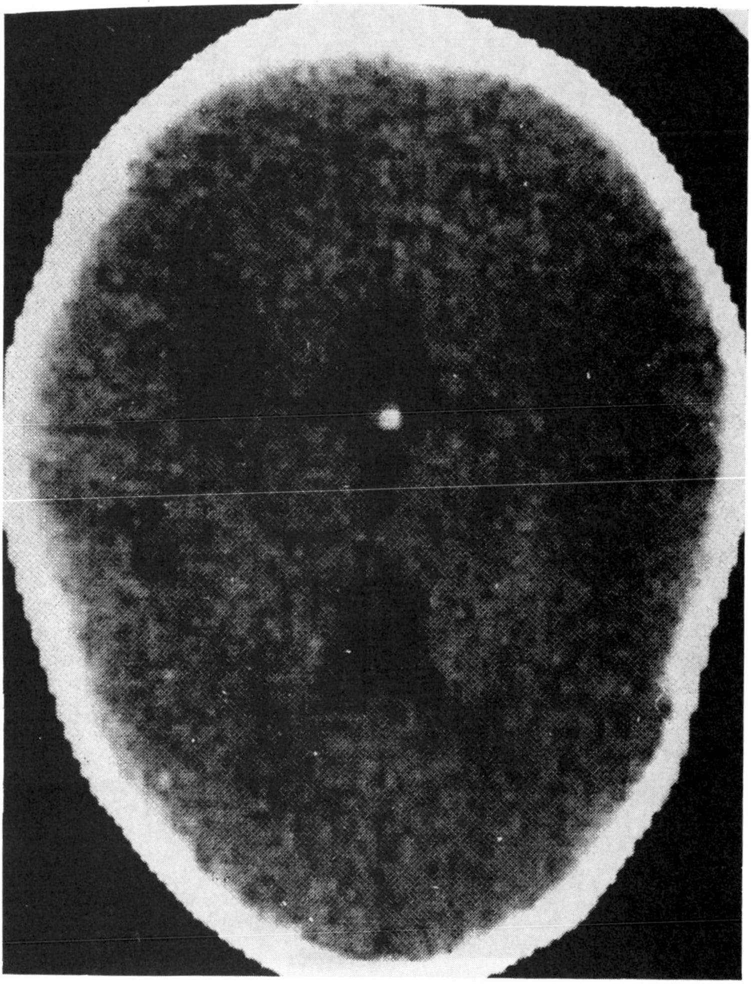

Fig. 7.2 CAT (computerized axial tomography) scan of a 3-year-old girl showing precocious calcification. (Reproduced with permission from Zimmerman, R.A. and Bilaniuk, L.T.; published by Plenum, 1982.)

Pineal tumors though rare provide the only situation where extirpation of the gland is performed although for specific types of tumor radiation therapy is the method of choice. Surgery of the pineal is a major undertaking, given its position deep within the brain, and assessment post-surgery always has to consider damage to surrounding brain tissue.

Pinealectomy in humans removes virtually all plasma melatonin (Figure 7.3). Other consequences of the operation consist of diffuse neurological problems which do not add up to a consistent functional effect as yet and may be more related to non-specific effects of operation. As in other species it is almost certainly necessary to challenge the organism by, for example, abrupt shifting of the light–dark cycle or major changes in applied artificial daylength, in order to show pineal-mediated effects. So far there are no such reports

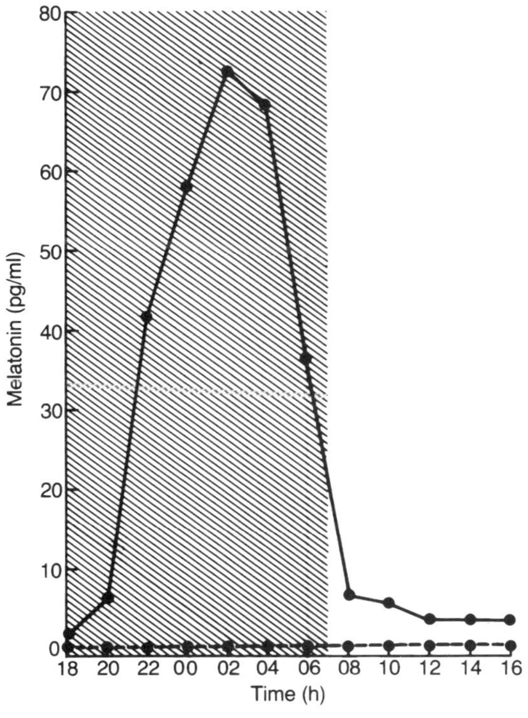

Fig. 7.3 Human plasma levels of melatonin before and after (- - -) removal of a tumor involving the pineal gland as measured by gas chromatography–mass spectrometry (limit of detection <1 pg/ml). (Reprinted, by permission of the *New England Journal of Medicine*, **19**, 1132–35, 1983. Neuwelt E.A. and Lewy, A.J.)

although the author is aware of attempts to maintain pinealectomized humans in environmental isolation in order to study circadian function. One potentially useful approach is suggested by the sleep and temperature response to very short and very long days in healthy human volunteers (see p. 81) which is probably pineal mediated.

Tumors of the pineal region in children are frequently associated with abnormal pubertal development whether delayed or advanced. Marburg's original hypothesis (Marburg 1907, 1930; Heubner 1898) to explain precocious puberty in boys with pineal tumors was that the tumors destroyed the inhibitory capacity of the pineal on sexual development. In fact there is much evidence that precocity is due to the production of human chorionic gonadotropin by germ cell tumors of the pineal. This relationship has nevertheless stimulated much work on the possible role of the pineal, through the secretion of melatonin, as a means of timing human puberty. Further discussion of pineal tumors is to be found on p. 230.

7.3 FUNCTIONS OF THE HUMAN PINEAL

It is only when the clear function of the pineal as a transducer of light–dark information is appreciated that a proper theoretical conception of the possible roles of the pineal in human physiology can be formulated. In the light of current knowledge we can state that these are:

(a) the transmission of information concerning daylength or photoperiod for the organization of seasonal responses;
(b) the transmission of information concerning phase and strength of the daily light–dark cycle for appropriate synchronization of the circadian system; and
(c) subsequent effects on physiological and behavioral rhythms such as sleep, temperature, reproduction, etc.

Clearly the importance of the pineal in human physiology depends on the importance of light in human physiology. This was considered in Chapter 4 and will also emerge in the course of this chapter.

7.4 HUMAN MELATONIN PRODUCTION – BASIC RHYTHM CHARACTERISTICS

The measurement of melatonin is critically important to human pineal studies. For several years after the development of the first assay methods the reported results were inconsistent between laboratories and greatly differing concentrations of melatonin were reported. Many of these early studies require re-evaluation for methodological reasons. There is now a reasonable consensus as to appropriate methodology: specific RIA or gas chromatography–mass spectrometry for body fluids, these, or high-performance liquid chromatography, for glandular estimations and for pharmacokinetics after exogenous melatonin.

The major urinary metabolite of melatonin in humans is 6-sulfatoxymelatonin (aMT6s). Both in plasma and in urine it shows a very close quantitative and qualitative relationship to melatonin itself (Figure 7.4 and see Figure 3.8). It has been extensively used in human studies, particularly in urine, as urine sampling is non-invasive and can be conducted over long periods of time. It has been used for circadian rhythm assessment, for clinical pathology and in field conditions. It has the added advantage of being very stable, present in much larger concentrations in urine than melatonin and is easy to measure by radioimmunoassay. In general it can be used as a reliable index of melatonin secretion albeit with some loss of subtle changes in urine measurements. Much information on the circadian characteristics of melatonin secretion has been derived from aMT6s measurements. A comprehensive review of the assay of melatonin and its metabolites can be found in Arendt (1986).

In a 'normal' environment melatonin is secreted during the night in healthy humans as in all other species. The average maximum levels attained in plasma in adults are of the order of 50–70 pg/ml when measured with highly specific assays. Mean maximum concentrations of aMT6s attain 80–100 pg/ml (in different mammalian species there is a relatively narrow range of circulating concentrations although birds have more). Minimum concentrations of both compounds are often below the limit of detection of radioimmunoassay in plasma and are usually below 10 pg/ml. The peak concentrations of melatonin in plasma occur between 0200 and 0400 hours.

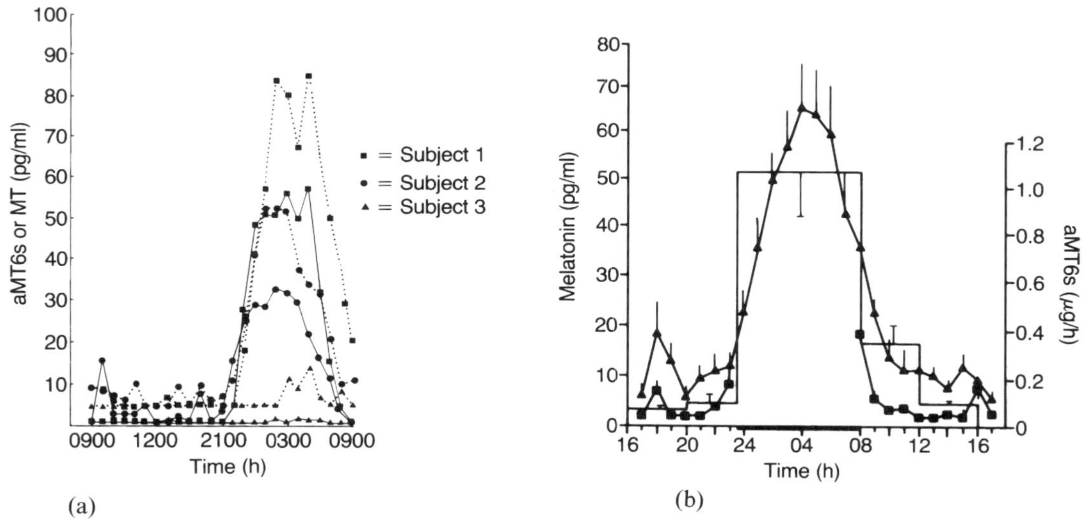

Fig. 7.4 (a) Plasma levels of 6-sulfatoxymelatonin (aMT6s) (......) and melatonin (———) in three healthy subjects sampled hourly over 24 hours. Note the large individual differences in amplitude and the undetectable plasma melatonin and extremely low aMT6s in one subject. The latter subject is a young man in his early twenties with no clinical problems but a tendency to sleep late. A small percentage of apparently normal people have undetectable plasma melatonin throughout 24 hours. (Reproduced with permission from Arendt, J. *et al.*, Immunoassay of 6- hydroxy-melatonin sulphate (aMT6s) in human plasma and urine : abolition of the urinary 24-hour rhythm with atenolol. *Journal of Clinical Endocrinology and Metabolism*, **60**, 1166–73, 1985.) (b) Plasma melatonin at hourly intervals by indwelling antecubital cannula, urinary 6-sulfatoxymelatonin (aMT6s) in sequential 4-hourly collections (8-hourly during the night) and saliva melatonin hourly during waking hours. Strong correlations are present between all three variables in amplitude and timing although for better definition using saliva and urine it is necessary to take samples through the night. For this subjects must be wakened, thus compromising any parallel sleep studies *inter alia*. (Reproduced with permission from Deacon, S.J. and Arendt, J., 1994; and see Vakkuri, Leppaluoto and Kauppila, 1985, and Nowak *et al.*, 1987.)

Grace Dean, RN, MSN

APPOINTMENTS

The onset of secretion is usually around 2100–2200 hours and the offset at 0700–0900 hours in temperate zones. The appearance and peak levels of aMT6s in plasma are delayed by 1–2 hours and the morning decline by 3–4 hours. In urine 70–80 % of aMT6s appears in the overnight sample (2400–0800 hours) and it is low but rarely undetectable in the afternoon and early evening.

There are some reports of episodic or pulsatile secretion of melatonin in humans. Two and sometimes three peaks of secretion during the night are frequently present but rapid pulses are not seen in peripheral blood using highly specific assays. However, there is good evidence for pulsatility when animals are sampled in the confluens sinus or the jugular vein. Thus secretion probably is pulsatile but through tissue diffusion and metabolism this is damped in the peripheral circulation. Posture may play a minor role in the amplitude of circulating melatonin concentration: recumbency induces a rapid but small (about 20 %) decline in levels.

The human melatonin rhythm is endogenous and free-runs in the absence of sufficient time cues, with a tau usually greater than 24 hours. Light of sufficient intensity is the major synchronizer (see Chapter 4) and serves to both entrain, and in long photoperiods, to suppress secretion (Chapter 4). According to Wever (1989) the only important 'masking' of the melatonin rhythm is suppression by light of sufficient intensity. This means that, providing that the light environment is controlled, melatonin can show the 'purest' rhythm characteristics of any circadian variable. Its parameters may be the best estimate of the phase of the hypothalamic clock (SCN), which is presumed to function as the major central rhythm generator in humans as in other mammals.

Possibly the most striking characteristic of the normal human melatonin rhythm is its reproducibility from day to day and from week to week in the same individual when this individual's behavior and environment remain relatively constant. Both the amplitude and the details of the circadian profile are intra-individually consistent, rather like an hormonal fingerprint (Figure 7.5). As noted with regard to masking it is remarkably resistant to factors such as nutrition, metabolic changes and to stress. These are precisely the characteristics that would be required of a daily time cue and coordinator of rhythm physiology. In view of this stability small changes in phase, and more rarely amplitude, can be highly significant.

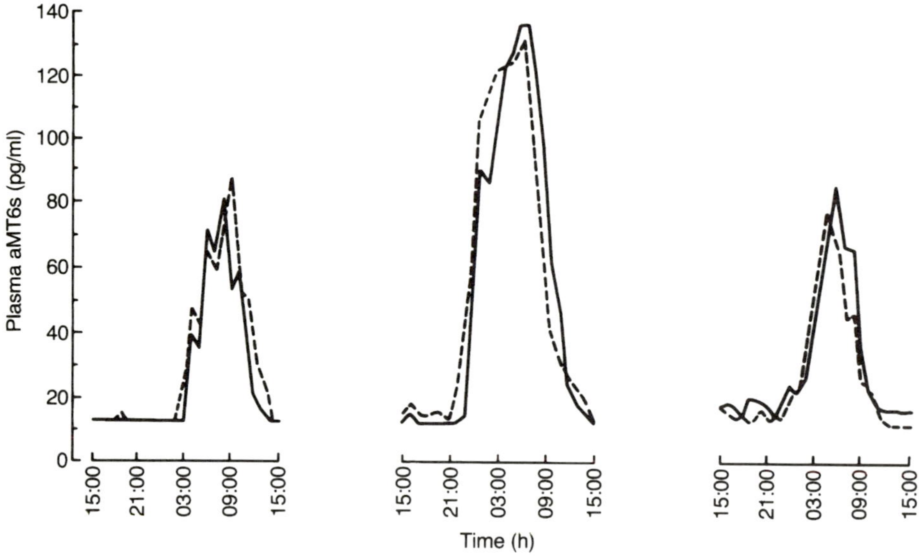

Fig. 7.5 Plasma concentrations of 6-sulfatoxymelatonin in three subjects sampled at hourly intervals by indwelling cannulae for 24 hours on two occasions, approximately 4–6 weeks apart during autumn in Antarctica. Note the reproducibility of amplitude and detailed pattern of secretion within individuals, rather like an hormonal fingerprint. (Reproduced from Arendt, 1988, and Broadway, Arendt and Folkard, 1987.)

In spite of intra-individual stability there is very large variability in amplitude of the rhythm between subjects. A small number of apparently normal individuals have no detectable melatonin in plasma at all times of day and in one subject at five different times of the year (Figure 7.4a). This does not mean that they have no melatonin, as all assay methods have limits of detection, and it may be particularly rapidly metabolized in some people. It does suggest that if melatonin is in any sense an essential hormone, low circulating levels are not pathological and that individuals may adapt to their 'low melatonin environment' through changes in sensitivity. However an alternative explanation is that there is much redundancy in rhythm control mechanisms and in the absence of one component others may assume the same function.

In a structured environment there is a small phase delay of the melatonin rhythm in winter compared to summer of approximately 1–2 hours and no change in amplitude (Figure 7.6). Basal (daytime)

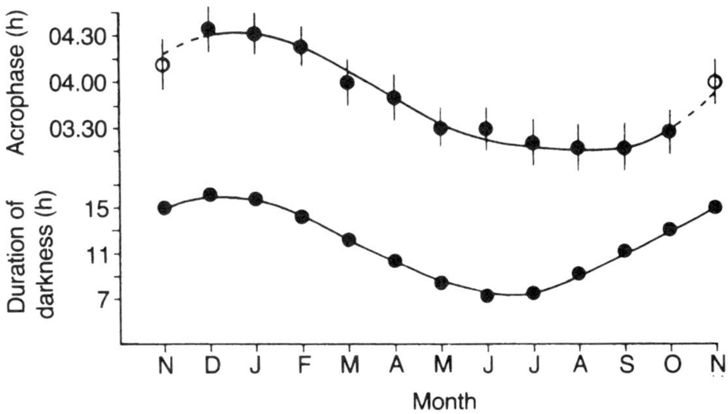

Fig. 7.6 Change in timing of the urinary aMT6s rhythm in the course of a year at 52 °N. Urine was sampled over 48 hours at monthly intervals and the monthly acrophases (calculated peak times) were smoothed by a 3-month moving average. The rhythm is phase advanced in summer compared to winter. (Reproduced with permission from Bojkowski, C. and Arendt, J.; published by Rhodos International, 1988; and see Illnerova, Zvolsky and Vanecek, 1985 and Broadway, Arendt and Folkard, 1987.)

concentrations of both melatonin (in plasma) and aMT6s (in urine) are reported to have a bimodal pattern with peaks in summer and winter. In artificial short (8L 16D) and long (16L 8D) photoperiods the melatonin rhythm expands and contracts to fill the dark phase as it does in seasonal mammals in natural or artificial light (see Chapter 4). This has rarely been observed in humans in uncontrolled conditions with the exception of polar regions.

7.5 ASSOCIATION WITH TEMPERATURE

The pineal is clearly concerned with circadian and seasonal thermoregulatory functions in animals. Many associations with temperature exist in humans. Since the initial assessments of human melatonin production it has been evident that there is a reciprocal relationship in circadian profiles whereby the temperature nadir

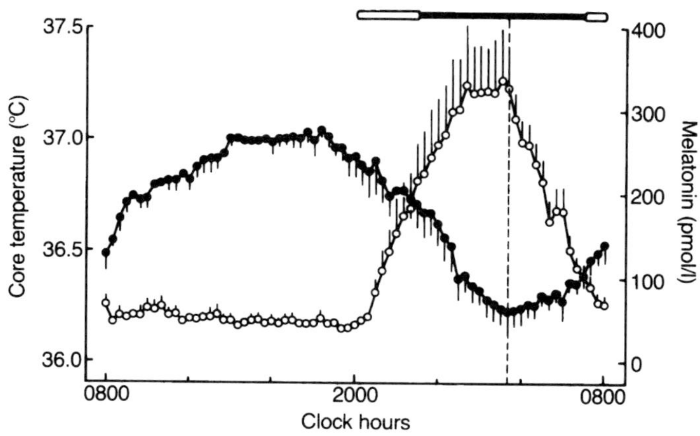

Fig. 7.7 Time course and 24-hour patterns of mean (±SEM) serum melatonin levels (O) and core body temperature (●) in 12 early follicular phase women. (▭) Duration of nocturnal melatonin secretion; (▬), sleep time. Close temporal correlations are evident between the onset or offset of melatonin secretion and the decline or rise in body temperature. (Reproduced with permission from Cagnacci, A. Elliott, J. and Yen, S. Melatonin: a major regulator of the circadian rhythm of core temperature in humans. *J. Clin. Endocrinol. Metab.* **75**, 447–52, 1992 © The Endocrine Society.)

correlates closely with the peak of melatonin (Figure 7.7). There is a possible causal relationship as exogenous melatonin can acutely depress body temperature in humans and bright light, which increases body temperature, suppresses melatonin. The ovulatory rise in temperature during the menstrual cycle is associated with a reported decline in amplitude of melatonin but the decline in melatonin is not a consistent observation.

7.6 ASSOCIATION WITH SLEEP

There are obvious correlations between melatonin production at night and sleep, and again some specific causal relationships may exist (see Chapter 8). However, sleep deprivation does not abolish the melatonin rhythm and in dim light does not affect secretion, unlike its effect on sleep-related hormones such as prolactin and growth hormone. Both nocturnal and diurnal animals secrete melatonin at night, thus it cannot be directly related to sleep induction as a general rule. It may have an endogenous function in the timing of sleep, however. As yet, suppression of melatonin secretion with adrenergic antagonists has not been shown to have specific major effects on sleep. During sleep deprivation self-rated fatigue ('sleepiness') exhibits a circadian rhythm which is closely positively correlated with plasma melatonin and negatively correlated with performance measures (Figure 7.8). Many studies have attempted to relate sleep stages to detailed melatonin profiles with no striking associations emerging.

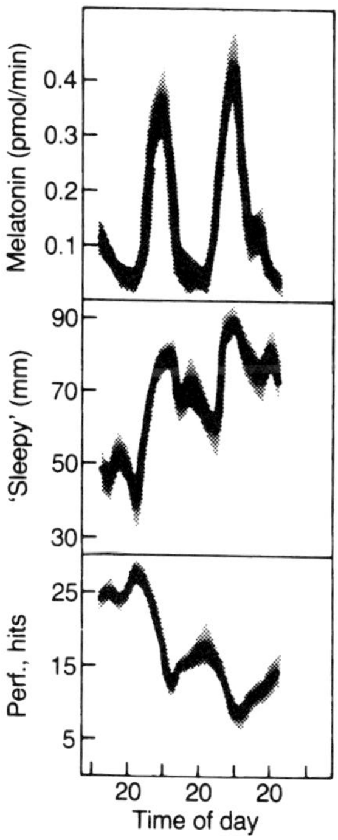

Fig. 7.8 Covariance of plasma melatonin, fatigue ('sleepy') and performance (perf., hits) during sleep deprivation in normal subjects. These studies also showed for the first time the negative correlation of melatonin production with body temperature. (Reprinted from *Psychoneuroendocrinology*, **4**, Akerstedt, T. *et al.* Melatonin excretion, body temperature and subjective arousal during 64 hours of sleep deprivation, pp. 219–25, copyright 1979, with kind permission from Pergamon Press Ltd, Headington Hill Hall, Oxford OX3 0BW, UK.)

7.7 OTHER ASSOCIATIONS

Prolactin is a sleep-related hormone in humans and its interaction with melatonin is described later. Many aspects of the immune system show striking circadian and seasonal variations. In some animals these are photoperiod dependent with evident links to melatonin (see p. 226). Obviously any variable with a circadian rhythm will show correlations with melatonin, if necessary displaced in time. The most important are cited earlier.

7.8 CONTROL OF HUMAN MELATONIN SECRETION

The neural and biochemical pathways known to control pineal function in animals are presumed to be similar in humans (Figure 7.9). There is good evidence for most of the components and retinally mediated light–dark control of human melatonin secretion is considered in Chapter 3. Evidence for the importance of the SCN as the master pacemaker in humans is rare and circumstantial; however, the retinohypothalamic pathway is present and in at least one case where pathological lesions of the SCN area were present, circadian rhythmicity was lost. Patients with degenerative diseases of the sympathetic nervous system lose the rhythm of melatonin. For example, it is abolished in peripheral diabetic neuropathy with sympathetic degeneration and in the presence of cervical cord section. The sympathetic innervation of the pineal is clearly essential for the control of human melatonin secretion.

Acute administration of clinical antihypertensive doses of β-adrenergic antagonists such as propanolol and atenolol in the afternoon completely abolishes the melatonin rhythm during the subsequent night; thus melatonin production is under adrenergic control (Figure 7.10). Atenolol acts peripherally in the clinical dose range and its effect indicates that the human pineal is outside the blood–brain barrier. Stimulation of melatonin production directly by β-agonists encounters cardiovascular problems in humans. Indirect stimulation is possible by noradrenergic uptake inhibitors such as desmethylimipramine (DMI) and oxaprotiline. DMI specifically advances the onset of secretion in the evening. The

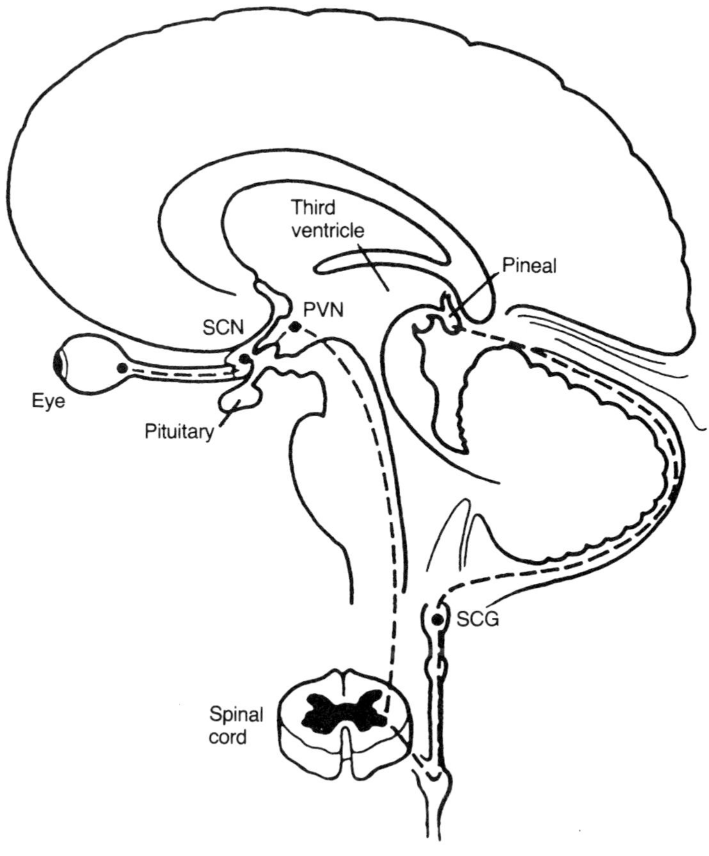

Fig. 7.9 Neural connections of the human pineal. SCN=suprachiasmatic nucleus, PVN=paraventricular nucleus, SCG=superior cervical ganglion. (Redrawn from Tamarkin, K., Baird, C.J. and Almeida, O.F.X. Melatonin: a coordinating signal for mammalian reproduction *Science*, **227**, 714–20, copyright 1985 by the AAAS.)

serotonin uptake inhibitor fluvoxamine induces a large increase in amplitude of secretion without changing the timing (see Chapter 4). Many inhibitors of catecholamine and indoleamine metabolism or uptake will increase human melatonin production, as will the serotonin precursor 5HTP. It is possible that $\alpha 2$ presynaptic

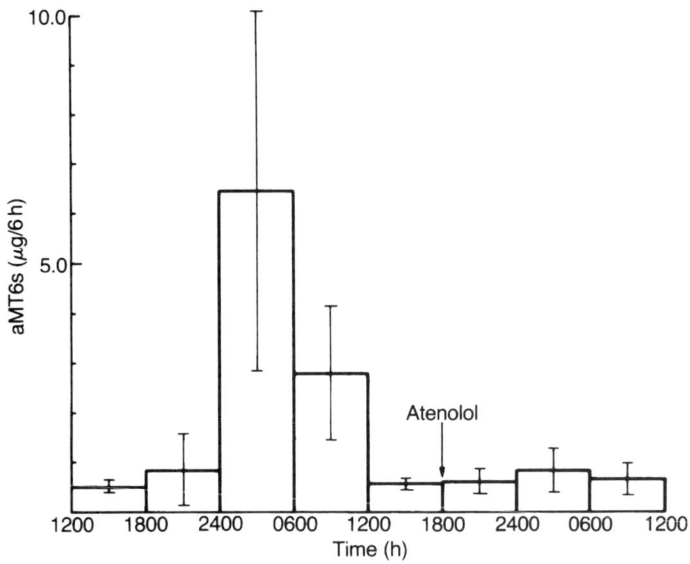

Fig. 7.10 Complete suppression of aMT6s production in healthy human subjects by 100 mg atenolol, a peripheral β-antagonist, at 1800 h. (Reproduced with permission from Arendt, J. *et al.*, Immunoassay of 6-hydroxy-melatonin sulphate (aMT6s) in human plasma and urine: abolition of the urinary 24-hour rhythm with atenolol. *Journal of Clinical Endocrinology and Metabolism*, **60**, 1166–73, 1985, and see Wetterberg, 1978, and Cowen *et al.*, 1983.)

receptors are involved in the control of human melatonin secretion.

Taken together these observations indicate that human melatonin production is controlled by noradrenalin acting primarily via β1-receptors in the pineal. Serotonin availability is probably important for the amplitude of secretion whilst noradrenalin is more concerned with timing.

In addition to drugs affecting catecholamines and indoleamines a number of other substances reportedly stimulate or inhibit human melatonin secretion. The class of compounds known as psoralens are used to sensitize the skin to UV light in the treatment of psoriasis. They induce large increases in melatonin production. One explanation is that they sensitize the retina to light thereby increasing the amplitude of circadian rhythms. It is more likely that they inhibit the metabolism of melatonin. The latter explanation also applies to the stimulatory effects of chlorpromazine. Some non-steroidal antiinflammatory drugs (e.g. ibuprofen) and some benzodiazepines (e.g. alprazolam) appear to exert transitory, minor inhibition of night-time secretion. The precursor of dopamine, L-dihydroxyphenylalanine, with or without peripheral decarboxylase inhibition does not affect melatonin in Parkinson patients although the theoretical possibility of dopamine stimulation exists.

Changing posture from standing to recumbent reduces plasma melatonin concentrations during the rising phase of secretion by 20–30 %. The effect is reversed on returning to a standing position. This phenomenon is probably due to hemoconcentration of weakly protein-bound (albumin) melatonin following diffusion of small molecules into the extracellular space during standing (Deacon and Arendt, 1994). The effects of stress, exercise and some other non-pharmacological interventions on melatonin production are controversial and do not appear to play a major role in humans. Insufficient data are available to make further judgements.

7.9 POST-MORTEM PINEAL STUDIES

The use of post-mortem tissue is never ideal due to the time lag between death and tissue extraction. However the stability of pineal melatonin synthesizing enzymes, at least HIOMT, is apparently good in post-mortem tissue. Early work demonstrated the presence of HIOMT activity in tissue even from elderly calcified pineals. There may be a bimodal annual rhythm in synthetic capacity similar to basal melatonin secretion. It is possible to show that the melatonin content of human pineals is related to the time of death with, as expected, higher values at night.

7.10 EVOLUTION OF MELATONIN SECRETION IN NORMAL HUMANS

The evolution of melatonin secretion from birth to death follows a reasonably well-established course. Shortly after birth very little

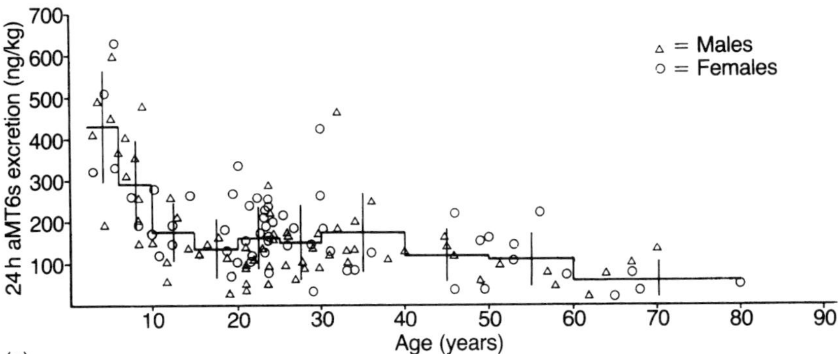

700
600
500
400
300
200
100
24 h aMT6s excretion (ng/kg)
10
20
30
40
50
60
70
80
90
Age (years)
△ = Males
○ = Females

(a)

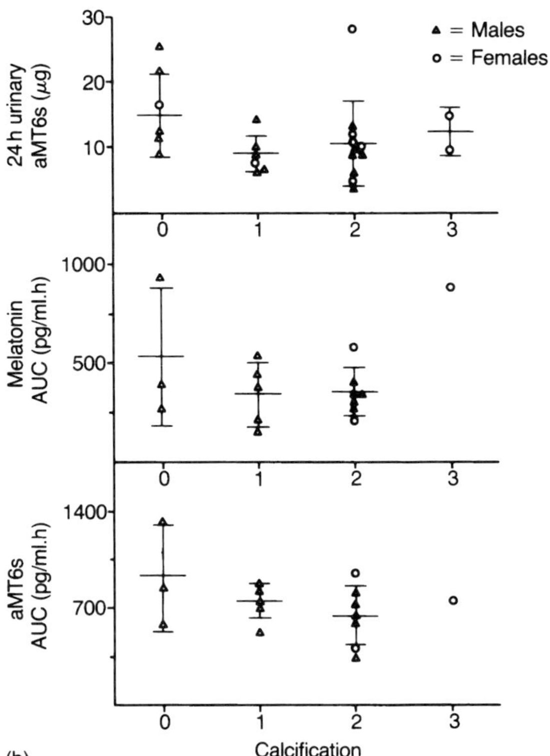

▲ = Males
o = Females
30
20
10
24 h urinary aMT6s (μg)
0
1
2
3
1000
500
Melatonin AUC (pg/ml.h)
1400
700
aMT6s AUC (pg/ml.h)
Calcification

(b)

melatonin, or aMT6s, is detectable in body fluids. Rhythmic secretion appears at approximately 2 months of age. The plasma concentration of melatonin increases rapidly thereafter and reaches a lifetime peak on average at 3–5 years old. Both daytime and night-time plasma concentrations are higher at this time but the increment is greater at night. Subsequently a steady decrease is seen, reaching mean adult concentrations in mid-to-late teenage with the major decline occurring before puberty. Values remain relatively unchanged until 35–40 years and a final decline in amplitude then takes place until very low levels are seen in old age (Figure 7.11a). Individual differences are of course superimposed on this general trend. Calcification in the human pineal does not relate to melatonin production within the age range 20–50 years (Figure 7.11b) although clearly as calcification appears during development it must relate to age within this time scale.

7.11 SEX, WEIGHT AND HEIGHT

Reports of differences in secretion in adults with sex, height or body weight are not consistent. However the measured plasma concentrations of melatonin in children are clearly negatively correlated to body weight. aMT6s in urine provides an integrated measure of melatonin production and in this case there are no changes in absolute amounts excreted per 24 hours until the decline in old age. If expressed as a function of body weight a similar pattern to plasma melatonin is seen from infants to adult life. It would appear that after the onset of secretion in very early life the pineal produces a constant

Fig. 7.11 (a) Total 24-h 6-sulfatoxymelatonin (aMT6s) excretion in normal subjects of different ages (n=130) expressed as a function of body weight (ng/kg). Note the decline during development and again during old age, corresponding to similar observations on plasma melatonin (Waldhauser *et al.*, 1984, Iguchi, Kato and Ibayashi, 1982). (Reproduced with permission from Bojkowski, C. and Arendt, J.; published by Blackwell Scientific Publications, 1990.) (b) There is no relationship between extent of pineal calcification, radiographically determined, and aMT6s (24-h excretion), plasma melatonin (area under the 24-h profile) or plasma aMT6s (area under the 24-h profile) in young to middle-aged adults. (Reproduced with permission from Bojkowski, C. and Arendt, J.; published by Blackwell Scientific Publications, 1990.)

amount of melatonin which appears to decline in plasma due to increasing weight. In old age body weight appears not to influence circulating melatonin.

7.12 PUBERTY

Melatonin blood concentrations clearly decline during pubertal development (Figure 7.12). Body weight is of considerable importance in the timing of puberty. The observations on melatonin in children suggest that programmed growth coupled with constant pineal function leads to passively changing melatonin concentrations. If melatonin has a role in human pubertal development, it might thereby achieve a critical level for the theoretical induction, or removal of inhibition, of one or more physiological responses. Many if not all physiological factors change in the course of development however, and these circumstantial arguments in no way prove that melatonin is involved in human puberty.

Photoperiod via melatonin secretion determines the timing of puberty in some species, provided that a sufficient degree of physical maturity has been reached. Interestingly, photoperiod perception by the fetus is present before birth in rodents and ungulates and ensures a rate of development appropriate to environmental conditions. Melatonin crosses the placenta and injections to the mother can dictate postnatal reproductive development. The effects of photoperiod on human puberty are entirely unknown although circumstantial evidence for a role of light exists. For example, blind subjects may have abnormally timed pubertal development and there may be seasonal influences on the timing of menarche.

Studies in precocious and delayed puberty have not yet proved very rewarding probably due to the large interindividual differences in melatonin amplitude and insufficient attention to the phase and duration of secretion. Lower melatonin has, however, been reported in precocious puberty compared to age-matched controls, although according to some authors it may be more related to skeletal age. Ovarian suppression with LHRH analog in precocious girls is not accompanied by changes in melatonin secretion. Several reports of increased melatonin in delayed puberty exist. One 15-year-old female

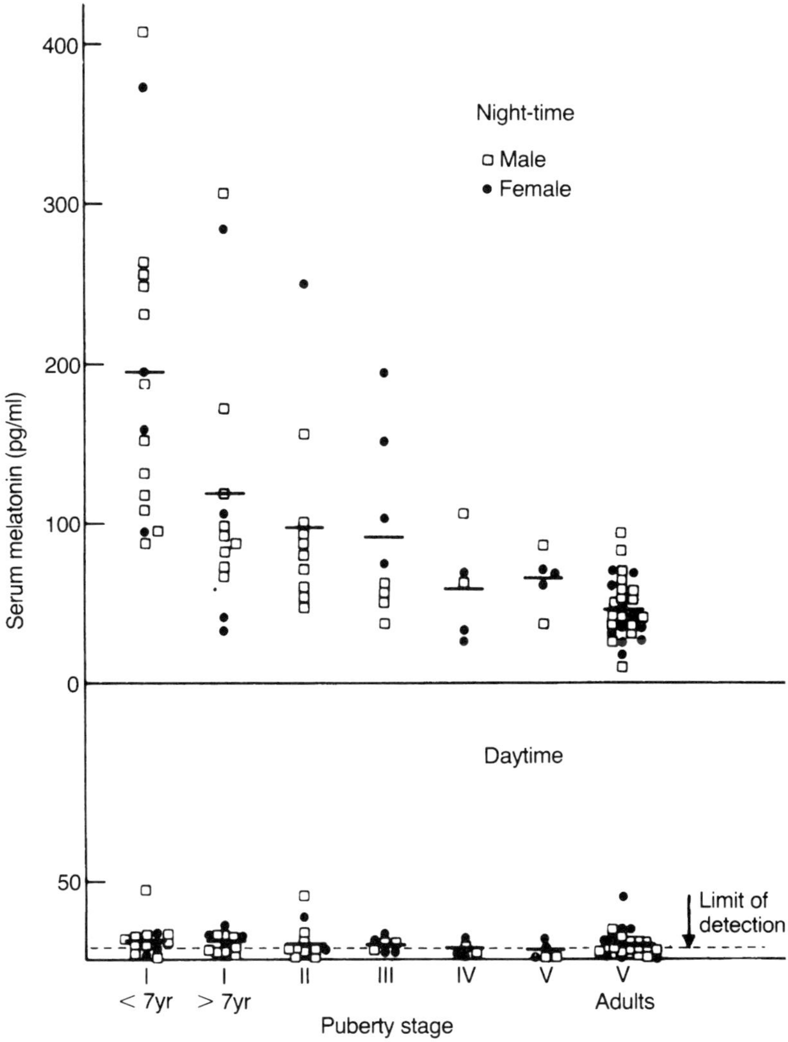

Fig. 7.12 Daytime and night-time serum melatonin in children, adolescents and young adults grouped according to their stage of sexual maturation. (Reproduced with permission from Waldhauser, F. *et al.*; published by Elsevier, 1984.)

subject with delayed puberty related to craniopharyngioma displayed aMT6s concentrations that were characteristic of 4–5-year-old children. A substantial decrease was seen in aMT6s production during estrogen treatment for the induction of secondary sexual characteristics. More recently delayed puberty in a male subject with a pineal tumor was accompanied by greatly elevated plasma melatonin concentrations. In the course of treatment with testosterone these concentrations declined and the patient became fully reproductively active. More longitudinal case studies of this sort are required.

7.13 DEVELOPMENT OF THE MELATONIN RHYTHM

The onset of circadian rhythmicity in human infants occurs several weeks after birth with individual variations. Initially arrhythmic, human infants develop free-running activity–rest cycles which subsequently entrain to the prevailing 24-hour time cues. These facts evidently explain the problems of human parents with night active offspring. It is likely that the maturation of the central pacemaker occurs first, followed by entrainment pathways such as the retinohypothalamic projection. In concert with maturation of entrainment pathways a 24-hour melatonin rhythm appears at around 6–8 weeks of life (Figure 7.13). Whether in specific individuals this corresponds to the organization and synchronization of other circadian variables such as sleep remains open and a question of very considerable interest.

Studies in rodents suggest that melatonin is the most important humoral signal *in utero* for the organization of the fetal circadian system. If human maternal melatonin acts in this way pre- and postnatally, we may speculate that the setting of circadian phase in early development depends on the way human infants are nourished. Melatonin is present in milk (at least in animals such as goats) with much greater quantities in morning milk than in evening samples; hence breast-fed babies will be exposed to rhythmic melatonin. We may safely assume that commercial cows' milk or other artificial feeds are incapable of providing rhythmic melatonin. Thus if melatonin serves to set circadian phase in humans it is quite probable that breast-fed babies will develop circadian organization before bottle-

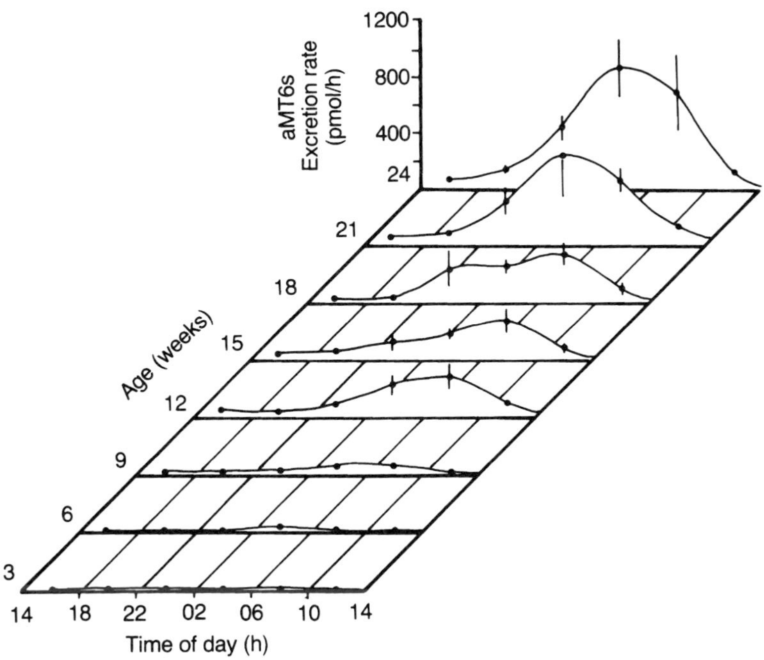

Fig. 7.13 Development of the circadian rhythm of 6-sulfatoxymelatonin (aMT6s) excretion in healthy full-term infants. Infants using at least eight diapers during 24 h and with at least one diaper change between 2200 h and 0600 h (n=11), were used to construct this figure. aMT6s excretion rates were calculated for six 4-h periods and the mean (±SEM) pmol/h plotted. Absence of error bars indicates the line was of similar size to the symbol. Premature infants excreted less aMT6s than full-term infants even at 52 weeks after birth (not shown). (Reproduced with permission from Kennaway, D.J., Stamp, G.E. and Goble, F.C. Development of melatonin production in infants and the impact of prematurity. *J. Clin. Endocrinol. Metab.* **75**, 367–9, 1992 © The Endocrine Society.)

fed babies. This may well have long-term consequences. At present the advantages of early circadian organization are unknown with regard to the offspring, but parents might well have cause to be grateful for day-active infants.

7.14 AGING

There is a striking decline in the amplitude of melatonin production with age in humans (Figure 7.14). This has led to speculation that circadian abnormalities present in old age may be secondary to loss of the melatonin rhythm. Furthermore it is suggested that exogenous melatonin may restore circadian organization and even improve longevity. At the time of writing there is no information on the effects of exogenous melatonin in old age in humans (see Chapter 8).

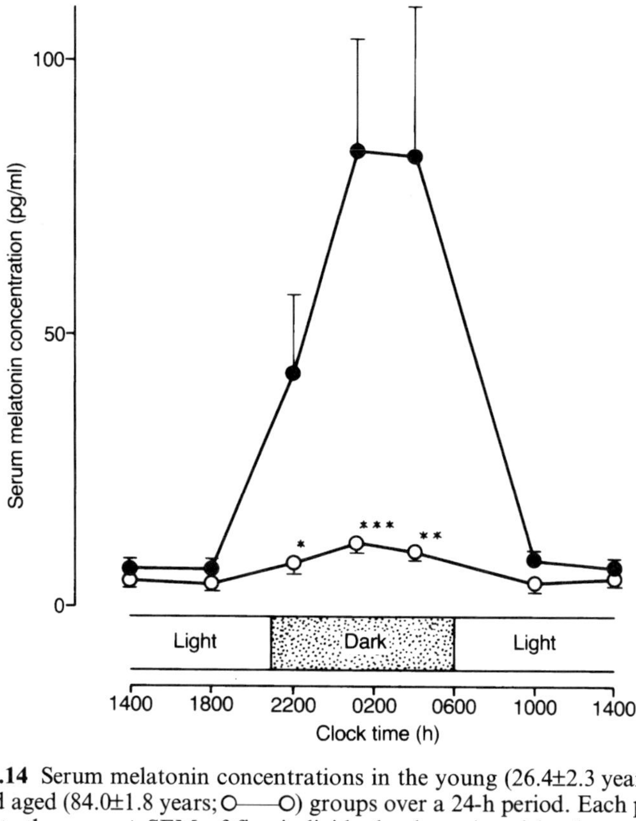

Fig. 7.14 Serum melatonin concentrations in the young (26.4±2.3 years; ●——●) and aged (84.0±1.8 years; ○——○) groups over a 24-h period. Each point represents the mean ± SEM of five individual values. Asterisks show a significant difference in serum melatonin concentrations at 2200 h, 0100 h, and 0400 h between the two groups. (Reproduced with permission from Iguch, H., Kato, K. and Ibayashi, H. Age-dependent reduction in serum melatonin concentrations in healthy human subjects. *J. Clin. Endocrinol. Metab.* **55**, 27, 1982 © The Endocrine Society.)

7.15 MENSTRUAL CYCLE

Some of the very earliest reports on human melatonin described low pre-ovulatory concentrations in morning plasma and suggested that low melatonin was facilitatory to the pre-ovulatory LH peak. This is, however, an inconsistent observation and the most recent work indicates that neither the amplitude nor the phase of melatonin is altered in the course of the cycle. Total aMT6s production in urine over 24 hours was not altered in a small number of women during the menstrual cycle.

Phase is difficult to assess carefully over a relatively long period such as the menstrual cycle. Longitudinal studies in the same individual are essential in small numbers of subjects in view of the interindividual variability. It is necessary to take 'snapshots' of plasma melatonin at intervals to avoid unacceptable loss of blood and so far there are no reports of saliva melatonin estimation in the course of the cycle. It is not possible so far to be certain that there are no small phase or amplitude changes which may be physiologically significant.

It is difficult to eliminate altogether a possible function of melatonin in the human cycle. There are reports that bright light given during mid-cycle at night may modify the timing of menstruation and there is a circadian basis for the timing of the ovulatory LH peak which is also influenced by season. All of these phenomena may be influenced by melatonin (see Chapter 8 for effects of melatonin on human reproductive function).

In primary hypothalamic amenorrhea the amplitude of melatonin secretion is increased in most reports but so far this is a correlative not a causal association.

7.16 HUMAN SEASONAL REPRODUCTIVE PHYSIOLOGY

The relevance of the pineal and melatonin to human reproduction is related to seasonal as well as circadian function. Seasonal patterns of conception rate are present in large data series and considered attempts have been made to relate these to environmental factors such as photoperiod. Roenneberg and Aschoff (1990) have concluded that both photoperiod and external temperature influence human reproduction but are overlaid with many social, behavioral and cultural factors. In western Europe there tend to be two peaks of conception, one in May–June and a minor peak in December; however, over many years the patterns may shift and in comparable latitudes in the USA the pattern may be reversed.

There are also seasonal influences on the onset of menarche with a maximum number of first menstruations in summer. Conception of dizygotic twin pregnancies is more common in the summer months during continuous light in Finland and increased endometrial hyperplasia and anovulatory cycles are reported in winter, associated with an increase in daytime melatonin secretion. Patients suffering from seasonal affective disorder (SAD, winter depression) are more seasonal than the general population in terms of reproduction (conception time), body weight, etc. There is evidence that they are supersensitive to light in terms of melatonin suppression and once again a photoperiodic effect may be involved.

Many factors other than reproduction show seasonal variations in humans. The evidence of residual human photoperiodism is accumulating. There is scope for the therapeutic use of bright light and/or melatonin for the manipulation of human seasonality.

7.17 POSSIBLE RELATIONSHIPS OF THE PINEAL AND MELATONIN TO THE IMMUNE SYSTEM

Whether the pineal gland and melatonin are closely related to immune function or not is a subject of some considerable topical interest. Pinealectomy is reported to suppress various aspects of immune function and in many (but not all) studies enhances tumor growth. Thus it is a reasonable deduction that the pineal does indeed influence the immune system. Which of the many biologically active compounds produced by the pineal is/are responsible for these effects remains an interesting question. Recent work by Maestroni and colleagues (1981,1988) indicates that suppression of melatonin production using the β-adrenoceptor antagonist propanolol or the inhibition of serotonin (and thus by implication, melatonin) biosynthesis by *para*-chlorophenylalanine, impairs humoral and cellular immune responses and that this impairment may be reversed by melatonin administration. These observations give credence to the likely role of melatonin in mediating at least some pineal effects on the immune system. The physiological function of melatonin in relation to the immune system has not been properly addressed in either these or other related studies. Nevertheless, in general, melatonin has been found to have immunostimulating effects, with some notable exceptions in the literature (e.g. Lewinski *et al.*, 1989). This brief account will consider the possible relation of melatonin to the rhythmicity of the immune system.

7.17.1 Circadian rhythms

In humans 24-hour rhythms are present in most, if not all, immune functions. For a recent extensive review of immune system rhythmicity see Reinberg and coworkers (Levi, Reinberg and Canon, 1989). Whether or not these 24-hour (sometimes 12 h and shorter periods are observed) periodicities are endogenous, i.e. truly circadian, persisting in a time-free environment, appears to be unknown. If both endogenous and exogenous components are present in immune system rhythmicity, as is the case for many 24-hour rhythms, there exist a number of conceivable ways in which melatonin may interact. If immune-system rhythmicity is endogenous and generated, like many other circadian rhythms, in the SCN, then any effect of melatonin on the entrainment and synchronization of central rhythm generating systems (see Chapter 8) can be extended in principle to include the immune system. So far, however, there is little, if any, evidence for the effects of melatonin on the spontaneous temporal variations of the immune system.

In some cases the peak time of immune system variables (e.g. total circulating lymphocytes, natural killer (NK) cell activity) corresponds closely to that of melatonin (Figure 7.15). In other cases, however, peak times are far apart. Any such correlations are, of course, not necessarily meaningful. There is very little information to date on entrainment of rhythms in the immune system to the light–dark cycle or other potential zeitgebers. Given the important effects of bright light on winter depression in humans, and the

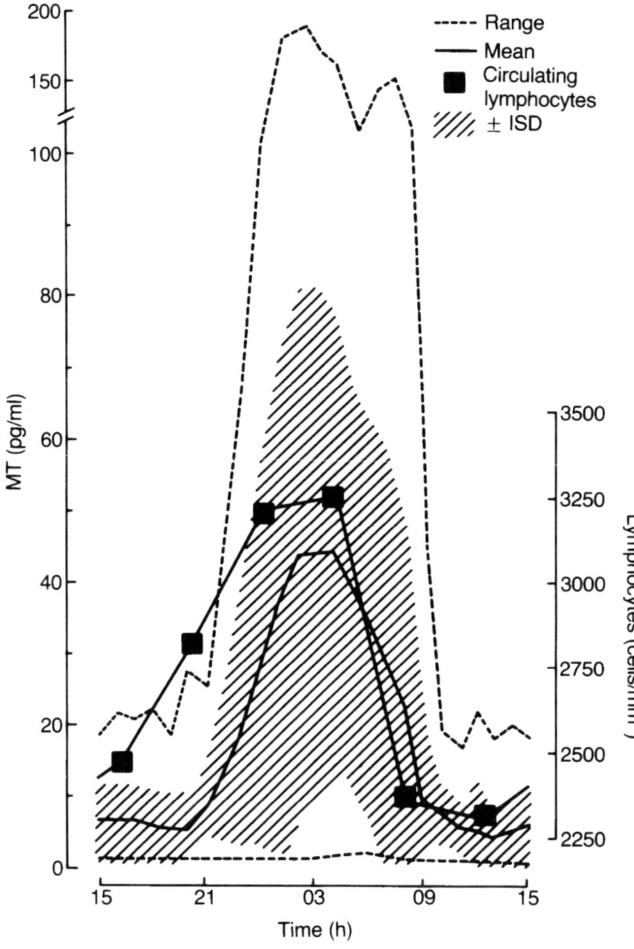

Fig. 7.15 Normal values of human plasma melatonin as compared to total circulating lymphocytes. A causal relationship has not been demonstrated. (Reproduced with permission from Arendt, J.; published by Plenum Publishing, 1991.)

relationship between depression, stress and immune response, such information would be of considerable interest.

Before proceeding to consider annual rhythmicity it is worth noting that 12-hour periodicity has been reported in lymphocyte activity (Levi, Reinberg and Canon, 1989) but no 12-hour rhythmicity has ever been reported for melatonin. Moreover, a 7-day rhythm in rejection of kidney grafts, and possibly other variables, may exist. Although one publication reports 7-day rhythmicity in serotonin-*N*-acetyltransferase activity in rats with peak values at the weekends, it would not be surprising to find such variations given the marked environmental (in laboratories) and social (in humans) changes during Saturday and Sunday compared with the working week. It seems doubtful that 7-day variations are endogenous. If they indeed exist, masking by changing environmental conditions is likely to be the cause.

The decline in amplitude of melatonin production in old age has been extensively discussed as the reason for practically every deficit seen in old age from nocturnal enuresis to Alzheimer's disease. Naturally the immune system enters into these speculations together with the possible decrease of melatonin in depression (see p. 235) and its conceivable direct or indirect links with immunology. All told, however, in the opinion of this author, there is little evidence to suggest causal relationships between melatonin and spontaneous temporal circadian variations of the immune system.

7.17.2 Circannual variations and photoperiodism

Photoperiodic influences on the immune system promise to be a rich field of enquiry. In rodents photoperiod influences splenic weight, lymphocyte and macrophage counts with clear evidence for stimulation by short photoperiod in hamsters. Likewise there are many reported effects of photoperiod on tumor growth (see p. 277). Suffice it to say that as melatonin is the major photoperiodic messenger molecule, photoperiod effects on the immune system and cancer are likely to be transduced by melatonin.

In contrast to the clear and careful work reported concerning photoperiodism and the immune system, annual rhythmicity in human immune functions, although frequently observed, has not been characterized in great detail and comparison with melatonin production is difficult.

7.18 PATHOLOGY

7.18.1 Pineal hyper- and hypoplasia

There are a number of reports of variations in pineal weight post-mortem as a function of cause of death which have been summarized by Tapp (1979). Of the most interesting, hypoplasia of the pineal in association with retinal disease may be causally interrelated. Tapp has reported that pineals in patients dying from carcinoma of the breast or melanoma are heavier than those from patients with other cancers. Giant pineals, a very rare familial condition known as Mendenhall's syndrome, is associated with enormous development of the pineal, which may weigh as much as 1–2 g compared to the normal range of 50–200 mg. A single case report has described high concentrations of 6-sulfatoxymelatonin in urine.

Sudden infant death syndrome (SIDS) is associated with small pineals and decreased melatonin production. SIDS deaths usually occur at night and may be associated with abnormalities of sleep. If melatonin serves to coordinate circadian organization in the developing infant its underproduction may contribute to the pathology.

7.18.2 Pineal tumors

Pineal tumors are heterogeneous and may arise from germ cells (teratomas, germinomas, choriocarcinomas, endodermal sinus tumor, mixed germ cell tumor), pineal parenchymal cells (pineoblastoma and pineocytoma) and from the supporting stroma (gliomas). All are rare (less than 1 % of intracranial space occupying lesions) and tend to occur during childhood below 20 years of age with the exception of parenchymal cell tumors which occur equally in adults and children. Germinomas respond well to radiotherapy whereas primary surgery is more frequently the treatment of choice in other types. The tumor markers alpha-fetoprotein and β-hCG together with CSF cytology and computerized tomography (CT) serve for differential diagnosis. The most common symptoms are secondary to hydrocephalus (headache, vomiting and drowsiness) together with the triad of visual problems, diabetes insipidus and reproductive abnormalities. Germinomas and teratomas occur predominantly in males. Both precocious and delayed puberty have been associated with pineal tumors. Precocious puberty is more commonly associated with teratoma. Cohen and coworkers (1991) have recently reviewed the occurrence of precocious puberty in parallel with β-hCG secreting pineal tumors. As β-hCG is identical to β-LH they conclude that pubertal development can be directly attributed to ectopic β-hCG production in many cases. Moreover the predominance in boys may be explained on the basis that LH alone can stimulate testosterone production whereas in girls both LH and FSH are required for ovarian follicular development and estrogen production.

There is no consistent information on over- or underproduction of melatonin with specific types of tumor. However pinealectomy removes circulating melatonin and radiation therapy appears to greatly suppress melatonin. At present long-term evaluation of patients after surgery is under way in a number of centers. Early reports suggest that pinealectomy is associated with migraine, cluster headache and a variety of other neurological symptoms.

7.18.3 Association with other solid tumors

The relationship of the pineal and melatonin to cancer is an area which arouses much interest following early work suggesting that the pineal contains oncostatic activity. Cohen and coworkers proposed, in 1978, that human breast cancer was a melatonin deficient disease. Since this time many clinical studies have been performed to assess melatonin secretion in cancer patients. The consensus would indicate that low levels are associated at least with breast and prostatic cancer. However, one study which compared women with malignant and benign breast growths and carefully controlled for age found no associations other than a decline with age. Early work is plagued by uncontrolled conditions and often sampling only in the daytime. A number of broad studies which included various oncological conditions report significant differences, both increases and decreases, in plasma melatonin between types of cancer and from control populations. At present these are uninterpretable.

There is some evidence that melatonin is oncostatic *in vitro*, but possibly more important with regard to the control of cancer initiation and proliferation are the generally stimulatory effects of melatonin on the immune system (see Chapter 8).

There are many problems with clinical observations on melatonin. Subtle changes in individual phase and amplitude may be of real importance, but such is the general nature of clinical research that cross-sectional populations are used rather than longitudinal studies which are much more time consuming. Indeed it is usually impossible to acquire data before patients become ill and thus the critical observations may be missing. The timing and amplitude of melatonin secretion is dependent on behavior (e.g. outdoor or indoor work, early or late rising) by extrapolation from animal work, and in general such considerations are ignored. It is particularly pertinent that hospitalized patients are kept in a light–dark cycle which may differ substantially from non-hospitalized controls and are required to adhere to a particular routine, usually involving very early wake-up and bedtime, and frequently recumbency. Only when the light–dark input to the pineal and melatonin are appreciated does clinical research provide useful insights.

7.18.4 Cardiovascular system

Major circadian variations exist in the cardiovascular system from blood pressure (highest in the morning) to average time of death from heart disease (very early morning). Pinealectomy of rats leads to hypertension and melatonin administration can induce transient hypotension. These relationships require further exploration.

7.18.5 General endocrine relationships

Many clinical attempts have been made to relate circulating melatonin to endocrine pathology. The results on the whole are difficult to interpret and inconsistent. The most frequent observations concern cortisol and prolactin secretion in view of their marked circadian rhythm and dependence on the light–dark cycle. Wetterberg *et al.* (1982) have proposed an inverse quantitative relationship between melatonin and cortisol in depression (see p. 235). Reports of changed melatonin secretion in hyperprolactinemia are inconsistent. In panhypopituitarism melatonin is very low, corresponding to observations in animals after hypophysectomy. Administration of anterior pituitary releasing hormones and neurotransmitter precursors (except possibly 5HTP) does not appear to consistently affect melatonin. Opioid peptides and opiates may have some stimulatory effect during the day. Pineal–thyroid inter-relationships exist but as with all other observations, factors with circadian rhythms will inevitably be found to correlate with melatonin. Like the reproductive and the immune system the thyroid is clearly photoperiod dependent in animals, and thus almost certainly responsive to melatonin as a coordinator of rhythm physiology.

7.18.6 Anorexia, amenorrhea

The problem of anorexia nervosa lies within the domain of psychoneuroendocrinology but the cause is essentially unknown. Patients starve themselves and usually over-exercise. Their reproductive function can regress to prepubertal status with loss of menstrual periods. Reproductive function is impaired by a gross loss of body weight and excessive exercise. However, a primary hypothalamic dysfunction is suggested by the fact that menstrual cycling may cease before weight loss and not return until after regain of normal body weight. It is extremely difficult to sort out cause and effect.

Melatonin is fairly consistently of high amplitude in anorexia (Figure 7.16) and in some other forms of hypothalamic amenorrhea. In some but not all cases this may relate to body weight as in prepubertal children. As gonadectomy (in sheep) can at least transiently enhance melatonin, it is possible that low circulating steroid concentrations are responsible for increased melatonin. Even exercise according to some but not all reports increases the amplitude of secretion. Whether it is important to the course of the disease remains an open question.

7.18.7 Liver disease

Liver disease, such as cirrhosis, which impairs metabolic function leads to higher than normal plasma concentrations of melatonin.

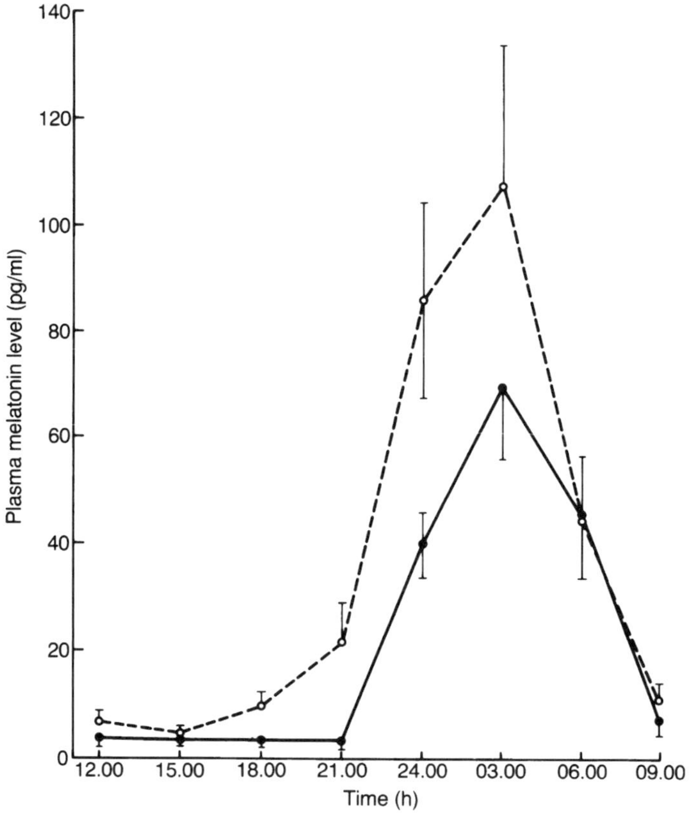

Fig. 7.16 Mean plasma melatonin levels (± SEM) in 11 patients with anorexia nervosa (○– – –○) and 9 healthy women (●——●) (pg/ml × 4.31=nmol/l). Subjects were in darkness between 2100 h and 0600 h. (Reproduced with permission from Arendt, J. *et al.*; published by the Royal College of Psychiatrists, 1992, and see Brown *et al.*, 1989.)

Likewise drugs which stimulate or suppress hydroxylation and conjugation mechanisms or which compete for metabolic pathways can be expected to affect circulating melatonin.

7.18.8 Disorders of the sympathetic nervous system

Disruption of the innervation of the human pineal abolishes the melatonin rhythm. This condition has been reported in cases of cervical cord lesion, diabetic autonomic neuropathy, Shy-Drager syndrome and dysautonomia.

7.18.9 Sleep disorder

Surprisingly, there is little evidence for a disturbance of melatonin secretion in major sleep disorders such as narcolepsy and Klein Levine syndrome, and in delayed sleep phase insomnia only a small delay in the rhythm is found. Rhythm abnormalities have been reported in non-specific insomnia.

7.18.10 Parkinson's disease

Melatonin secretion was normal in untreated Parkinson patients and was not modified by treatment with L-DOPA (dihydroxyphenylalanine) and/or aromatic amino acid decarboxylase inhibitors.

7.19 MELATONIN IN PSYCHIATRIC DISORDER

Abnormalities of circadian function have been postulated in depression and mania. Melatonin is arguably the best index of biological clock function: only bright light 'masks' the expression of the endogenous rhythm, and it has been extensively used in psychiatry and in other fields to assess biological clock status. There is evidence for a decline in amplitude of the melatonin rhythm in depression and in schizophrenia. In depression this is associated with an increase in cortisol. There is possibly an increase in amplitude in mania although not all studies are consistent.

There is little evidence for abnormal timing of melatonin, although Lewy (Lewy *et al.*, 1987) has reported exceptionally delayed melatonin rhythms in winter in patients with seasonal affective disorder (SAD) compared to the small delay seen in normals. This observation remains to be confirmed but is of considerable theoretical interest. The treatment originally proposed for SAD patients was the creation of an artificial summer day length using 3 hours of bright full spectrum light (Vitalite, 2500 lux) morning and evening. The 'melatonin hypothesis' predicted that such light treatment would shorten the duration of melatonin secretion thus generating a summer day length signal by analogy with animal work. The light treatment appears to be efficient but it does not appear to work through melatonin (Figure 7.17). Bipolar (manic depressive) patients are more sensitive to light suppression of melatonin than normals. This observation may prove to be of both diagnostic and therapeutic importance.

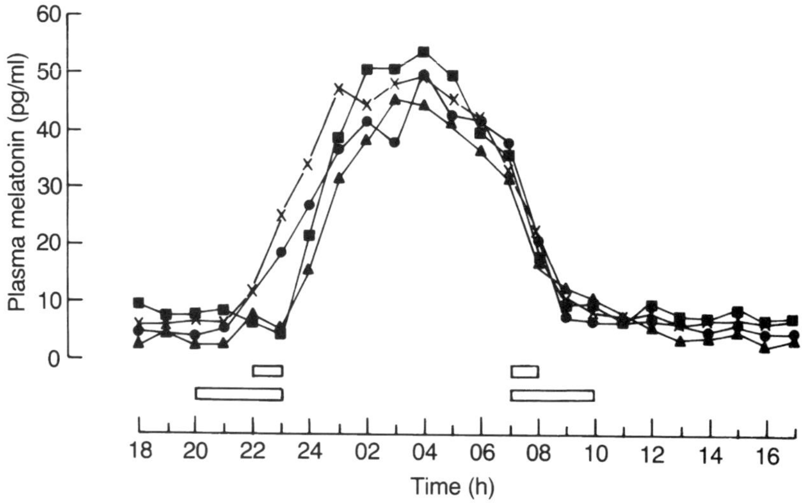

Fig. 7.17 Mean plasma melatonin concentrations in seven patients with seasonal affective disorder (SAD) before treatment (×) and after treatment with either 3-h (▲) or 1-h (■) bright white (2500 lux) light, or 3-h dim (400 lux) red light (●) morning and evening as shown by the rectangles. Only 3-h white light was effective in the alleviation of symptoms, however both 1-h and 3-h of white light delayed the onset of the melatonin rhythm. As with other data (e.g. Rosenthal *et al.*, 1988) this argues against a role for melatonin in the causality of seasonal affective disorder. (Reproduced with permission from Checkley, S.; published by Butterworth-Heinemann, 1989.) Lewy *et al.*, (1987) report delayed melatonin rhythms in SAD patients and that early morning light, in a phase advancing position, is most effective in inducing remission.

Most pharmacological antidepressant treatments will stimulate melatonin secretion, acting through increased availability of the precursors tryptophan and serotonin and the major pineal neurotransmitter noradrenalin, or by direct action on serotonin and catecholamine receptors. There may be a link between an increase in melatonin production and efficacy of treatment and this possibility merits exploration.

A more extended review of melatonin in psychiatric disorder can be found in Arendt (1989).

7.20 BLINDNESS

This subject is covered in Chapter 4.

REFERENCES

Akerstedt, T., Froberg, J.E., Friberg, W. and Wetterberg, L. (1979) Melatonin excretion, body temperature and subjective arousal during 64 hours of sleep deprivation. *Psychoneuroendocrinology*, **4**, 219.

Arendt, J. (1986) Assay of melatonin and its metabolites: results in normal and unusual environments, in *Melatonin in Humans* (eds R.J. Wurtman and F. Waldhauser), *Journal of Neural Transmission*, Suppl. 21, 11–33.

Arendt, J. (1988) Melatonin. *Clinical Endocrinology*, **29**, 205–29.

Arendt, J. (1989) Melatonin – a new probe in psychiatric investigation. *British Journal of Psychiatry*, **155**, 585–90.

Arendt, J. (1991) Significance of melatonin to chronobiology: immunological correlations, in *Proceedings of the Nato Workshop: Role of melatonin and pineal peptides in Neuroimmunomodulation* (eds F. Fraschini and R.J. Reiter), Plenum Press, NY, pp. 281–920.

Arendt, J., Bhanji, S., Franey, C. and Mattingley, D. (1992) Plasma melatonin levels in anorexia nervosa. *British Journal of Psychiatry*, **161**, 361–4.

Arendt, J., Bojkowski, C., Franey, C. *et al.* (1985) Immunoassay of 6-hydroxy-melatonin sulphate (aMT6s) in human plasma and urine: abolition of the urinary 24-hour rhythm with atenolol. *Journal of Clinical Endocrinology and Metabolism*, **60**, 1166–73.

Bojkowski, C. and Arendt, J. (1988) Annual changes in 6-sulphatoxymelatonin excretion in man. *Acta Endocrinology*, **117**, 470–6.

Bojkowski, C. and Arendt, J. (1990) Factors influencing urinary 6-sulphatoxymelatonin, a major melatonin metabolite, in normal human subjects. *Clinical Endocrinology*, **33**, 435–44.

Broadway, J., Arendt, J. and Folkard, S. (1987) Bright light phase shifts the human melatonin rhythm during the Antarctic winter. *Neuroscience Letters*, **79**, 185–9.

Brown, G.M., Chik, C.L., Ho, A.K. *et al.* (1989) Effects of food restriction on pineal function, in *Advances in Pineal Research* (eds R.J. Reiter and S.F. Pang), John Libbey, London, **3**, 87–92.

Cagnacci, A., Elliott, J. and Yen, S. (1992) Melatonin: a major regulator of the circadian rhythm of core temperature in humans. *Journal of Clinical Endocrinology and Metabolism*, **75**, 447–52.

Checkley, S. (1989) The relationship between biological rhythms and the affective disorders, in *Biological Rhythms in Clinical Practice* (eds J. Arendt, D.S. Minors and J. Waterhouse), Butterworth, London.

Cowen, P.J., Fraser, S., Sammons, R. *et al.* (1983) Atenolol reduces plasma melatonin concentrations in man. *British Journal of Clinical Pharmacology*, **15**, 579–81.

Cohen, M., Lippman, M. and Chabner, B. (1978) Role of pineal gland in aetiology and treatment of breast cancer. *Lancet*, **ii,** 814–6.

Cohen, A.R., Wilson, J.A. and Sadeghi-Nejad, A. (1991) Gonadotrophin-secreting pineal teratoma causing precocious puberty. *Neurosurgery*, **28**, 597–602.

Deacon, S. and Arendt, J. (1994) Posture influences melatonin secretion in humans. *Neuroscience Letters*, **167**, 191–4.

Iguchi, H., Kato, K. and Ibayashi, H. (1982) Age-dependent reduction in serum melatonin concentrations in healthy human subjects. *Journal of Endocrinology and Metabolism*, **55**, 27.

Illnerova, H., Zvolsky, P. and Vanecek, J. (1985) The circadian rhythm in plasma melatonin concentration of the urbanized man, the effect of summer and winter time. *Brain Research*, **328**, 186–9.

Kennaway, D.J., Stamp, G.E. and Goble, F.C. (1992) Development of melatonin production in infants and the impact of prematurity. *Journal of Clinical Endocrinology and Metabolism*, **75**, 367–9.

Levi, F., Reinberg, A. and Canon, C. (1989) Clinical immunology and allergy, in *Biological Rhythms in Clinical Practice* (eds J. Arendt, D.S.M. Minors and J.M. Waterhouse), Wright, London, pp. 99–135.

Lewinski, A., Zelazowski, P., Sewerynek, E. *et al.* (1989) Melatonin-induced suppression of human lymphocyte natural killer activity *in vitro*. *Journal of Pineal Research*, **7**, 153–64.

Lewy, A.J., Sack, R.L., Miller, L.S. *et al.* (1987) Anti-depressant and circadian phase-shifting effects of light. *Science*, **235**, 352–4.

Maestroni, G.J., Conti, A. and Pierpaoli, W. (1988) Role of the pineal gland in immunity. III. Melatonin antagonises the immunosuppressive effect of acute stress via an opiatergic mechanism. *Immunology*, **63**, 465–9.

Maestroni, G.J.M. and Pierpaoli, W. (1981) Pharmacologic control of the hormonally mediated immune response, in *Psychoneuroimmunology* (ed. R. Ader), Academic Press, New York, p.405.

Marburg, O. (1907) Zur Kenntnis der normalen and pathologischen Histologie der Zirbeldruse. *Arb. Wien, Neurol. Inst.*, **17**, 217–49.

Marburg, O. (1930) Die physiologie der zirbeldruse (*Glandula pinealis*), in *Handbuch der normalen und Pathologischen Physiologie*, Volume 13, pp. 493–590.

Neuwelt, E.A. and Lewy, A.J. (1983) Disappearance of plasma melatonin after removal of a neoplastic pineal gland. *New England Journal of Medicine*, **19**, 1132–5.

Nowak, R., McMillen, C., Redman, J. and Short, R.V. (1987) The correlation between serum and salivary melatonin concentrations and urinary 6-hydroxymelatonin sulphate: two non invasive techniques for monitoring human circadian rhythmicity. *Clinical Endocrinology*, **27**, 445–52.

Roenneberg, T. and Aschoff, J. (1990) Annual rhythms in human reproduction: I. Biology, sociology or both? *Journal of Biological Rhythms,* **5**, 195–216.

Rosenthal, N.E., Jacobsen, F.M., Sack, D.A. *et al.* (1988) Atenolol in seasonal affective disorder: a test of the melatonin hypothesis. *American Journal of Psychiatry*, **145**, 52–6.

Tamarkin, K., Baird, C.J. and Almeida, O.F.X. (1985) Melatonin: a coordinating signal for mammalian reproduction. *Science*, **227**, 714–20.

Tapp, E. (1979) The histology and pathology of the human pineal gland. *Progress in Brain Research*, **52**, 481–500.

Tapp, E. and Huxley, M. (1972) The histological appearance of the human pineal gland from puberty to old age. *Journal of Pathology*, **108**, 137–44.

Vakkuri, O., Leppaluoto, J. and Kauppila, A. (1985) Oral administration and distribution of melatonin in human serum, saliva and urine. *Life Sciences*, **37**, 489–95.

Waldhauser, F., Frisch, H., Waldhauser, M. *et al.* (1984) Fall in nocturnal serum melatonin during prepuberty and pubescence. *Lancet*, **i**, 362–365.

Wetterberg, L. (1978) Melatonin in humans physiological and clinical studies. *Journal of Neural Transmission*, (suppl.) **13**, 389–410.

Wetterberg, L., Aperia, B., Beck-Friis, J. *et al.* (1982) Melatonin and cortisol levels in psychiatric illness. *Lancet*, **ii**, 100.

Wever, R.A. (1989) Light effects on human circadian rhythms: a review of recent Andechs experiments. *Journal of Biological Rhythms*, **4**, 161–86.

Zimmerman, R.A. and Bilaniuk, L.T. (1982) Age-related incidence of pineal calcification detected by computed tomography. *Radiology/Neuroradiology*, **142**, 659–62.

FURTHER READING

Arendt, J. (1978) Melatonin assays in body fluids. *Journal of Neural Transmission*, (Suppl) **13**, 265–78.

Arendt, J. (1985) Mammalian pineal rhythms. *Pineal Research Reviews*, **3**, 161–213.

Arendt, J., Labib, M.H., Bojkowski, C. *et al.* (1989) Rapid decrease in melatonin production during treatment of delayed puberty with oestradiol in a case of craniopharyngioma. *Lancet*, **i**, 1326.

Attanasio, A., Borrelli, P., Marini, R. *et al.* (1983) Serum melatonin in children with early and delayed puberty. *Neuroendocrinology Letters*, **5**, 387–95.

Axelrod, L. (1977) Endocrine dysfunction in patients with tumours of the pineal region, in *Pineal Tumours* (ed. H.H. Schmidek), Masson Publishing, USA, pp. 61–77.

Badia, P., Myers, B. and Murphy, P. (1992) Melatonin and thermoregulation, in *Melatonin: Biosynthesis, Physiological Effects, and Clinical Applications* (eds R.J. Reiter and H.S. Yu), CRC Press, Boca Raton, Florida, pp. 349 – 64 .

Bartsch, C., Bartsch, H. and Fluchter, S.H. (1985) Evidence for modulation of melatonin secretion in men with benign and malignant tumours of the prostate: relationship with the pituitary hormones. *Journal of Pineal Research*, **2**, 121–32.

Bartsch, C., Bartsch, H. and Fuchs, U. (1989) Stage-dependent depression of melatonin in patients with primary breast cancer. *Cancer*, **64**, 426–33.

Berga, S.L., Jones, K.L. and Kaufmann, S. (1989) Nocturnal melatonin levels are unaltered by ovarian suppression in girls with central precocious puberty. *Fertility and Sterilisation*, **52**, 937–41.

Berga, S.L., Mortola, J.F. and Yen, S.S.C. (1988) Amplification of nocturnal melatonin secretion in women with functional hypothalamic amenorrhea. *Journal of Clinical Endocrinology and Metabolism*, **66**, 242–4.

Bojkowski, C., Arendt, J., Shih, M. *et al.* (1987) Assessment of melatonin secretion in man by measurement of its metabolite: 6-sulfatoxymelatonin. *Clinical Chemistry*, **33**, 1343–8.

Brainard, G., Knobler, R.L., Podolin, P.L. *et al.* (1987) Neuroimmunology: modulation of the hamster immune system by photoperiod. *Life Sciences*, **40**, 1319–26.

Brzezinski, A. and Wurtman, R.J. (1988) The pineal gland: its possible roles in human reproduction. *Obstetrics and Gynecology Survey*, **43**, 197–207.

Cagnacci, A., Elliot, J.A. and Yen, S.S.C. (1991) Amplification of pulsatile LH secretion by exogenous melatonin in women. *Journal of Clinical Endocrinology and Metabolism*, **73**, 210–2.

Claustrat, B., Chazot, G. and Brun, J. (1984) A chronobiological study of melatonin and cortisol secretion in depressed subjects: plasma melatonin, a biochemical marker in major depression. *Biological Psychiatry*, **19**, 1215–28.

Cozzi, B., Ravault, J.P., Ferraudi, B. *et al.* (1988) Melatonin concentration in the cerebral vascular sinuses of sheep and evidence for its episodic release. *Journal of Pineal Research,* **5**, 535–44.

Critchley, P.H.S., Malcolm, P., Malcolm, G. *et al.* (1991) Plasma melatonin levels are normal in Parkinson's disease: lack of effect of benserazide. *Journal of Neurology and Neuropsychiatry*, **54**, 91–2.

Dahlitz, M., Alvarez, B., Vignau, J., *et al.* (1991) Delayed sleep phase syndrome response to melatonin. *Lancet*, i, **337**, 1121–4.

Das Gupta, D., Riedel, L., Frick, J.H. *et al.* (1983) Circulating melatonin in children: in relation to puberty, endocrine disorders, functional tests and racial origin. *Neuroendocrinology Letters*, **5**, 63–78.

Deacon, S.J. and Arendt, J. (1994) Phase shifts in melatonin, 6-sulphatoxymelatonin and alertness rhythms after treatment with moderately bright light at night. *Clinical Endocrinology*, **40**, 413–20.

Demisch, K., Demisch, L., Bodinik, H.J. *et al.* (1986) Melatonin and cortisol increase after fluoxamine. *British Journal of Clinical Pharmacology*, **22**, 620–2.

Eastwood, M.R. and Peacocke, J. (1976) Seasonal patterns of suicide depression and electroconvulsive therapy. *British Journal of Psychiatry*, **129**, 472–5.

Fellenberg, A.J., Phillipou, G. and Seamark, R.F. (1981) Urinary 6-sulphatoxymelatonin excretion and melatonin production rate: studies in sheep and man, in *Pineal Function* (eds C.D. Matthews and R.F. Seamark), Elsevier, Amsterdam, pp. 143–9.

Franey, C., Burton, S., Checkley, S. *et al.* (1986) Acute desipramine stimulates melatonin and 6-hydroxymelatonin sulphate production in man. *British Journal of Pharmacology*, **22**, 73–9.

Fraschini, F., Esposti, D., Esposti, G. *et al.* (1987) Clinical and experimental data on the relationships between pineal endocrine function and the opioid system. *Advances in Pineal Research*, **2**, 171–80.

Frazer, A., Brown, R. and Kocsis, J. (1986) Patterns of melatonin rhythms in depression. *Journal of Neural Transmission*, **21**, 269–90.

Gupta, D. and Attanasio, A. (1988) Pathophysiology of pineal function in health and disease in children. *Pineal Research Reviews,* **6**, 261–300.

Hill, S.M. and Blask, D.E. (1988) Effects of the pineal hormone melatonin on the proliferation and morphological characteristics of human breast cancer cells (MCF-7) in culture. *Cancer Research*, **48**, 6121–5.

Holmes, S.W. and Sugden, D. (1975) The effect of melatonin on pinealectomy-induced hypertension in the rat. *Proceedings of the British Pharmacological Society*, 17–19th December.

Horowitz, M.B. (1991) Central nervous system germinomas, a review. *Archives of Neurology*, **48**, 652–7.

Heubner, O. (1898) Tumor der *Glandula pinealis*. *Dtsch. Med. Wschr.*, **24**, 214.

Illnerova, H. (1988) Entrainment of mammalian circadian rhythms in melatonin production by light. *Pineal Research Reviews*, **6**, 173–217.

Kauppila, A., Kivela, A., Pakarinen, A. *et al.* (1987) Inverse seasonal relationship between melatonin and ovarian activity in humans in a region with a strong seasonal contrast in luminosity. *Journal of Clinical Endocrinology and Metabolism*, **65**, 823–8.

Lewy, A.J., Sack, R.L. and Singer, C.M. (1984) Assessment and treatment of chronobiologic disorders using plasma melatonin levels and bright light exposure, the clock-gate model and the phase response curve. *Psychopharmacology Bulletin*, **20**, 561–5.

Lewy, A.J., Sack, R.L. and Singer, C.M. (1985) Treating phase typed chronobiologic sleep and mood disorders using appropriately timed bright artificial light. *Psychopharmacology Bulletin*, **21**, 368–72.

Lewy, A.J., Wehr, T.A., Gold, P.W. *et al.* (1978) Plasma melatonin in manic-depressives illness, in *Catecholamines: Basic and Clinical Frontiers*, vol. 2 (eds E. Usdin, I.J. Kipin and J. Barchas), Pergamon, Oxford, pp. 1173–5.

Lewy, A.J., Wehr, T.A., Goodwin, F.K. *et al.* (1981) Manic depressive patients may be supersensitive to light. *Lancet*, **i,** 383–4.

Lewy, J.J., Kern, H.E., Rosenthal, N.E. *et al.* (1982) Bright artificial light treatment of a manic-depressive patient with a seasonal mood cycle. *American Journal of Psychiatry*, **139**, 1496–8.

Maestroni, G.J.M., Conti, A. and Pierpaoli, W. (1989) Melatonin, stress and the immune system. *Pineal Research Reviews*, **7**, 268.

Minors, D.S., Waterhouse, J.M. and Wirz-Justice, A. (1992) A human phase response curve to light. *Neuroscience Letters*, **133**, 36–40.

Neuwelt, E.A., Mickey, B. and Lewy, A.J. (1986) The importance of melatonin and tumour markers in pineal tumours, in *Melatonin in Humans* (eds R.J. Wurtman and F. Waldhauser), *Journal of Neural Transmission*, (Suppl 21), 397–413.

Nurnberger, J.I., Berrettini, W., Tamarkin, L. *et al.* (1988) Supersensitivity to melatonin suppression by light in young people at high risk for affective disorder. *Neuropsychopharmacology*, **1**, 217–23.

Papousek, M. (1975) Chronobiologische aspekte der zyklothymie. *Fortschr. Neurology and Psychiatry*, **43**, 381–440.

Parkes, A.S. (1976) *Patterns of Sexuality and Reproduction.* Oxford University Press, Oxford.

Puig-Domingo, M., Webb, S., Serrano, J. *et al.* (1992) Brief report: melatonin related hypogonadotropic hypogonadism. *New England Journal of Medicine*, **327**, 1356–9.

Rosenthal, N.E., Sack, D.A., Gillin, J.C. *et al.* (1984) Seasonal affective disorder. A description of the syndrome and preliminary findings with light therapy. *Archives of General Psychiatry*, **41**, 72–79.

Skene, D.J., Bojkowski, C.J., Currie, J.E. *et al.* (1990) 6-Sulphatoxymelatonin production in breast cancer patients. *Journal of Pineal Research*, **8**, 269–76.

Souetre, E.F., Salvati, E., Tamarkin, L. *et al.* (1988) Stimulating effect of 5-methoxypsoralen on the secretion of the melatonin by the pineal gland. *Annual Review of Chromopharmacology*, **5**, 107–110.

Sparks, D.L. and Hunsaker, J.C. (1988) III: The pineal gland in sudden infant death syndrome: preliminary observations. *Journal of Pineal Research*, **5**, 111–8.

Sturner, W.Q., Lynch, H.J., Deng, M.H. *et al.* (1990) Melatonin levels in the sudden infant death syndrome. *Forensic Science International*, **45**, 171–80.

Tamarkin, L., Cohen, M., Roselle, D. *et al.* (1981) Melatonin inhibition and pinealectomy enhancement of 7,12-dimethylbenz-(a) anthracene-induced mammary tumours in the rat. *Cancer Research*, **41**, 4432–6.

Tamarkin, L., Danforth, D., Lichter, A. *et al.* (1982) Decreased nocturnal plasma melatonin peak in patients with estrogen receptor positive breast cancer. *Science*, **216**, 1003–5.

Testart, J., Frydman, R. and Roger, M. (1982) Seasonal influence of diurnal rhythms in the onset of the plasma luteinizing hormone surge in women. *Journal of Clinical Endocrinology and Metabolism*, **55**, 374–7.

Thompson, C., Franey, C., Arendt, J. *et al.* (1988) A comparison of melatonin secretion in depressed patients and normal subjects. *British Journal of Psychiatry*, **152**, 260–5.

Thompson, C., Obrecht, R., Franey, C. *et al.* (1985) Neuroendocrine rhythms in a patient with the Kleine–Levin syndrome. *British Journal of Psychiatry*, **147**, 440–3.

Touitou, Y., Fevre, M., Bogdan, A. *et al.* (1984) Patterns of plasma melatonin with aging and mental condition: stability of nyctohemeral rhythms and differences in seasonal variation. *Acta Endocrinologica*, **106**, 145.

Valcavi, R., Dieguez, C., Azzarito, C. *et al.* (1987) Effect of oral administration of melatonin on GH responses to GRF_{144} in normal subjects. *Clinical Endocrinology*, **26**, 453–8.

Vaughan, G.M. (1984) Melatonin in humans. *Pineal Research Reviews*, **2**, 141–201.

Voordow, B.C.G., Euser, R., Verdonk, R.E.R. et al. (1992) Melatonin and melatonin-progestin combinations alter pituitary-ovarian function in women and can inhibit ovulation. Journal of Clinical Endocrinology Metabolism, 74, 108–17.

Vriend, J. (1983) Evidence for pineal gland modulation of the neuroendocrine-thyroid axis. *Neuroendocrinology*, **36**, 68–78.

Waldhauser, F., Boepple, P., Schemper, M. *et al.* (1991) Serum melatonin in central precocious puberty is lower than in age matched pre-pubertal children. *Journal of Clinical Endocrinology Metabolism*, **73**, 793–6.

Waldhauser, F., Steger, H. and Vorkapic, P. (1987) Melatonin secretion in man and the influence of exogenous melatonin on some physiological and behavioural variables. *Advances in Pineal Research*, **2**, 207–23.

Wass, J.A.L., Jones, A.E., Rees, L.H. *et al.* (1982) hCGβ producing pineal choriocarcinoma. *Clinical Endocrinology*, **17**, 423–31.

Wehr, T.A. (1991) The durations of human melatonin secretion and sleep respond to changes in daylength (photoperiod). *Journal of Clinical Endocrinology and Metabolism*, **73**, 1276–80.

Wehr, T.A., Jacobsen, F.M., Sack, D.A. *et al.* (1986) The efficacy of phototherapy in seasonal affective disorder appears not to depend on its timing or its effect on melatonin secretion. *Archives of General Psychiatry*, **43**, 870–5.

Weitzman, E.D., Moline, M.L., Czeisler, C.A. *et al.*(1982) Chronobiology of ageing: temperature, sleep–wake rhythms and entrainment. *Neurobiology of Ageing*, **3**, 299.

West, R.J., Leonard, J.V., (1980) Familial insulin resistance with pineal hyperplasia: metabolic studies and effect of hypophysectomy. *Archives of Diseases in Childhood*, **55**, 619–21.

Wever, R.A. (1986) Characteristics of circadian rhythms in human functions, in *Melatonin in humans* (eds R.J. Wurtman and F. Waldhauser) *Journal of Neural Transmission*, Suppl, **21**, 323–74.

Young, I.M., Francis, P.L., Leone, A.M. *et al.* (1986) Night/day urinary 6-hydroxymelatonin production as a function of age, bodymass and urinary creatinine levels: a population study in 110 subjects aged 3–80. *Journal of Endocrinology*, Suppl. Abstract No. 32, 111.

Effects of melatonin: therapeutic potential and significance to human health

8

Contents

8.1 CIRCADIAN RHYTHM DISORDERS

There is no doubt that melatonin has profound effects on light dependent rhythmic function in animals, acting either as a synchronizer, a coupling agent or a combination of both (see Chapters 5 and 6). A very large number of people would benefit from the ability to manipulate rhythms at will. For example, circadian rhythm disturbance is associated with occupational health conditions where rapid phase shifts are imposed, such as shift work and jet lag: about 20–25% of the working population of developed countries works shifts. In some blind people free-running circadian rhythms lead to intermittent mismatches between the internal and external environment such that, for example, there may be an overpowering desire to sleep in the middle of the day and wakefulness at night. The elderly may display disorganized rhythm physiology. It is also probable that insufficient exposure to strong time cues such as bright light in the general population is conducive to abnormal phase relationships between the internal and external environment and consequent problems of health and performance.

The pathology of affective disorder is replete with reports of abnormal rhythms, particularly advanced or delayed phase and/or lowered amplitude – although the precise relationship to the development of the disease has yet to be clarified. Insomnia is clearly a disturbed sleep–wake cycle and in the cases of advanced and delayed sleep phase insomnia an underlying disturbance of the central biological clock or pacemaker, rather than a downstream change in a 'hand of the clock' is one possible explanation.

Situations where rhythm abnormalities are associated with lack of well being and/or poor performance include:

- shift work
- jet lag
- delayed and advanced sleep phase insomnia
- affective disorder
- blindness
- old age.

These conditions represent a very large proportion of the population. The more we behave as though the daily and seasonal cycles do not exist, the more problematic our general health is likely to become, as long as our biology still responds to environmental signals and cannot maintain itself with a lack of signals. It is conceivable that sensitization to a more appropriate zeitgeber than natural bright light, such as clock time, would improve our efficiency – but the quality of life would be unimaginably diminished if this were to require avoidance of natural light.

At present a search for a general chronobiotic – a compound able rapidly to shift the biological clock in all or some of its manifestations – occupies much scientific time and effort. To date bright light is the only treatment that in suitable intensity and duration is able to do this. A number of pharmaceutical products and steroid hormones have been shown to shift aspects of the circadian system in animals – usually the activity–rest cycle. In some circumstances the ability to manipulate one specific rhythm without affecting others may be more useful than affecting all rhythms. The possible use of melatonin in this area became evident in the course of investigating its functions in seasonal breeders and lower vertebrates (see Chapters 5 and 6).

Early experiments on the effects of melatonin in humans were designed in the absence of a confirmed role for the pineal in rhythmic function and in retrospect appear crude. However some important insights can be gained from a review of early work.

8.2 EARLY WORK

Aaron Lerner, the discoverer of melatonin, was the first to administer it to humans and one has the impression that like all good scientists he tried it on himself first. In 1978 he and J.J. Nordlund reviewed the current clinical pharmacology of melatonin and stated that 'The first person to ever receive 200 mg melatonin intravenously daily for 5 days has no evidence of delayed toxicity 18 years later'. This dose is the equivalent of, at a conservative estimate, the melatonin content of about a million human pineals daily. Then and now there is good evidence for very low toxicity of melatonin. Sugden (1983), for example, in one of the very few publications on the subject was unable to find an oral LD_{50} for melatonin given to rats and it is not mutagenic in standard *in vitro* tests. It is likely that much

Table 8.1 Acute toxicity of melatonin. This table compares the LD_{50} values of melatonin by different routes of administration in the mouse and rat. The numbers in parenthesis indicate the 95% confidence limits of each LD_{50} value. Five mice or three rats were used at a minimum of four to five dose levels

Route of administration	*LD_{50} (mg/kg)*	
	Mouse	*Rat*
i.p.	1168 (987–1382)	1131 (800–1600)
oral	1250 (1090–1434)	>3200
s.c.	>>1600	>>1600
i.v.	472 (400–564)	356 (115–1105)

Source: Sugden, D., Psychopharmacological effects of melatonin in mouse and rat. *J. Pharmacol. Exp. Ther.* **222**, 587–91; published by Waverly, 1983.

unpublished data on the lack of melatonin toxicity exists, as clinical trials certificates have been issued to cover long-term administration of up to 300 mg per day. Since Lerner's report it has been reasonable to conclude that melatonin can be used in clinical trials with safety.

Sugden (1983) described for the first time the acute toxicology and pharmacology of melatonin. These pioneering observations are given in Tables 8.1–8.5.

Thus melatonin in pharmacological doses potentiated barbitone sleeping time, had significant anticonvulsant and analgesic activity and very low toxicity. It almost completely abolished locomotor activity in mice at 100 mg/kg. These experiments were performed in the light phase of the day without regard to specific timing. They are

Table 8.2 Potentiation of pentobarbitone sleeping time. Melatonin was given to rats and mice 15 (i.p. and i.v.) or 30 min (oral) before pentobarbitone administration (35 mg/kg i.v.) The duration of the loss of the righting reflex was measured and the ED_{50} values and 95% confidence limits (in parenthesis) were calculated. Five animals were used at a minimum of four dose levels in each experiment

Route of administration	*ED_{50} (mg/kg)*	
	Mouse	*Rat*
i.p.	6.1 (1.9–9.6)	27.4 (22.5–31.8)
oral	20.6 (13.2–33.6)	109.5 (72.5–186.6)
i.v.	13.4 (5.3–19.3)	24.6 (20.0–30.2)

Source: Sugden, D., Psychopharmacological effects of melatonin in mouse and rat. *J. Pharmacol. Exp. Ther.* **222**, 587–91; published by Waverly, 1983.

Table 8.3 Potentiation of barbitone sleeping time. Mice were given barbitone (200 mg/kg i.v.) 15 min before melatonin (20 mg/kg i.p.) or its vehicle. The time from barbitone administration to the loss of the righting reflex and the duration of the loss of righting reflex (sleep time) were measured. Ten mice were used in each group

Treatment	*Time to loss of righting reflex (min)*	*Sleep time (min)*
Vehicle	30.4 ± 1.4	57.0 ± 7.9
Melatonin	26.9 ± 1.7	113.6 ± 11.5*

*p<0.001 compared with vehicle-treated group using Student's t test.
Source: Sugden, D., Psychopharmacological effects of melatonin in mouse and rat. *J. Pharmacol. Exp. Ther.* **222**, 587–91; published by Waverly, 1983.

nevertheless of considerable importance and paved the way for subsequent human experimentation.

Enormous doses of melatonin (50 mg to 6.6 g = 33 million pineals!) in the daytime had no beneficial effects on Parkinson patients, Huntington's chorea, depression (in fact depression was worsened) and schizophrenia. Skin pigmentation was not affected: human pigment cells do not resemble amphibian melanophores in pigment migration phenomena. Small decreases in plasma LH and FSH were observed. Large amounts of melatonin such as these may produce

Table 8.4 Anticonvulsant effects of melatonin. Mice were dosed i.p. with vehicle or melatonin 30 min before challenge with the convulsant; ECS (916 mA, 1 msec pulse width at 100 Hz for 0.2 s), 3-MPA (45 mg/kg i.p.) and PTZ (pentylenetetrazole; i.v. infusion of 12 mg/ml at 0.175 ml/min). For ECS and 3-MPA the figures refer to the number of mice showing tonic extension of the hindlimbs. For PTZ, the seizure threshold indicates the mean ± SEM dose of PTZ infused at the time the first running convulsion occurred. Eight mice were used at each dose level of melatonin

Treatment	*Dose (mg/kg)*	*Anticonvulsant test*		
		ECS (number of responders)	*3-MPA*	*PTZ (seizure threshold) (mg/kg)*
Vehicle		10/10	6/6	54.5 ± 3.0
Melatonin	50	9/10	5/6	NT
	100	8/10	4/6	50.0 ± 2.5
	200	4/10	1/6	66.5 ± 4.5*
	400	0/10	0/6	122.5 ± 19.5†

*p<0.05; †p<0.01 compared with the vehicle-treated group using Student's t test.
NT, not tested.

Table 8.5 Analgesic effects of melatonin. Mice were dosed with melatonin i.p. 30 min before the analgesic test. In the hot-plate test the mean ± SEM reaction time (s) of 10 mice in each group is given. In the writhing test the number of mice showing a pain response in each group of 10 mice is given. Naloxone (1 mg/kg s.c.) was given 15 min before the analgesic test

Treatment	*Dose (mg/kg)*	*Analgesic hot-plate (reaction time, s)*	*Writhing test (number of responders)*
Vehicle		8.3 ± 0.9	10/10
Melatonin	100	11.6 ± 1.4	9/10
	200	23.3 ± 4.9*	5/10
	400	43.3 ± 6.9†	0/10
Melatonin+	200		
Naloxone	1	30.1 ± 5.8*	4/10

*$p<0.01$; †$p<0.001$ compared with the vehicle-treated group using Student's *t* test. Details of the hot-plate and writhing test are given in the 'Materials and Methods' section in Sugden (1983).
Source: Sugden, D., Psychopharmacological effects of melatonin in mouse and rat. *J. Pharmacol. Exp. Ther.* **222**, 587–91; published by Waverly, 1983.

headache and abdominal cramps. This may possibly be due to a false transmitter effect on serotoninergic systems.

The most consistent action of melatonin was to induce sedation and/or sleep. This remains the most consistent effect of high doses of melatonin in humans and animals; however it is interesting that 1–6 g of melatonin given to humans in the daytime did not induce a reversal of the sleep–wake cycle. Patients were reportedly 'somnolent' which suggests that the effect is not that of a classical hypnotic. It is also important to note that much lower (2 mg intranasally, 1–80 mg by oral or intravenous route) but still pharmacological doses of melatonin can induce transient sleepiness or sleep. These transient effects of lower doses may be due to the rapid metabolism of intranasal, intravenous or oral melatonin. The discrepancies between individuals may be due to large variations in both metabolism and absorption.

Even with rapid clearance, the administration of very large doses of melatonin means that supraphysiological concentrations may be present in the plasma throughout 24 hours. Such a situation is present in animals implanted with slow-release melatonin. Here the initial effect of the implant depends firstly on photoperiodic history as we have seen. For example, in the short day breeder implanted in autumn the first response is a short day response. However if the implant remains in place and continues to supply high levels of melatonin the animal enters a refractory state and fails subsequently to respond to photoperiod change in the presence of the implant. This situation is in some respects like the effect of pinealectomy when the animal is functionally blind to photoperiod change. It is possible that daily high dose melatonin has a similar effect in humans. The circadian timing mechanisms of sleep related to the light–dark cycle and melatonin may no longer be functional and one manifestation of this may be somnolence during the daytime.

As melatonin is effective as a seasonal time cue when physiological night-time levels are created artificially, it is evident that the early work on melatonin in humans essentially investigated its pharmacological properties. Recently observed pharmacological properties of melatonin in animals include sedation, hypothermia, anxioloysis, muscle hypotonia, decrease in locomotor activity with a rebound increase on increasing the dose, slight analgesia and slight protection against electroconvulsive shock. These have been summed up by Guardiola-Lemaitre (1991) as a sedative profile with an anxiolytic component, but with a behavioral profile dissimilar to benzodiazepines.

8.3 TIMED ADMINISTRATION

Only when the importance of timing in the physiology of melatonin was appreciated, rather than the pharmacological emphasis on dose, did the design of human experiments produce coherent and interpretable results.

The administration of amounts of melatonin that can be cleared within 24 hours can be conceived in several different ways. If, for example, the circadian dusk signal is the first onset of secretion in the evening, then rising plasma melatonin from an endogenous or exogenous source could be a time cue for sleep onset (diurnal animals) and the time of onset of sleep should respond to exogenous melatonin. Likewise simulation of declining morning melatonin could act as a dawn signal. If the maintained night-time levels of melatonin are important for sleep maintenance then a slow-release preparation given in the evening may prove of greater benefit to sleep duration than a rapidly metabolized dose. Good slow-release preparations of melatonin for human use have only been developed recently and there is a lack of information in this area (see Chapter 4).

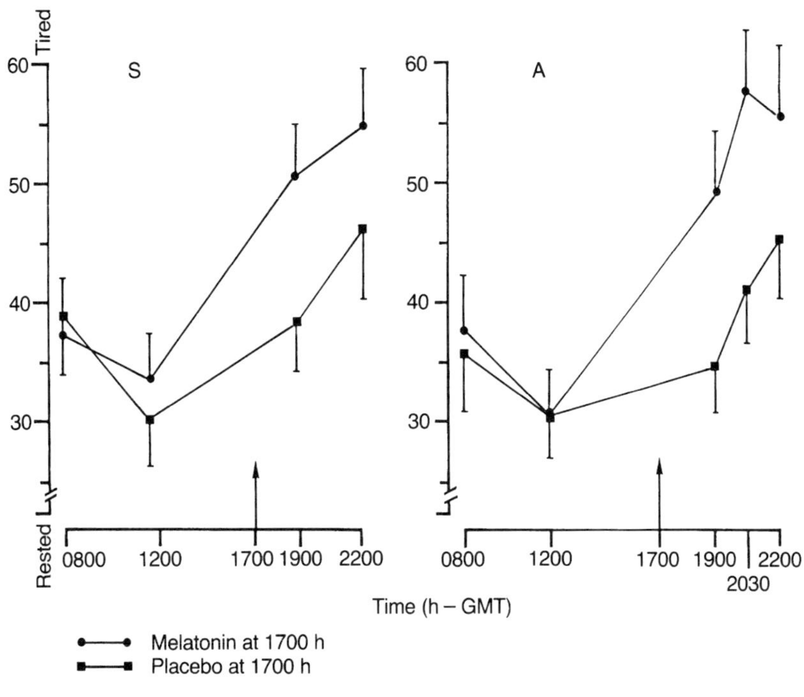

Fig. 8.1(a) Melatonin (2 mg) in the late afternoon (1700 h) taken daily for one month in spring (S) or autumn (A), advances the onset of evening self-rated fatigue or sleep, in 12 normal subjects compared to placebo (double blind cross-over design). Group significance was achieved within 4 days. Fatigue was rated on a visual analog scale (VAS) rested–tired There were no significant effects on mood, LH, FSH, T4, testosterone, GH or cortisol but the timing of the prolactin rhythm was advanced. (Reproduced with permission from Arendt *et al.*, 1984 and 1985.)

8.3.1 Phase shifts induced by melatonin

Daily feeding of low dose (2 mg) melatonin in the late afternoon advances the timing of evening self-rated fatigue, the endogenous melatonin rhythm and the morning decline of prolactin compared to placebo (Figures 8.1–8.3). The duration and quality of sleep were not significantly altered but there was a non-significant advance of morning wake-up time. The evening rise of melatonin was advanced more than the morning decline. There were no significant effects on self-rated mood, or on LH, FSH, testosterone, cortisol, growth hormone (Figure 8.4) or T4. The effect on fatigue was not immediately apparent but became significant in the group as a whole after 4 days. No deleterious effects were reported by the subjects. This study, conducted twice over 4 weeks on two different occasions in spring and autumn showed that in low doses melatonin has chronobiotic effects in humans at least with respect to sleep, its own rhythm and prolactin. It indicated that such a daily dose is well

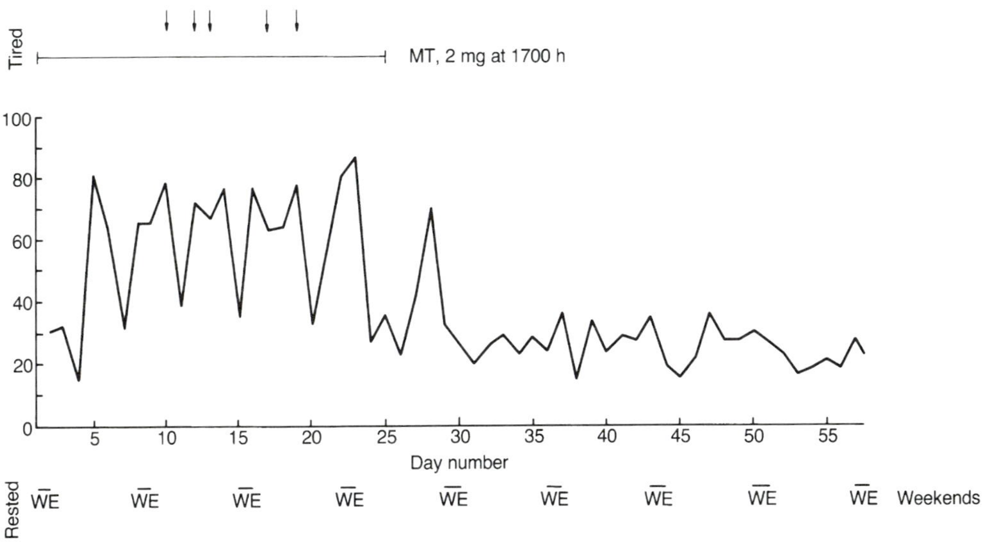

Fig. 8.1(b) Fatigue ratings [see (a)] in one individual at 1900 h from the group in (a) taking melatonin, then placebo. Note the periodic nature of the increase in fatigue in this subject: relative coordination, characteristic of a weak zeitgeber. (Reproduced with permission from Arendt, J., The human circadian system, in *Melatonin: Clinical Perspectives*, (eds A. Miles, D.R.S. Philbrick and C. Thompson); published by Oxford University Press, 1988.)

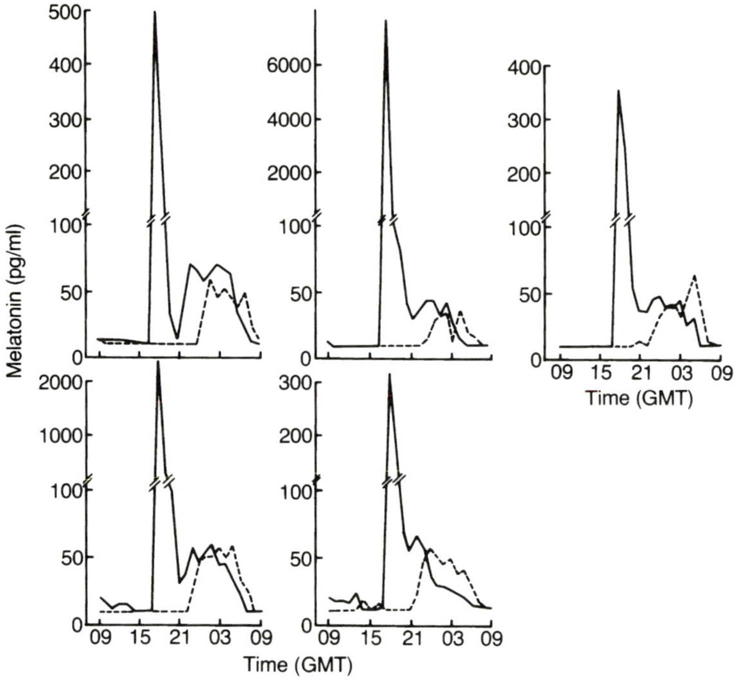

Fig. 8.2 Melatonin (2 mg) in the late afternoon [1700 h see Figure 8.1(a)] phase advances the onset of its own endogenous secretion (——) compared to placebo (– – –) in those subjects where the endogenous and exogenous plasma melatonin can be distinguished. Note the supraphysiological peak of exogenous melatonin followed by the endogenous night-time secretion. (Reproduced from Arendt *et al.*, 1985.)

tolerated over 4 weeks. Importantly, it showed no significant reproductive effects and indeed the dose is well below that needed to influence human reproduction as a combined contraceptive (see p. 273). It clearly suggested that melatonin would be useful for phase shifting and synchronizing strategies at least for phase advances.

Since this initial study a substantial volume of work has served to confirm this prediction.

Acute administration of melatonin (1–5 mg) in the morning or the early evening (at 1700 h) acutely lowers body temperature and after evening administration, the following day the endogenous melatonin

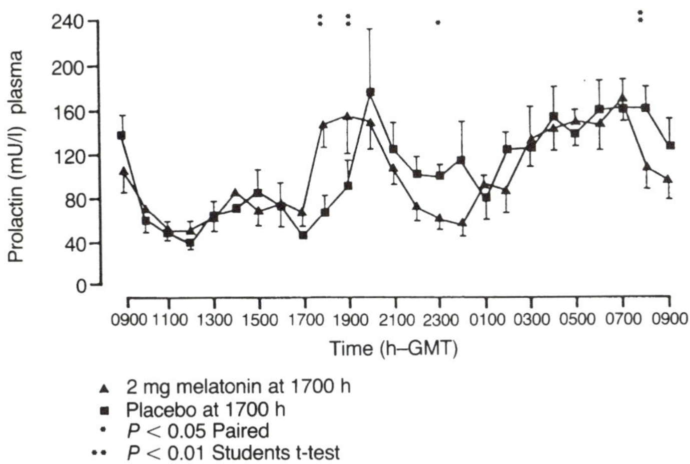

Fig. 8.3 Melatonin (2 mg) in the late afternoon [1700 h see Figure 8.1(a)] advances the timing of the prolactin rhythm compared to placebo. (Reproduced with permission from Wright, J. *et al.*, The effects of exogenous melatonin on endocrine function in man. *Clinical Endocrinology*, **24**, 375–82; published by Blackwell Scientific Publications Ltd, Oxford, 1986.)

rhythm is advanced by approximately 1 hour, together with a trend in the body temperature rhythm (Figure 8.5).

One elegant approach to the time dependent effects of melatonin on sleep is the measurement of 'sleep propensity'. Here subjects attempt to sleep for 7 minutes every 20 minutes during 24 hours. The latency to sleep is an index of sleep propensity. Small (5 mg) doses of melatonin have striking time-dependent effects on the timing and magnitude of sleep propensity, with notably abolition of a 'forbidden sleep zone' in the early evening following late afternoon administration (Figure 8.6).

8.4 A PRC FOR MELATONIN

Following the phase response curve (PRC) reported for the effect of melatonin on rest–activity cycles in rats and the evident phase advancing effects in humans, a PRC in humans was the next step. This will hopefully improve predictions as to the optimum time of

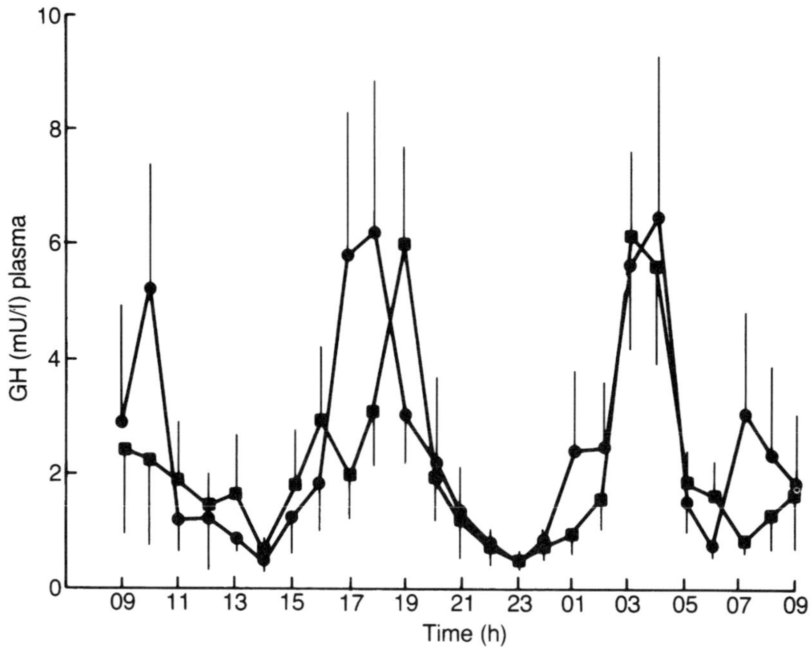

Fig. 8.4 Melatonin (2 mg) in the late afternoon [1700 h see Figure 8.1(a)] does not significantly affect the timing of the GH rhythm compared to placebo. ● = after melatonin treatment; ■ = after placebo treatment; mean ± SEM. (Arendt, English and Wright, unpublished.)

administration for a desired response. Lewy and coworkers (1992) have described a PRC for melatonin in both blind and sighted people with the endogenous melatonin rhythm as a circadian marker. Using 2 × 0.5 mg spaced by 2 hours they report phase advances in the late subjective day, with a maximum 6 hours before the onset of melatonin secretion (on average about 1600–1700 h). In contrast to rats, where only phase advances are seen, they also found small phase delays during the early morning. These data are approximately 12 hours out of phase with the PRC to light and are consistent with the long held view that melatonin acts like darkness in the timing of both circadian and seasonal rhythms (Figure 8.7).

There is other evidence for the phase delaying effects of melatonin in humans in that there is subjective improvement of jet lag when

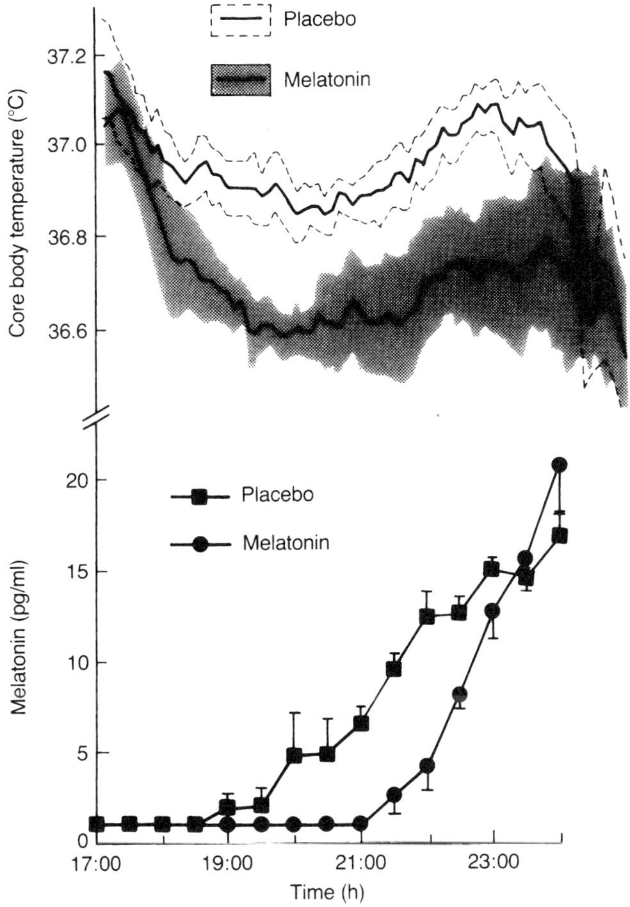

Fig. 8.5 Phase shift of melatonin and suppression of core body temperature using 5 mg melatonin compared to placebo. A single treatment of melatonin (5 mg) in the late afternoon (1700 h) acutely suppresses temperature and phase advances the endogenous melatonin rhythm (in saliva) the following day, when subjects are kept seated and in dim (<200 lux) light from 1700 h until bedtime (2300 h). (From data in Deacon, English and Arendt, 1994, unpublished figure.)

travelling westwards, with melatonin treatment during 'circadian early morning' and in preliminary shift work studies (see section 8.5). Also, when absorbed in the morning there are significant effects on morning sleepiness similar to extension of darkness into the morning hours over several days.

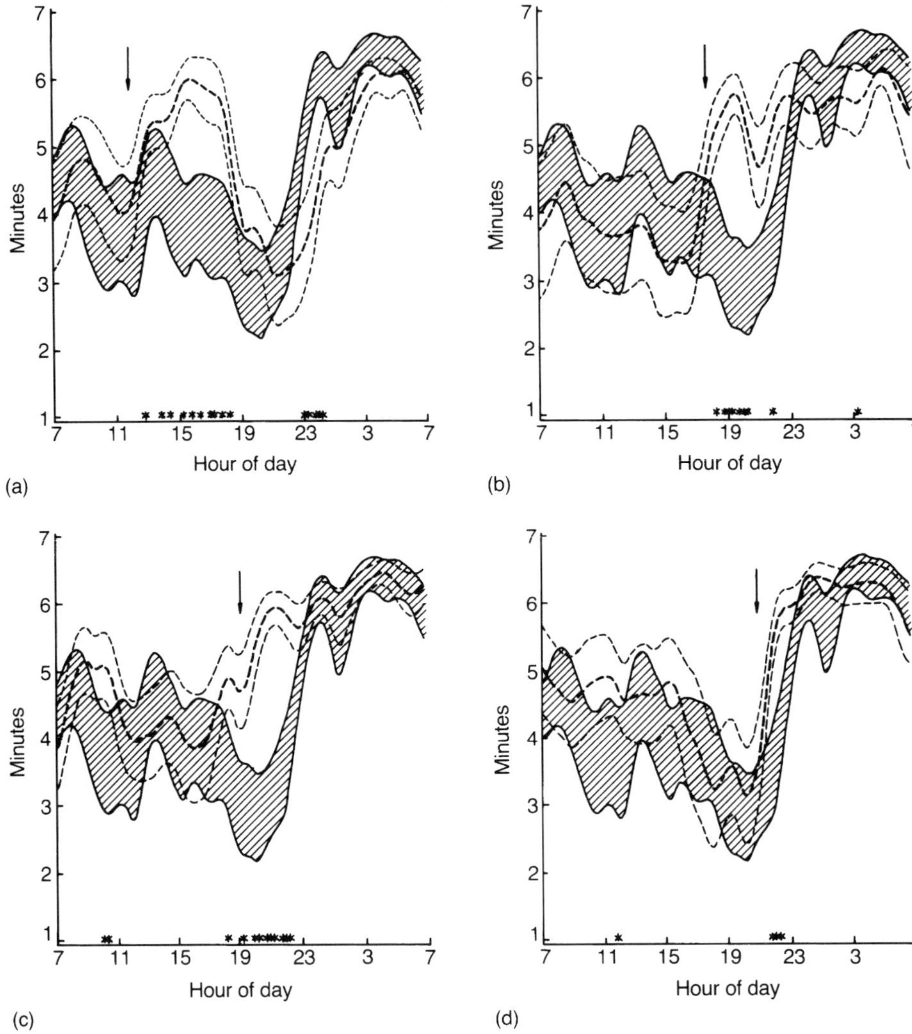

Fig. 8.6 The sleep propensity function was measured for a 24-hour period over 5 weeks using the 7/13 sleep–wake paradigm developed by Lavie (1993). In this paradigm every 20 min subjects are instructed to lie in bed and to attempt to fall asleep. Electrophysiological recordings are carried out during the 7-min sleep attempts to determine the amount of sleep. At the end of each nap, regardless of

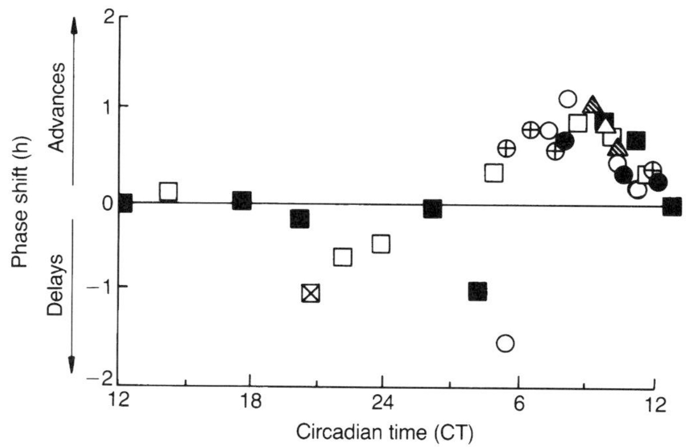

Fig. 8.7(a) Phase response curve of the effect of melatonin on the endogenous melatonin rhythm. The onset of melatonin secretion in evening dim light (the so-called dim light melatonin onset or DLMO, designated CT 14) was used as a phase marker. Exogenous melatonin was administered at various times with respect to the time of endogenous melatonin production (CT-circadian time). Translated roughly to clock time, maximum phase advances were found around 1700 h and delays around 0600 h. (Reproduced from Lewy, A.J. *et al.* Melatonin shifts human circadian rhythms according to a phase response curve. *Chronobiology International*, **9**, 380–92, by permission of Raven Press, 1992.)

Fig. 8.6 cont the amount of sleep obtained, subjects are requested to leave the bedroom. They remain awake under supervision for 13 min. Each point represents total sleep (mean ± SD) for a given 7-min attempt across all subjects (n=8). Sleep propensity is operationally defined as total sleep during each nap. Each figure presents the sleep propensity function (±SD) in the placebo condition (hatched area) and in one of the treatment conditions (dotted line). The times of melatonin administration (5 mg) are indicated by arrows: (a) 1200 h; (b) 1700 h; (c) 1900 h and (d) 2100 h. Asterisks indicate significant placebo/melatonin difference. Note the time- dependent delayed increase in sleep propensity by melatonin. (From Lavie *et al.*, 1993, personal communication.)

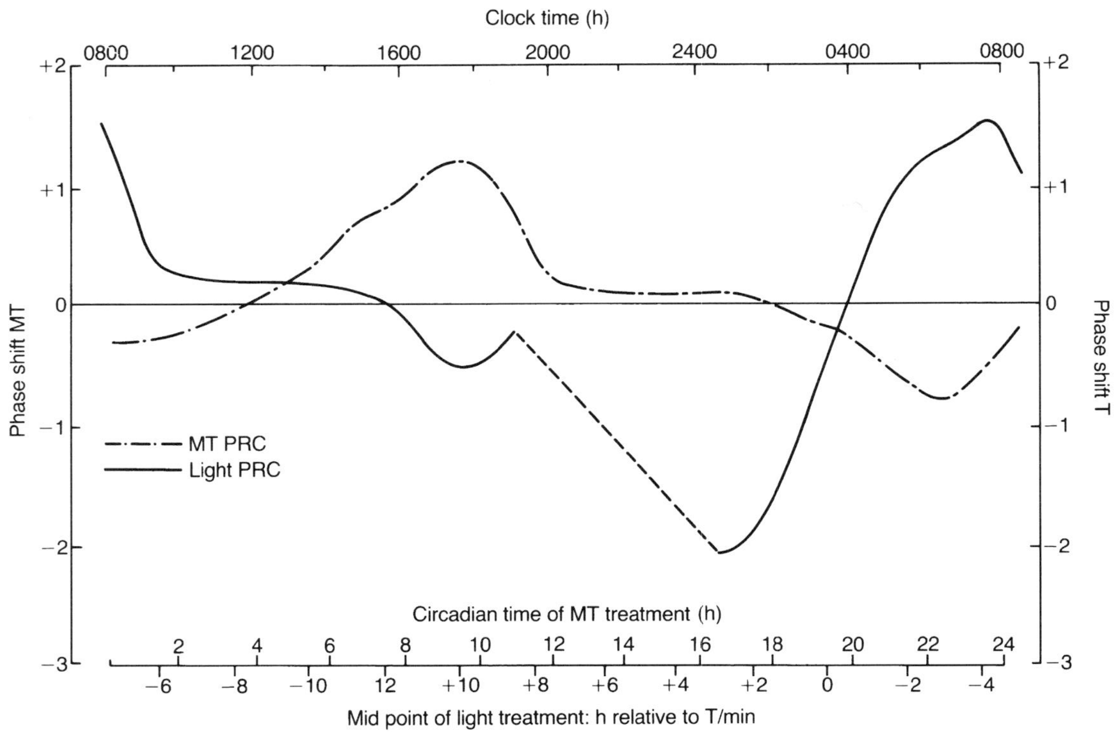

Fig. 8.7(b) A comparison of the phase response curves of core body temperature to bright white light redrawn from Minors, Waterhouse and Wirz-Justice (1992). and of the endogenous melatonin rhythm to exogenous melatonin [redrawn from Figure 8.7(a)]. Note that the responses mirror each other to some extent, e.g. morning bright light and evening melatonin phase advance, evening bright light and morning melatonin phase delay. MT = melatonin, T = core body temperature, PRC = phase response curve. Approximate clock time is indicated. (Adapted with permission from Minors, D.S., Waterhouse, J.M. and Wirz-Justice, A.; A human phase response curve to light. *Neuroscience Letters*, **133**, 36–40; published by Elsevier Science, Amsterdam, 1992.)

8.5 JET LAG AND SHIFT WORK

Melatonin has been used in the alleviation of jet lag in both real life and simulation conditions. Field studies show that self-rated jet lag can be reduced on average by 50% with appropriately timed treatment both westwards and eastwards (Figures 8.8, 8.9). The improvement is greater with larger numbers of time zones, and is greater in an eastwards direction compared to westwards. Using advance shifts over eight time zones with pre-shift administration of melatonin in the late subjective day and post-shift administration at the desired bedtime, the subjective impressions are reinforced by improved latency and quality of sleep, greater daytime alertness and more

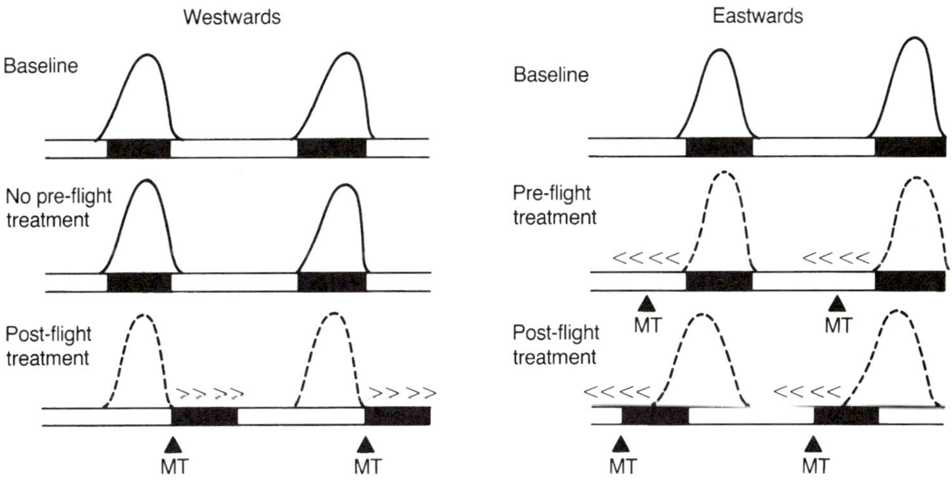

Fig. 8.8 Theoretical basis for the treatment of jet lag by melatonin, based on the predicted phase position of the endogenous melatonin rhythm as a function of the local light–dark cycle (dark bar = darkness). Preflight melatonin in solid lines, postflight predicted melatonin in dotted lines. Arrowheads indicate the time of melatonin (MT) treatment. Westwards postflight melatonin (4 days) at bedtime should initiate phase delays, Eastwards pre-(1–2 days) and post-(4 days) flight melatonin as indicated should initiate and maintain phase advances. Using this approach more than 400 subjects have participated in jet lag studies during the last 7 years. Melatonin reduces jet lag by 60% eastwards and 40% westwards, the benefit being greater over larger numbers of time zones (not shown). (Arendt, English and Wright, unpublished.)

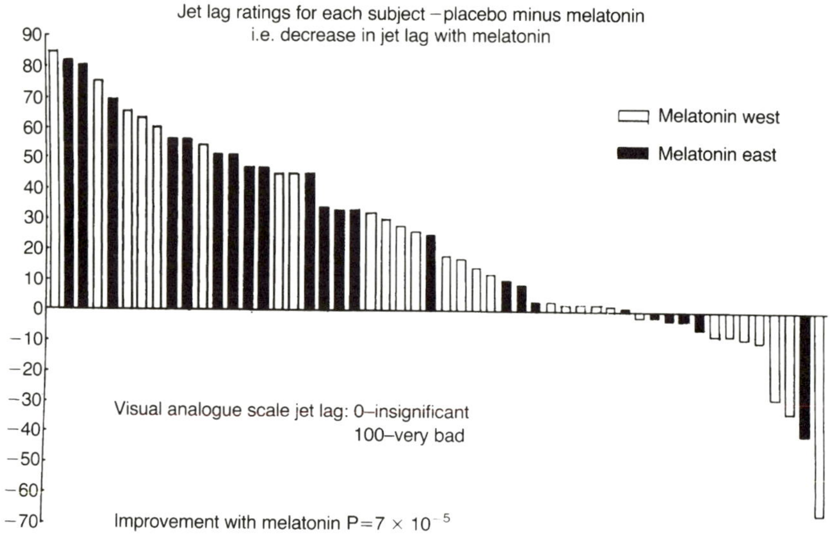

Fig. 8.9 Double blind crossover study of the alleviation of jet lag by melatonin (5 mg, see Figure 8.8) in individual subjects traveling over 10–12 time zones to and from Australia and New Zealand with adaptation of at least 14 days between flights. Subjects received a different treatment outward and inward and the visual analog scale ratings (VAS) of jet lag (negligible–very bad) after melatonin are subtracted from those on placebo. Note that the vast majority of subjects benefit from melatonin. Four subjects were adversely affected and it is likely that if it were possible to time treatment according to individual circadian phase there would be no adverse effects. (Reproduced with permission from Skene *et al*; published by Gustav Fischer Verlag, Stuttgart, 1989.)

rapid resynchronization of melatonin (Figure 8.10) and cortisol rhythms. In simulation studies using advanced phase shift and a similar administration protocol, effects on sleep are not significant but there is a marked increase in the rate of re-entrainment of both hormonal and electrolyte rhythms and an immediate effect of lowering body temperature which persists as more rapid re-entrainment. Neither the dose nor the timing of melatonin administration has been fully optimized for different flight times and time zone shifts although Figure 8.8 and the phase response curve for melatonin can serve to indicate preferred treatment times.

A minority (less than 10%) of subjects feel worse after melatonin.

In these cases the individual circadian status is probably a major determinant of response in that undesirable changes in direction of entrainment may occur with inappropriate timing of treatment. Moreover unpredictable exposure to bright light can theoretically act in opposition to the desired result. However, recent results from the author's laboratory (Deacon and Arendt, unpublished) suggest that melatonin alleviates the problems related to abrupt phase shifts even in the presence of conflicting bright light.

There is very little published work on the use of melatonin in shift work although exposure to bright light during the night is clearly beneficial to night shift workers. One preliminary report describes improved sleep and increased daytime alertness in night shift workers receiving melatonin at the desired bedtime during a night shift week compared to placebo and baseline conditions (Figure 8.11). Melatonin was taken at a time calculated to induce a phase delay as workers moved from day shift to night shift. There was a suggestion that one performance task (visual search speed) was adversely affected by melatonin in this study. This is a very important consideration if melatonin is to be used on the workforce. There are no reported deleterious effects on performance in jet lag studies but acute studies using much higher doses (240 mg) describe a brief decline in visual reaction time, accompanied by an increase in fatigue ratings but with no effect on memory. Recently, bright light treatment during the night has been used to force phase delays in human volunteers kept otherwise in a normal environment. Melatonin was used in the post-phase shift adaptation to induce phase advances. In the course of this study no deleterious effects on performance were seen (Deveson and Arendt, unpublished). It will be essential to assess thoroughly performance effects of melatonin in the work place and on specific tasks before its use can be recommended fully.

8.6 BLINDNESS

Some blind people, in spite of living in normal society, have free-running rhythms of sleep and wakefulness which are accompanied in the few reported studies by other free-running variables such as melatonin and cortisol (see Chapter 4, Figure 8.12). Others exhibit an

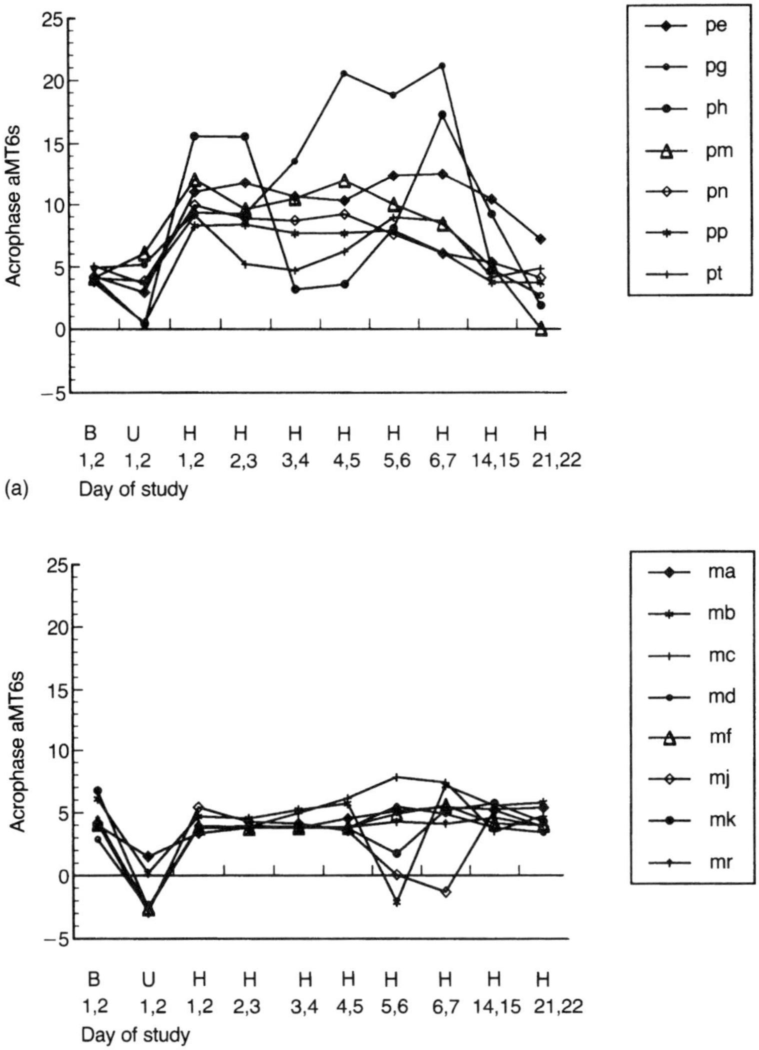

Fig. 8.10 Adaptation of the acrophase of the endogenous melatonin rhythm in individuals taking melatonin (5 mg, see Figure 8.8 (a)) for travel over eight time zones eastwards compared to placebo. B = baseline acrophase; U = acrophase after adaptation to an eight time zone change westward; H = acrophase during sequential days (1–7, 14, 15, 21, 22) adaptation to an eight time zone change eastward, using data over 48 h (e.g. days 1, 2; 2, 3, etc.) for the calculations. In the the melatonin group (b) the acrophase for days 1–5 represents exogenous melatonin, days 6–22 are endogenous melatonin. Melatonin group: ma, mb etc., placebo group: pe, pg etc. Adaptation is significantly faster in the melatonin-treated group, with return to baseline by days 6, 7. (Previously unpublished figure with data from Arendt, Aldhous and Marks, 1987.)

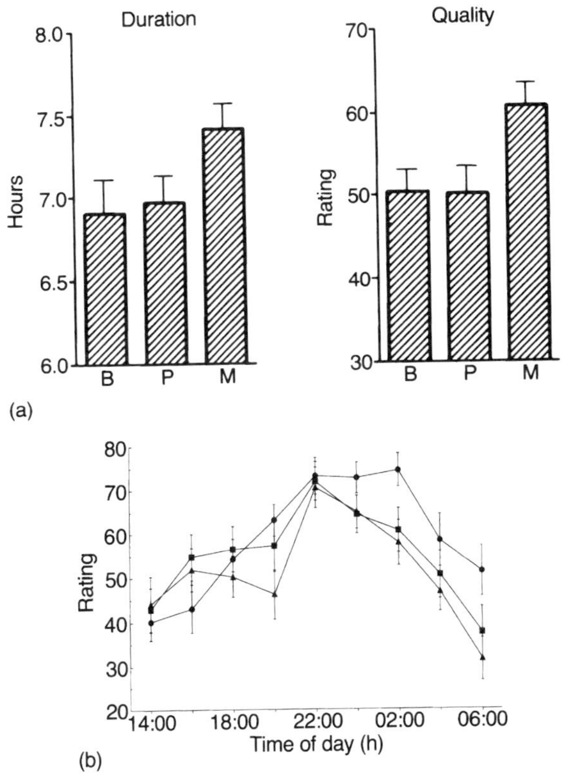

Fig. 8.11(a) Melatonin taken at desired bedtime (0600–0700 h) during 7 days of night shift improves sleep duration and quality in studies of night shift workers compared to placebo and no treatment. B = no treatment, P = placebo, M = melatonin. (b) Melatonin taken at desired bedtime (0600–0700 h) during 7 days of night shift improves on-shift alertness especially in the early hours of the morning in studies of night shift workers compared to placebo and no treatment. Combined alertness ratings for the 3rd, 5th and 7th night shift, VAS: drowsy–alert test completed 2-hourly, (■) = no treatment, (▲) = placebo, (●) = melatonin. (Reproduced from Folkard, S., Arendt, J. and Clark, M. Can melatonin improve shift workers' tolerance of the night shift? Some preliminary findings. *Chronobiology International*, **10**, 315–20, by permission of Raven Press, 1993.)

abnormal phase delay. To what extent these phenomena are present in the blind population as a whole is unknown. Afflicted free-running subjects often find the condition difficult to tolerate, with intermittent problems particularly of sleepiness and poor performance during the day and poor night-time sleep, when their pacemaker is out of phase with the environment.

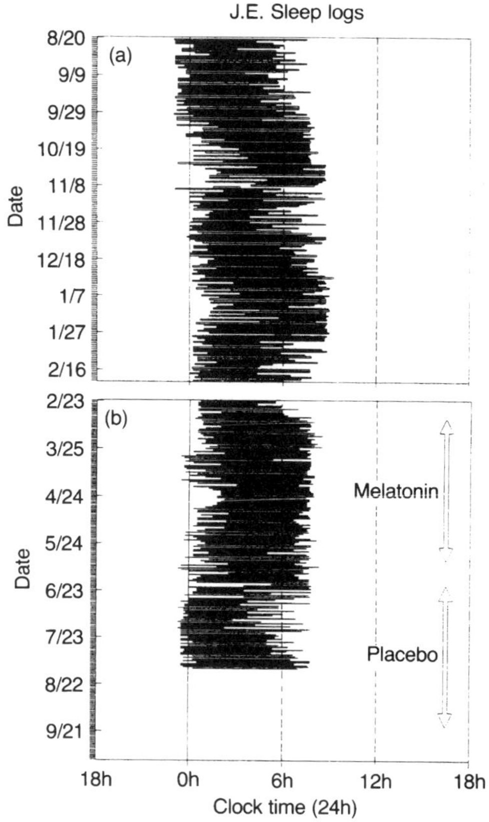

Fig. 8.12 Sleep record (horizontal bars are sleep times) of a blind woman whose endogenous free-running melatonin rhythm is delayed by 14 min per day (not shown) with (a) no treatment, (b) 5 mg melatonin at 22–2300 h, placebo at 22–2300 h. Treatment design single blind crossover. Melatonin stabilized the sleep period. (Aldhous and Arendt, unpublished.) Similar observations have been made by others but note that in limited studies, not all blind subjects suffering from sleep disturbance benefit from melatonin ingestion.

Only a small number of studies have been reported, but so far about one-third of blind subjects reporting sleep problems derive benefit from melatonin ingestion at desired bedtime. The major effect is a stabilization of sleep onset, sometimes with a significant improvement of some other sleep parameters such as quality and duration. In sufficiently long-term experiments it is clear that a synchronization of the sleep–wake cycle has occurred under melatonin (Figure 8.12). The

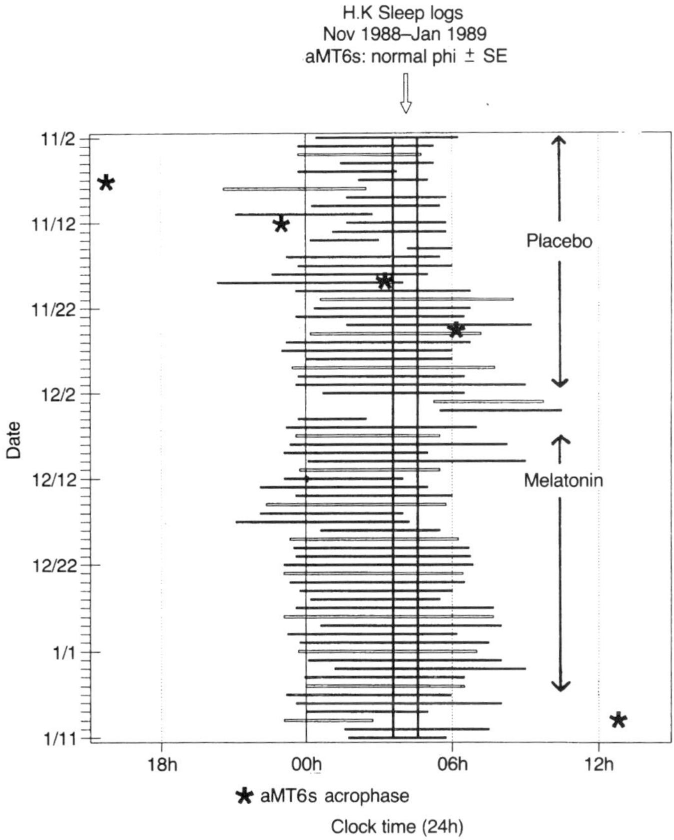

Fig. 8.13 Sleep record (horizontal bars are sleep times) of a blind man whose melatonin rhythm delayed by 41 min per day, treated with either 5 mg melatonin at 22–2300 h or placebo. The acrophases of the urinary aMT6s rhythm are indicated by asterisks. Note that sleep is most disrupted during the antiphase (acrophase in the afternoon) of the aMT6s rhythm. Melatonin treatment stabilized sleep onset, improved mood and alertness but did not synchronize the core body temperature, endogenous melatonin or cortisol rhythms, although temperature showed relative coordination. (From Arendt *et al.*, 1988, and Folkard *et al.*, 1990, previously unpublished figure.)

successful treatments have included both free-running subjects with tau greater than 24 hours and others with abnormal phase delay. Insufficient data are available to determine in which blind people melatonin is likely to be useful as yet. As melatonin may have sites of action within the visual system the integrity of such targets may be compromised in some blind people.

Whether synchronization of sleep–wake is accompanied by synchronization of other circadian variables in these circumstances is a question of much interest. One subject with a previously free-running sleep–wake cycle and whose sleep has been stabilized for a number of years showed a persistent free-run in melatonin, cortisol and temperature rhythms (Figure 8.13). In spite of this his well-being was restored with improvement of mood and daytime alertness and from this one case it appears that if the sleep problem is allayed, the behavior of other major circadian variables is irrelevant. It also suggests that behavioral changes do not necessarily entrain all of the pacemaker in the absence of any light–dark cycle. There may be enormous individual differences however. In the work of Lewy and colleagues (Sack *et al.*, 1992) other blind subjects show clear phase shifts and occasionally entrainment of the endogenous melatonin rhythm during melatonin treatment.

8.7 DELAYED SLEEP PHASE INSOMNIA

Another situation in which phase delay forms an integral part of a clinical problem is delayed sleep phase insomnia (DSPI). Patients cannot sleep at the socially acceptable time of night and delay sleep onset until the early hours of the morning, sleeping through much of the day. This condition is endemic in teenagers and in the student population where it is usually self limiting. The obligation to earn a living provides sufficient motivation to overcome the problem. In true DSPI subjects the aberrant phase position cannot easily be overcome, with consequent inability to work and socialize at the usual times. This condition has been successfully treated with bright light in the early morning to induce phase advances of the clock. In others evening melatonin (5 mg at 2200 h) will also advance sleep time significantly.

8.8 LIGHT AND MELATONIN

The interest in melatonin as a thermoregulatory factor has been reinforced by studies designed to uncover the acute effects of melatonin and light in humans. There is an inverse relationship between acute suppression of melatonin and acute elevation of body temperature by bright light. Volunteers, sleep deprived in bright light, received melatonin by infusion. The elevation of body temperature found in bright light was reduced by melatonin to values found in dim light. Moreover administration of melatonin in close to physiological concentrations in the morning to healthy female volunteers acutely suppresses body temperature (Figure 8.14). It is possible that suppression by endogenous melatonin can account for 30–40% of the amplitude of the body temperature rhythm in entrained conditions. The acute and opposite effects of bright light and melatonin may represent the primary manifestation of a phase shift in the making, in terms of brain metabolic events.

The physiological role of melatonin in circadian organization in humans can be considered in the same way as in the previous discussion concerning animals (Chapter 6). With a tau greater than 24 hours the phase advancing effects of morning light and evening melatonin may combine with other zeitgebers as yet incompletely defined to maintain 24 hour synchrony. The reverse situation can evidently be postulated for the rare individual with a tau less than 24 hours. With this in mind it would obviously be advantageous to combine controlled bright light (artificial or natural) and melatonin treatments in the management of rhythm disorder.

8.9 SIGNIFICANCE OF RHYTHM OPTIMIZATION TO INDUSTRY

The diurnal variation in performance and alertness has major implications for industrial performance. Night shift workers may operate a number of different scheduling systems, all of which involve some circadian disruption. In the worst case a desynchronized system can lead to the trough of performance occurring at a time of maximum required efficiency. It is probably no coincidence that many major disasters have occurred during the night shift (for example, the nuclear disasters of Chernobyl and Three Mile Island). Transport operatives have a clear peak in accident rate in the middle of the night and numerous other examples can be found in the literature. There are several useful publications advising on coping strategies for such situations and hopefully the worst scenarios can

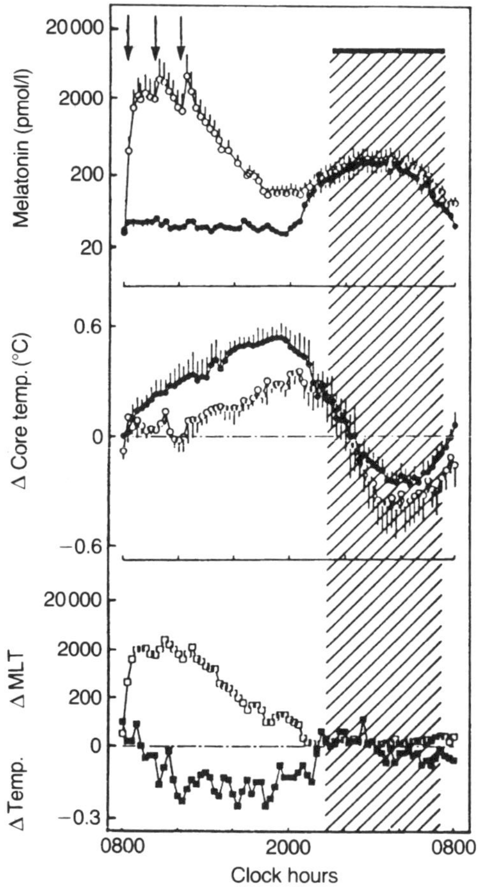

Fig. 8.14 The effect of exogenous melatonin in the morning (MLT; 1 mg at 0800 h; 0.75 mg at 1000 h; 0.75 mg at 1200 h) and placebo on body temperature. Top panel, increase in daytime melatonin after administration (○) compared to placebo (●). Middle panel, body temperature during and after melatonin (○) compared to placebo (p<0.01, 2-way ANOVA). Bottom panel, the striking inverse temporal correlation between high melatonin (□) and low temperature (■) is more apparent after the elimination of circadian 'masking' by subtracting for all points the mean melatonin or temperature value of the placebo day from the corresponding value of the melatonin day. The closed bar above the shaded area represents sleep time. Arrows indicate times of melatonin administration. Bright light has opposite effects on temperature. (Reproduced with permission from Cagnacci, A. Elliott, J. and Yen, S. Melatonin: a major regulator of the circadian rhythm of core temperature in humans. *J. Clin. Endocrinol. Metab.* **75**, 447–52, 1992 © The Endocrine Society.)

be avoided by appropriate behavior and exposure to time cues. Melatonin used to reinforce the circadian system has an obvious place in the hierarchy of therapeutic approaches.

8.10 RELATION BETWEEN CIRCADIAN AND PHOTOPERIODIC EFFECTS OF MELATONIN

Early experiments on the effects of low-dose sub-chronic melatonin administration in humans were designed using the principle that the duration of melatonin secretion is the critical parameter in the timing of seasonal functions. Late afternoon exogenous melatonin (2–5 mg) generates a sharp supraphysiological plasma peak which subsequently declines to night-time levels over the next 4–5 hours, thereby extending the duration of high levels. If humans show residual photoperiodic responses (for example in reproduction, prolactin secretion and the occurrence of winter depression) to long duration (short photoperiod) melatonin, this protocol would in principle unveil any significant effects. None were found other than the changes in circadian timing of sleep, melatonin and prolactin.

These experiments were performed in a normal environment and subject to the masking effects of light during the daytime, thus some masked changes may have been missed. Nevertheless they suggest that a primary photoperiodic response may be a change of phase within the circadian system.

These questions may be resolved with the development of slow-release melatonin preparations. Recently melatonin has been given to humans by infusion with a specific duration. This approach generates phase response curves assessed through endogenous melatonin secretion similar to those obtained using oral preparations. They have the advantage that much lower doses are required and supraphysiological concentrations are not generated.

8.11 EFFECTS OF MELATONIN ON HUMAN REPRODUCTIVE FUNCTION

Much of the early work on the effects of melatonin was directed towards an investigation of its possible role in non-seasonal reproductive function. This was primarily due to initial observations of antigonadal activity in rats and hamsters in the 1960s. Subsequently it was found that *in vitro* physiological concentrations of melatonin inhibit LHRH-induced LH release in pituitary cultures from prepubertal rats, probably via a calcium dependent mechanism, and *in vivo* will delay pubertal development in both male and female rats. In pharmacological amounts it will significantly suppress LH in pre- and postmenopausal women.

More recent work suggests that oral melatonin taken in the morning at 0800 h amplifies LH pulses in premenopausal healthy women in the early follicular phase (Figure 8.15). No effect on FSH or ovarian steroids was noted in this latter study, neither was the frequency of pulses affected. In ewes, high-amplitude, low-frequency pulses characterize anestrus, i.e. reproductive involution in long days. These results partially support the contention that in large doses, suitably administered, melatonin can inhibit human reproductive activity. It may act in long day seasonal breeders and humans to modify feedback sensitivity to gonadal steroids at pituitary and hypothalamic levels. It also has the potential to act at gonadal level by modification of steroid synthesis and metabolism. High concentrations are found in human follicular fluid.

Chronic administration of large amounts (up to 300 mg daily) of melatonin to cycling women will significantly suppress plasma LH but not estradiol after 4 months, treatment. The combination of these large daily doses with norethistrone (a synthetic progestin minipill) will inhibit ovulation in women. In the absence of side effects, yet to be fully assessed, especially over the long term, this preparation may prove to be clinically useful. In the author's opinion, low, timed, doses of melatonin, used to reinforce circadian organization, are likely to improve fertility in humans.

There are few reports of the administration of melatonin to young children, although it has been used to alleviate sleep problems (Jan *et al.*, 1994). By analogy with animal work (Chapter 6) melatonin may prove to have a primary role in the organization of physiological rhythms in the pre- and postnatal period in humans. Melatonin receptors are present in the human fetus and neonate. The ontogenesis of melatonin receptors in rats and the striking effects on control of LH release from the prepubertal pituitary reinforce the possible

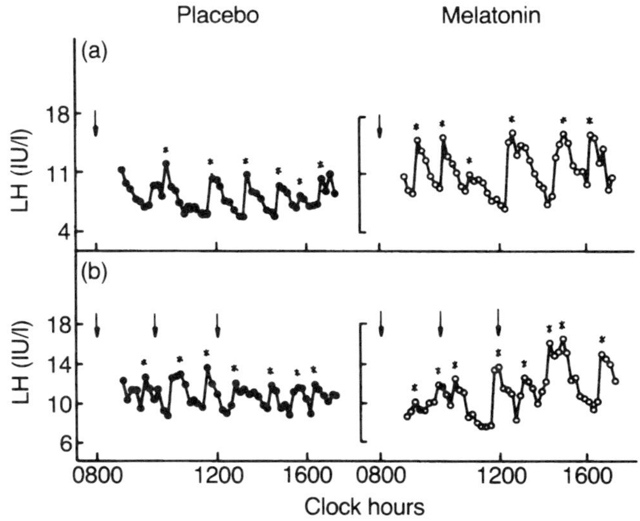

Fig. 8.15 Pulsatile pattern of LH secretion between 0900 and 1700 h in two representative early follicular phase women after the administration of placebo (●) or melatonin (○) at the dose of 100 mg (panel a) or of 2.5 mg (panel b). In both subjects an increased LH pulse amplitude after melatonin administration is evident. *= significant pulses. Arrows indicate time of treatment. (Reproduced with permission from Cagnacci, A., Elliott, J.A. and Yen S. Amplification of pulsatile LH secretion by exogenous melatonin in women. *J. Clin. Endocrinol. Metab.* **73**, 210–2, 1991 © The Endocrine Society.)

influence of melatonin on human puberty. It is conceivable that it may be useful in the treatment of abnormal pubertal development and in the synchronization of sleep–wake cycles in early life. Clearly however, there are major ethical concerns with regard to investigations in this area.

8.12 CANCER

There is considerable interest in the ability of melatonin to act as an oncostatic agent. Most of the available information concerns animal work but in view of the therapeutic potential it is included in this chapter.

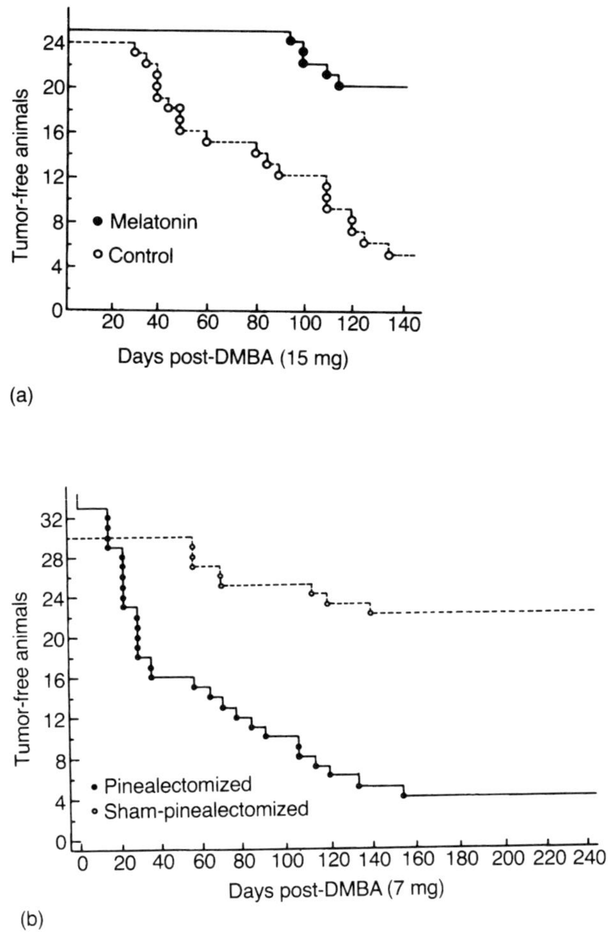

Fig. 8.16 (a) Incidence of breast carcinoma in Sprague-Dawley rats given 15 mg of DMBA at 50 days of age (day 0). From 0–90 days all animals received daily s.c. injections of melatonin (500 μg per animal) or vehicle (control) at 4 pm (lights on from 0600–0800 h). All animals were examined twice weekly for tumors until 140 days after DMBA treatment. (Reproduced with permission from Tamarkin, L. *et al.*, published by the American Association for Cancer Research, 1981.) (b) Incidence of breast carcinoma in Sprague-Dawley rats pinealectomized or sham operated at 20 days of age and given 7 mg DMBA at 50 days of age (day 0). All animals were examined twice weekly for tumors until 240 days after DMBA treatment. (Reproduced with permission from Tamarkin, L. *et al.*, published by the American Association for Cancer Research, 1981.)

There is good evidence for photoperiod dependency and/or melatonin responsiveness of the initiation and evolution of certain cancers, particularly hormone-dependent cancers, in animals. Early work has been summarized by Bartsch and Bartsch (1981). The most dramatic observations concern dimethylbenzanthracene (DMBA)-induced mammary tumors in rats (Figure 8.16). Here pinealectomy greatly increased the incidence of induced tumor growth and daily melatonin administration in the late light phase greatly decreased the incidence. Melatonin administration at this time of day mimics short day photoperiods and suppresses prolactin. Thus the antitumor effect of melatonin may act via suppression of prolactin, which is thought to promote mammary cancer.

One conclusion from this study might be that long days favour mammary tumor growth and short days suppress it. There is some preliminary evidence that breast cancer in humans is preferentially initiated in long days (prolactin is higher in the spring and summer than at other times of year in humans as in other species). However it is never possible to ascertain the precise time of cancer initiation in humans and these data should be treated with caution.

Melatonin also inhibits the growth of the Dunning R3327H1P tumor (an androgen insensitive rat prostatic adenocarcinoma), Yoshida sarcoma in rats, transplantable melanoma in hamsters, solid leukemia and B16 melanoma growth in athymic mice. Not all reports show positive results however. Photoperiodic effects of melatonin on reproduction are either stimulatory or inhibitory depending on conditions and this may apply to effects on cancer. Particularly good evidence for this exists with respect to hamster melanoma growth, fibrosarcoma ascites and Ehrlich solid tumors in albino mice; hence great care in experimental design is necessary.

One of the most interesting observations concerning, particularly, DMBA-induced mammary cancer in rats, is that blinding and underfeeding halts the development of breast adenocarcinomas and that this effect is pineal dependent (Blask, Hill and Pelletier, 1988). Underfeeding may be associated with high amplitude of the melatonin rhythm and blinding leads to free-running rhythms in rats and in some cases humans. In theory, therefore, we have a strong melatonin signal operating at the normal endogenous periodicity of the animal and hence, conceivably, greatly strengthened coupling of at least parts of the circadian system. Such reinforcement may well optimize physiological defence mechanisms. It is likely that poorly organized circadian rhythms are responsible for much more than jet lag.

Blask and coworkers (Blask, Hill and Pelletier, 1988; Hill and Blask, 1988) have shown dose dependent inhibition of the growth of the human ovarian cancer cell line MCF-7. Their effects were induced by physiological concentrations of melatonin and were dependent on the simultaneous presence of estradiol or prolactin. Others in this field reported growth inhibition with much larger doses.

Lissoni and colleagues (1992) have reported that the administration of melatonin (10 mg daily at 1900 h) to various cancer patients non-significantly prolonged the survival time of non-small cell lung cancer patients resistant to chemotherapy (Figure 8.17) and other reports of combined melatonin therapy in cancer patients have been optimistic.

8.13 EFFECTS ON THE IMMUNE SYSTEM

Melatonin, when appropriately administered, has generally stimulatory effects on aspects of the immune system. These are referred to briefly in Chapter 7. One interpretation of this phenomenon is that melatonin acts as an antistress hormone on the brain opioid system with consequent up-regulation of the immune system.

8.14 MISCELLANEOUS EFFECTS OF MELATONIN

Most effects of melatonin can be interpreted on a circadian or seasonal basis but not all. Previous reviews should be consulted for comprehensive lists of melatonin-induced effects. Some of the most interesting will be mentioned here.

Claustrat and coworkers have shown significant circadian phase abnormalities in migraine assessed by endogenous melatonin. Infu-

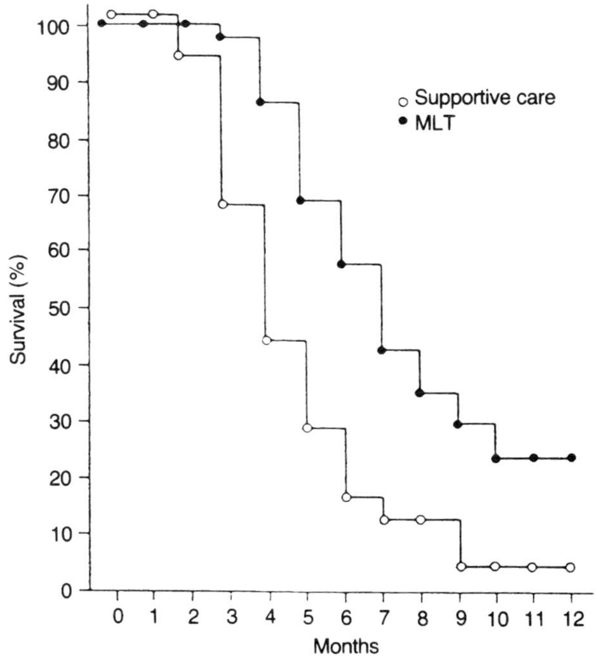

Fig. 8.17 Percentage of survival at 1 year from progression during first-line chemotherapy in advanced non-small cell lung cancer patients treated with melatonin (10 mg/day orally at 1900 h). MLT = melatonin. (Reproduced with permission from Lissoni, P. *et al.*, published by Karger, Basel, 1992.)

sion of melatonin such as to induce a phase advance led to remission (Claustrat, personal communication; Claustrat *et al.*, 1989).

As previously discussed in Chapter 7 there is evidence from animal work that pinealectomy shortens life span and daily melatonin administration retards the aging process, extending the life span in mice by 20% and the reproductive life of rats. If these observations prove to be consistent, they have obvious implications for human use. So far there are no reports of the controlled use of melatonin in human aging.

Large acute oral doses (100–200 mg) of melatonin stimulate prolactin secretion (Figure 8.18) in contrast to the suppression with long duration melatonin in photostimulated seasonal breeders, or change of phase seen after chronic late afternoon treatment in humans. The

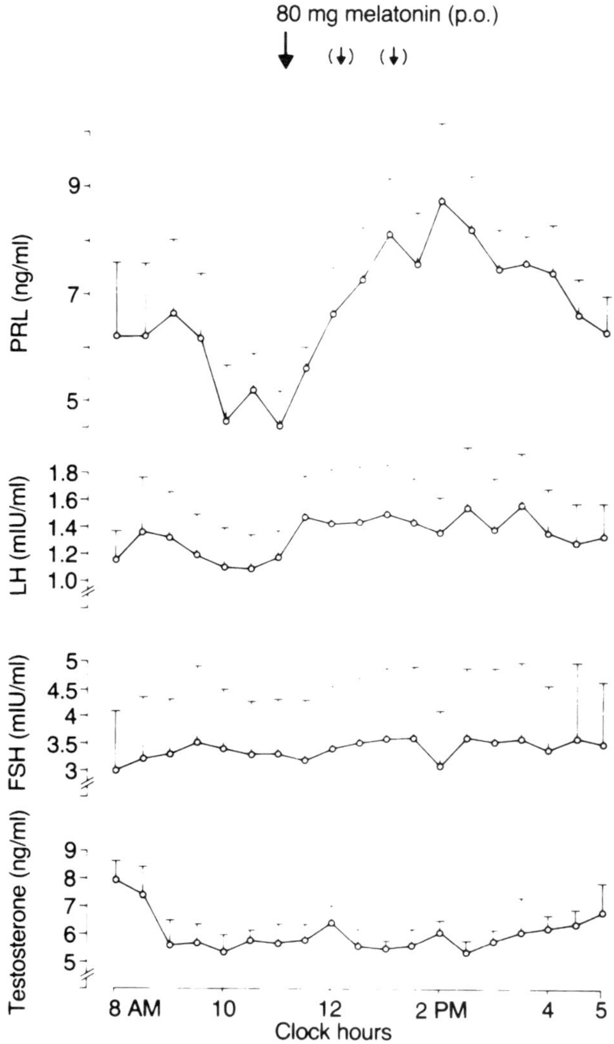

Fig. 8.18 Acute melatonin treatment increases prolactin secretion but does not change testosterone, FSH or LH. Serum levels of testosterone, FSH, LH and PRL (mean ± SEM) in eight male volunteers before, during and after administration of either one capsule (n=5) or three capsules (n=3) each containing 80 mg crystalline melatonin. Arrows indicate the administration times. (Reproduced with permission from Waldhauser, F. *et al.* published by Karger, Basel, 1987.)

reason for this acute effect may relate to the ability of melatonin in pharmacological amounts to inhibit some dopaminergic functions as dopamine is the main prolactin inhibitor *in vivo*. Doses in this range also acutely increase fatigue ratings and may have some deleterious effects on various performance tasks.

There is some evidence that melatonin may affect lipid metabolism in humans, however, in general its metabolic effects are diffuse and minor. In patients with seasonal affective disorder treated with bright light the concomitant administration of melatonin reinduces the symptoms of carbohydrate craving and increase in body weight without a return of depression. This suggests that the strictly somatic metabolic seasonal symptoms are melatonin dependent.

It is possible that melatonin modifies cardiovascular function. Hypotension has been described after acute administration of supraphysiological amounts. Cardiac rhythms were significantly altered in a single case report.

Melatonin or its kynurenamine metabolites can inhibit prostaglandin synthesis. They may influence clotting, but little published data are confirmed.

Recently Reiter and colleagues (1993) have reported a free radical scavenging ability of melatonin. They have evidence that melatonin is a very potent scavenger of hydroxyl radicals. If this is confirmed the therapeutic vistas for melatonin become even wider. Moreover some of the previously mentioned actions, e.g. anti-aging and antitumor effects may be explicable on this basis.

8.15 COMMERCIAL PREPARATIONS OF MELATONIN, ITS AGONISTS AND ANTAGONISTS

As a result of the various properties of melatonin described earlier a number of commercial preparations are under development together with agonists and antagonists. The latter will be considered in Chapter 9.

A transmucosal patch has been described which has been administered to dogs via the buccal mucosa. It is capable of creating artificial night-time levels of a duration dictated by the length of time of application within 24 hours. This promises to be of interest for both basic and therapeutic studies. Most human melatonin administration has been accomplished with crystalline melatonin incorporated into simple gelatine capsules with a filler or dissolved in a suitable vehicle.

At the time of writing melatonin has not yet received a product license at least in the UK. This is likely to be granted in the future and one can anticipate availability as a specifically formulated prescription medicine.

Problems with the therapeutic use of melatonin include the following:

- Its effects are time and dose dependent.
- The effects may be stimulatory, inhibitory, or ineffective depending on current and previous photoperiodic experience, circadian status, developmental stage, and previous exposure.
- There may be long-term consequences of treatment which are beneficial. On the other hand they may also be detrimental.
- There are no published full toxicity studies.
- There may be undesirable effects on cardiovascular function including clotting.
- Intermittent reported undesirable side effects include headache and nausea.
- Its interactions with other drug treatments are unknown.

REFERENCES

Arendt, J. (1988) Melatonin and the human circadian system, in *Melatonin, Clinical Perspectives* (eds C. Thompson, A. Miles and D. Philbrick), Oxford University Press, pp. 43–61.

Arendt, J., Aldhous, M. and Marks, M. (1987) Some effects of jet-lag and their treatment by melatonin. *Ergonomics*, **30**, 1379–93.

Arendt, J., Bojkowski, C., Folkard, S., Franey *et al.* (1985) Some effects of melatonin and the control of its secretion in man, in Ciba Foundation Symposium 117, *Photoperiodism, Melatonin and the Pineal* (eds D. Evered and S. Clark), Pitman, London, pp. 226–83.

Arendt, J., Borbely, A.A., Franey, C. *et al.* (1984) The effect of chronic, small doses of melatonin given in the late afternoon on fatigue in man: a preliminary study. *Neuroscience Letters*, **45**, 317–21.

Bartsch, H. and Bartsch, C. (1981) Effect of melatonin on experimental tumours under different photoperiods and times of administration. *Journal of Neural Transmission*, **52**, 269–79.

Blask, D.E., Hill, S.M. and Pelletier, D.B. (1988) Oncostatic signalling by the pineal gland and melatonin in the control of breast cancer, in *The Pineal Gland and Cancer* (eds D. Gupta, A. Attanasio and R.J. Reiter), Müller & Bass, Tubingen, Germany, pp 195–206.

Cagnacci, A., Elliott, J.A. and Yen, S.S.C. (1991) Amplification of pulsatile LH secretion by exogenous melatonin in women. *Journal of Clinical Endocrinology and Metabolism*, **73**, 210–2.

Cagnacci, A., Elliott, J.A., Yen, S.S.C. (1992) Melatonin: a major regulator of the circadian rhythm of core temperature in humans. *Journal of Clinical Endocrinology and Metabolism*, **75**, 447–52.

Claustrat, B., Loisy, C., Brun, J., Beorchia, S. *et al.* (1989) Nocturnal plasma melatonin levels in migraine: a preliminary report. *Headache*, **29**, 241–4.

Deacon, S., English, J. and Arendt, J. (1995) Acute phase-shifting effects of melatonin associated with suppression of core body temperature. *Neuroscience Letters* (in press).

Folkard, S., Arendt, J., Aldhous, M. *et al.* (1990) Melatonin stabilises sleep onset time in a blind man without entrainment of cortisol or temperature rhythms. *Neuroscience Letters*, **113**, 193–198.

Folkard, S., Arendt, J. and Clark, M. (1993) Can melatonin improve shift workers' tolerance of the night shift? Some preliminary findings. *Chronobiology International*, **10**, 315–20.

Guardiola-Lemaitre, B. (1991) Development of melatonin analogues, in *Advances in Pineal Research* (eds J. Arendt and P. Pevet). John Libbey, London, pp. 351–63.

Hill, S.M. and Blask, D.E. (1988) Effects of the pineal hormone melatonin on the proliferation and morphological characteristics of human breast cancer cells (MCF-7) in culture. *Cancer Research*, **48**, 6121–5.

Lavie, P. (1991) The 24-hour sleep propensity function (SPF): practical and theoretical considerations, in *Sleep, Sleepiness and Performance* (ed. T.H. Monk), John Wiley and Sons Ltd., ch.3.

Lerner, A.B. and Nordlund, J.J. (1978) Melatonin: clinical pharmacology. *Journal of Neural Transmission*, (Suppl. 13), 339–347.

Lewy, A.J., Saeeduddin, A., Latham-Jackson, J.M. *et al.* (1992) Melatonin shifts human circadian rhythms according to a phase response curve. *Chronobiology International*, **9**, 380–392.

Lissoni, P., Barni, S., Ardizzoia, A. *et al.* (1992) Randomized study with the pineal hormone melatonin versus supportive care alone in advanced nonsmall cell lung cancer resistant to a first line chemotherapy containing cisplatin. *Oncology*, **49**, 336–9.

Minors, D.S., Waterhouse, J.M. and Wirz-Justice, A. (1992) A human phase response curve to light. *Neuroscience Letters*, **133**, 36–40.

Sack, R.L., Lewy, A.J., Blood, M.L. *et al.* (1991) Melatonin administration to blind people: phase advances and entrainment. *Journal of Biological Rhythms,* **6**, 249–61.

Skene, D.J., Aldhous, M. and Arendt, J. (1989) Melatonin, jet-lag and the sleep-wake cycle, in *Sleep 1988* (ed. J. Horne), Proceedings of the European Sleep Congress, Jerusalem, 1988, Karger, Basel, pp. 39–41.

Sugden, D. (1983) Psychopharmacological effects of melatonin in mouse and rat. *Journal of Pharmacological Experimental Therapy*, **227**, 587–91.

Tamarkin, L., Cohen, M., Roselle, D. *et al.* (1981) Melatonin inhibition and pinealectomy enhancement of 7,12-dimethylbenz(a)anthracene-induced mammary tumours in the rat. *Cancer Research*, **41**, 4432–6.

Waldhauser, F., Lieberman, H., Lynch,H. *et al.* (1987) A pharmacological dose of melatonin increases PRL levels in males without altering those of GH, LH, FSH, TSH, testosterone or cortisol. *Neuroendocrinology*, **46**, 125–30.

Wright, J., Aldhous, M., Franey, C. *et al.* (1986) The effects of exogenous melatonin on endocrine function in man. *Clinical Endocrinology*, **24**, 375–82.

Zaidan, R., Geoffriau, M., Brun, J. *et al.* (1994) Melatonin is able to influence its secretion in humans: description of a phase-response curve. *Neuroendocrinology*, **60,** 105–12.

FURTHER READING

Akerstedt, T. (1985) Adjustment of physiological circadian rhythms and the sleep wake cycle to shift work, in *Hours of Work: Temporal Factors in Work Scheduling* (eds S. Folkard and T.H. Monk), John Wiley, New York, pp. 185–98.

Aldhous, M.E. and Arendt, J. (1991) Melatonin rhythms and the sleep wake cycle in blind subjects. Proceedings of the 7th Meeting of the European Society for Chronobiology, Marburg, Germany, 1991. *Journal of Interdisciplinary Cycle Research*, **22**, 84–5.

Angeli, A., Gatti, G., Sartori, M.L. *et al.* (1988) Effect of exogenous melatonin on human Natural Killer cell activity. An approach to the immunomodulatory role of the pineal gland, in *The Pineal Gland and Cancer* (eds D. Gupta, A. Attanasio and R.J. Reiter), Brain Research Promotion, Tubingen, pp. 145–56.

Arendt, J. and Aldhous, M. (1988) Synchronisation of a disturbed sleep-wake cycle in a blind man by melatonin treatment. *Lancet*, **i**, 772–3.

Arendt, J. and Aldhous, M. (1988) Further evaluation of the treatment of jet-lag by melatonin: a double blind crossover study. *Annual Review of Chronopharmacology*, **5**, 53–5.

Arendt, J., Aldhous, M. and Marks, V. (1986) Alleviation of jet-lag by melatonin: preliminary results of controlled double-blind trial. *British Medical Journal*, **292**, 1170.

Arendt, J. and Marks, V. (1982) Physiological changes underlying jet-lag. *British Medical Journal*, **284**, 144–6.

Aubert, C., Janiaud, P. and Lecalves, J. (1980) Effect of pinealectomy and melatonin on mammary tumour growth in Sprague-Dawley rats under different conditions of lighting. *Journal of Neural Transmission*, **47**, 121–30.

Badia, P., Myers, B. and Murphy, P. (1992) Melatonin and thermoregulation, in *Melatonin: Biosynthesis, Physiological Effects, and Clinical Applications* (eds R.J. Reiter and H.S. Yu), CRC Press, Boca Raton, Florida, pp. 349–64.

Blask, D.E. (1984) The pineal: an oncostatic gland? in *The Pineal Gland* (ed. R.J. Reiter), Raven, New York, pp. 253–84.

Cassone, V. (1990) Effects of melatonin on vertebrate circadian systems. *Trends in Neuroscience*, **13**, 457–63.

Cassone, V.M., Roberts, M.H. and Moore, R.Y. (1988) Effects of melatonin on 2-deoxy-(1-14-C)-glucose uptake within rat suprachiasmatic nucleus. *American Journal of Physiology*, **255**, R332.

Checkley, S.A. (1988) The relationship between biological rhythms and the affective disorders, in *Biological Rhythms in Clinical Practice* (eds J. Arendt, D.S. Minors and J. Waterhouse), Butterworth, London.

Cramer, H., Rudolph, J. and Consbruch, V. (1974) On the effects of melatonin on sleep and behaviour in man, in *Serotonin–New Vistas: Biochemistry and Behavioural and Clinical Studies Advances in Biochemical Psychopharmacology*, Volume 11 (eds E. Costa, G.I. Gessa and M. Sandler), Raven Press, New York, pp. 187–91.

Czeisler, C.A., Allan, J.S. and Strogatz, J.S. (1986) Bright light resets the human circadian pacemaker independent of the timing of the sleep–wake cycle. *Science*, **233**, 667–71.

Czeisler, C.A., Johnson, P.J., Duffy, J.F. *et al.* (1990) Exposure to bright light and darkness to treat physiologic maladaptation to night-work. *New England Journal of Medicine*, **322**, 1253–9.

Czeisler, C.A., Kronauer, R.E., Mooney, J.J. *et al.* (1987) Biologic rhythm disorders, depression, and phototherapy. The *Psychiatric Clinics of North America*, **10**, 687.

Dahlitz, M., Alvarez, B., Vignau, J. *et al.* (1991) Delayed sleep phase syndrome response to melatonin. *Lancet*, **337**, 1121–4.

Das Gupta, T.K. and Terz, J. (1967) Influence of pineal gland on the growth and spread of melanoma in the hamster. *Cancer Research*, **27**, 1306.

Folkard, S. and Monk, T.H. (1983) Chronopsychology: circadian rhythms and human performance, in *Physiological Correlates of Human Behaviour*, Academic Press, London.

Franchi, A.M., Gimeno, M.F., Cardinali, D.P. *et al.* (1987) Melatonin, 5-methoxytryptamine and some of their analogues as cyclooxygenase inhibitors in rat medial basal hypothalamus. *Brain Research*, **405**, 384–8.

Fraschini, F.O., Mess, B. and Martini, L. (1968) Pineal gland, melatonin and the control of luteinizing hormone secretion. *Endocrinology*, **82**, 919–24.

Hill, S., Spriggs, L., Simon, M. *et al.* (1992) The growth inhibitory action of melatonin on human breast cancer cells is linked to the estrogen response system. *Cancer Letters*, **64**, 249–56.

Holmes, S.W. and Sugden, D. (1975) The effect of melatonin on pinealectomy-induced hypertension in the rat. *Proceedings of the British Pharmacological Society*, 17–19th December, p. 360.

Karasek, M. and Fraschini, F. (1991) Is there a role for the pineal gland in neoplastic growth? in *Role of Melatonin and Pineal Peptides in Neuroimmunomodulation* (eds F. Fraschini and R.J. Reiter), Plenum Press, New York, pp. 243–51.

Klein, K.E. and Wegmann, H.M. (1979) Circadian rhythms in air operations, in *Sleep, Wakefulness and Circadian Rhythms* (ed. A.N. Nicholson), Neuilly-sur-Seine, NATO Advisory Group Aero R & D 10.1–10.25.

Krause, D.N. and Dubocovich, M.L. (1990) Regulatory sites in the melatonin system of mammals. *Trends in Neuroscience*, **13**, 464–70.

Lapin, V. (1979) Pineal influence on tumour. *Progress in Brain Research*, **52**, 523–33.

Lapin, V. and Ebels, I. (1981) The role of the pineal gland in neuroendocrine control mechanisms of neoplastic growth. *Journal of Neural Transmission*, **50**, 257–64.

Lewy, A.J. and Newsome, D.A. (1983) Different types of melatonin circadian secretory rhythms in some blind subjects. *Journal of Clinical Endocrinology and Metabolism*, **56**, 1103–7.

Lewy, A.J., Sack, R.L. and Latham, J.M. (1991) Melatonin and the acute suppressant effect of light may help regulate circadian rhythms in humans, in *Advances in Pineal Research* (eds J. Arendt and P. Pevet), **5**, 285–94.

Lewy, A.J., Sack, R.L., Miller, L.S. *et al.* (1987) Anti-depressant and circadian phase-shifting effects of light. *Science*, **235**, 352–4.

Lieberman, H.R., Waldhauser, F., Garfield, G. *et al.* (1984) Effects of melatonin on human mood and performance. *Brain Research*, **323**, 201–7.

Lissoni, P., Barni, S., Rovelli, F. *et al.* (1993) Neuroimmunotherapy of advanced solid neoplasms with single evening subcutaneous injection of low-dose interleukin-2 and melatonin. *European Journal of Cancer*, **29A,** 185–9.

Maestroni, G.J.M., Conti, A. and Pierpaoli, W. (1986) Role of the pineal gland in immunity. Circadian synthesis and release of melatonin modulates the antibody response and antagonises the immunosuppressive effect of corticosterone. *Journal of Neuroimmunology*, **13**, 19.

Maestroni, G.J.M., Conti, A. and Pierpaoli, W. (1989) Melatonin stress and the immune system. *Pineal Research Reviews*, **7**, 268.

Martin, J.E. and Klein, D.C. (1975) Melatonin inhibition of neonatal pituitary response to luteinising hormone-releasing factor. *Science*, **191**, 301–2.

Minors, D.S. and Waterhouse, J.M. (1981) *Circadian Rhythms and the Human*. Wright, Bristol, pp. 143–8.

Neville, S., Arendt, J. and Ioannides, C. (1988) A study of the mutagenicity of melatonin and 6-hydroxymelatonin. *Journal of Pineal Research*, **6**, 73–6.

Nickelsen, T., Lang, A. and Bergau, L. (1991) The effect of 6-, 9- and 11-hour time shifts on circadian rhythms: adaptation of sleep parameters and hormonal patterns following the intake of melatonin or placebo, in *Advanced Pineal Research* (eds J. Arendt and P. Pevet), **5**, 303–6.

Palm, L., Blennow, G. and Wetterberg, L. (1991) Correction of a non-24-hour sleep wake cycle by melatonin in a blind retarded boy. *Annals of Neurology*, **29**, 336–9.

Petrie, K., Conaglen, J.V., Thompson, L. *et al.* (1989) Effect of melatonin on jet-lag after long haul flights. *British Medical Journal*, **298**, 705–7.

Petterborg, L.J., Thalen, B.E., Kjellman, B.F. *et al.* (1991) Effect of melatonin replacement on serum hormone rhythms in a patient lacking endogenous melatonin. *Brain Research Bulletin*, **27**, 181–5.

Reiter, R.J., Poggeler, B., Tan, D.-X. *et al.* (1993) Antioxidant capacity of melatonin: a novel action not requiring a receptor. *Neuroendocrinology Letters*, **15**, 103–16.

Rodin, A.E. (1963) The growth and spread of Walker 256 carcinoma in pinealectomised rats. *Cancer Research*, **27**, 1545.

Rosenthal, N.E., Jacobsen, F.M., Sack, D.A. *et al.* (1988) Atenolol in seasonal affective disorder: a test of the melatonin hypothesis. *American Journal of Psychiatry*, **145**, 52–6.

Rosenthal, N.E., Sack, D.A., Gillin, J.C. *et al.* (1984) Seasonal affective disorder. A description of the syndrome and preliminary findings with light therapy. *Archives of General Psychiatry*, **41**, 72–9.

Sack, R.L., Lewy, A.J. and Hoban, T.M. (1987) Free-running melatonin rhythms in blind people: phase shifts with melatonin and Triazolam administration, in *Temporal Disorder in Human Oscillatory Systems* (eds L. Rensing, V. an der Heiden and M.C. Mackey), Springer-Verlag, Heidelberg, pp. 219–24.

Samel, A., Wegman, H.M., Vejvoda, M. *et al.* (1991) Influence of melatonin treatment on human circadian rhythmicity before and after a simulated 9 hour time shift. *Journal of Biological Rhythms*, **6**, 235–48.

Sarrafzadeh, A., Wirz-Justice, A., Arendt, J. *et al.* (1990) Melatonin stabilises sleep onset in a blind man, in *Sleep 1990* (ed. J. Horne), Proceedings of the European Sleep Conference, Strasbourg.

Shellard, S.A., Whelan, R.D.H. and Hill, B.T. (1989) Growth inhibitory and cytotoxic effects of melatonin and its metabolites on human tumour cell lines *in vitro*. *British Journal of Cancer*, **60**, 288–90.

Sizonenko, P.C., Lang, U., Rivest, R.W. *et al.* (1985) The pineal and pubertal development, in *Photoperiodism, Melatonin and the Pineal, Ciba Foundation Symposium 117*, (eds D. Evered and S. Clark), Pitman, London, pp. 208–30.

Smith, J.A., O'Hara, J. and Schiff, A.A. (1981) Altered diurnal serum melatonin rhythm in a blind man. *Lancet*, **ii**, 933.

Stanberry, R., Das Gupta, T.K. and Beattie, C.W. (1983) Photoperiodic control of melanoma growth in hamsters: influence of pinealectomy and melatonin. *Endocrinology*, **113**, 469–75.

Strassman, R.J., Qualls, C.R., Lisansky, E.J. *et al.* (1991) Elevated rectal temperature produced by all night bright light is reversed by melatonin infusion in man. *Journal of Applied Physiology*, **71**, 2178–81.

Trentini, G., Genazzani, A.R., Criscuolo, M. *et al.* (1992) Melatonin treatment delays reproductive aging of female rat via the opiatergic system. *Neuroendocrinology*, **56**, 364–70.

Trentini, G.P., Botticelli, A.R., Botticelli, C.S. *et al.* (1976) Decreased ovarian LH incorporation after melatonin treatment.

Hormone and Metabolic Research, **8**, 234–6.

Vollrath, L., Semm, P. and Gammel, G. (1981) Sleep induction by intranasal application of melatonin. *Advances in the Biosciences*, **29**, 327–9.

Voordow, B.C.G., Euser, R., Verdonk, R.E.R. *et al.* (1992) Melatonin and melatonin-progestin combinations alter pituitary-ovarian function in women and can inhibit ovulation. *Journal of Clinical Endocrinology and Metabolism*, **74**, 108–17.

Waldhauser, F., Steger, H. and Vorkapic, P. (1987) Melatonin secretion in man and the influence of exogenous melatonin on some physiological and behavioural variables. *Advances in Pineal Research*, **2**, 207–23.

Walker, R.F., McCamant, S. and Timiras, P.S. (1982) Melatonin and the influence of the pineal gland on timing of the LH surge in rats. *Neuroendocrinology*, **39**, 37–42.

Wehr, T.A. and Goodwin, F.K. (eds) (1983) *Circadian Rhythms in Psychiatry*, Boxwood Press, California.

Wever, R.A. (1989) Light effects on human circadian rhythms: a review of recent Andechs experiments. *Journal of Biological Rhythms*, **4**, 161.

Williams, L.M., Martinoli, M.G., Titchener, L.T. *et al.* (1991) The ontogeny of central melatonin binding sites in the rat. *Endocrinology*, **128**, 2083–90.

Wurtman, R.J., Axelrod, J. and Chu, E.W. (1963) Melatonin, a pineal substance: its effect on the rat ovary. *Science*, **141**, 277–80.

Wynn, V.T. and Arendt, J. (1988) The effect of melatonin on the human electrocardiogram and simple reaction time responses. *Journal of Pineal Research*, **5**, 427–36.

Target sites and mechanism of action of melatonin

9

Contents

9.1 INTRODUCTION

The concept presented in this book is that of the pineal gland, via melatonin, communicating light–dark information for the coordination of mammalian rhythms, with a similar but hierarchically more important role in lower vertebrates. The actions of melatonin from a current perspective are therefore multiple and must derive primarily from modification of events in the central nervous system. However peripheral actions are not excluded.

Possible sites of action in the control of photoperiodic seasonal functions are considered briefly in Chapter 5. Whether or not distinct sites exist for the separate control of seasonal and circadian functions is a question of fundamental importance which is by no means answered. In some hamsters for example, there is evidence that photoperiodic responses are independent of the SCN. But other work shows clearly a circadian basis for photoperiodism, through the presence of specific circadian windows of light sensitivity for photoperiodic induction. A resolution of this problem may lie in the concept that the circadian system is only important in photoperiodic responses because it is required to generate an appropriate melatonin rhythm. There appear to be substantial species differences in the reception and the interpretation of the melatonin signal.

9.2 TARGET SITES

9.2.1 Retina

Melatonin acts within the retina to affect the timing of disc shedding, pigment aggregation and calcium-activated dopamine release, all of which are light-dependent functions. It may act within photoreceptor cells to regulate their sensitivity to light. It also coordinates the pigment migration response to circadian light–dark cycles in dermal melanophores. All of these effects take place in cell types which are derived embryologically from neural crest tissue.

9.2.2 Brain

Lesions of brain areas thought to act as substrates for circadian and photoperiodic effects provide useful information on possible target tissues for melatonin but suffer from the disadvantage that other neural circuits may be inadvertently disrupted. This complicates the interpretation of findings. Nevertheless lesions of the SCN, and the anterior hypothalamic area can block photoperiodic and/or circadian effects of melatonin in some rodents and ungulates – but with a degree of disparity between laboratories.

Implants of melatonin in the hypothalamus suppress LH release in rats – a species whose photoperiodic responses are much curtailed – but where melatonin is clearly important for the timing of puberty. Implants or infusion of melatonin into the hypothalamus mimic or block photoperiodic responses in numerous species and are ineffective when placed within the brain, but at a distance from the hypothalamus.

Peripheral administration of melatonin elicits changes in neurotransmitters especially in the hypothalamus. Serotonin and opioid peptides in particular respond to melatonin in ways that can be related to changes in photoperiod. GABAergic transmission and norepinephrine turnover can both be enhanced by melatonin and a number of photoperiod-induced changes, such as LHRH and vasopressin immunoreactivity, are almost certainly melatonin dependent. These changes in neurotransmitters are of interest but to what extent they represent either events downstream of central rhythm generating systems or a focus of action remains inconclusive. Immediate early gene expression in the SCN can be induced by phase shifting application of light but there is little data as yet on induction of immediate early genes by melatonin.

9.2.3 Pituitary

In prepubertal rats melatonin inhibits LHRH-induced LH release in pituitary cultures at concentrations comparable to those circulating in the blood. This effect is not demonstrable in adult animals and may relate to the presence of binding sites in the pars distalis at an early age. Melatonin effects on the pars tuberalis of the pituitary are discussed on p. 307 *et seq.* and in Chapter 5.

9.2.4 Periphery

Melatonin has been shown to have biochemical effects in many peripheral tissues, of which the most interesting are related to reproductive function (see Chapter 5). *In vitro* changes in general are, however, difficult to relate to the relatively slow overall rhythm responses to melatonin.

9.2.5 Specificity of action

Melatonin is highly lipophilic and easily diffusable, thus specificity of action must be conferred by a discriminating system such as a receptor. A search for the immediate target sites of melatonin via uptake and binding studies has provided the most revealing information. We would with hindsight expect to find that the melanophore, the retina, the hypothalamus and the pituitary are target tissues for melatonin. If it acts via receptors, binding sites should be demonstrable and binding should elicit a relevant physiological response in these tissues.

9.2.6 Pharmacological effects

In addition to its function as an internal zeitgeber for the timing of seasonal breeding and circadian rhythmicity melatonin appears to have useful pharmacological effects of rather broad specificity such as its immediate sedative and anticonvulsant effects. These may depend on, for example, false transmitter function, and may therefore use the lower affinity responses of neurotransmitter receptors.

There is one major problem in the interpretation of data concerning the target tissues of melatonin and that is the fact that as yet its effective concentration at its target sites is not known with certainty.

9.3 UPTAKE STUDIES

The first indications as to a site of action, reported by Cardinali, Hyyppa and Wurtman (1973), were obtained by uptake and distribution studies in rodents using tritiated or carbon-14 labeled melatonin of low specific activity. Radiolabeled melatonin accumulated in the mid-brain and hypothalamus and indeed in the pineal gland itself, the pituitary and the retina. Two studies also suggested accumulation in the choroid plexus. Early radioimmunoassay also detected higher concentrations of endogenous melatonin in the hypothalamus compared to the rest of the brain.

Recently, with the availability of positron emission tomography carbon-11 labeled melatonin has been injected into one human male volunteer via the antecubital vein, and the distribution of radioactivity followed by external detection. Peak melatonin concentrations were found in the brain after 8.5 min whereas plasma concentrations peaked at 3.5 min. Disappointingly no accumulation in the brain structures could be demonstrated and radioactivity quickly disappeared from the brain after 8.5 min. It is likely that the specific activity of the ligand was insufficient to make small areas of concentrated accumulation detectable. Moreover the resolution of the method was not sufficient to delineate, for example, the suprachiasmatic nucleus.

9.4 BINDING STUDIES

After initial uptake work in rodents, subsequent studies designed to investigate binding to cell membranes, cytosol and nucleus formed the main thrust of efforts to uncover melatonin targets in mammals and other species. The available tritium and carbon-14 radiolabeled melatonin remained of relatively low specific activity. Binding could, however, be demonstrated to membrane and cytosolic fractions of brain and some peripheral tissues. Interestingly, Joss (1981) also reported nuclear binding. These pioneering studies served to focus attention on the brain as the most consistent area of high binding.

9.5 2-(^{125}I) IODOMELATONIN

The experimental approach to the determination of target sites was revolutionized by the development of high specific activity melatonin labeled in the 2-position with ^{125}I (iodomelatonin) initially by Vakkuri, Rintamaki and Leppaluoto (1985) for radioimmunoassay. It was subsequently used for binding studies by Laudon and Zisapel (1986) and Dubocovich and coworkers (e.g. Krause and Dubocovich, 1990) followed by an explosion of publications from

2-[^{125}I]-Iodomelatonin

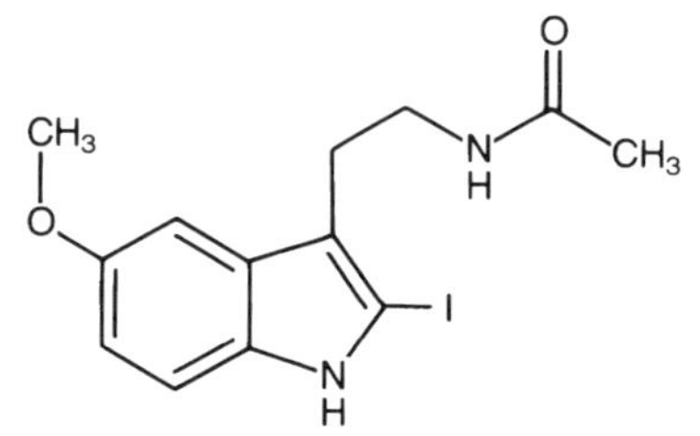

(a)

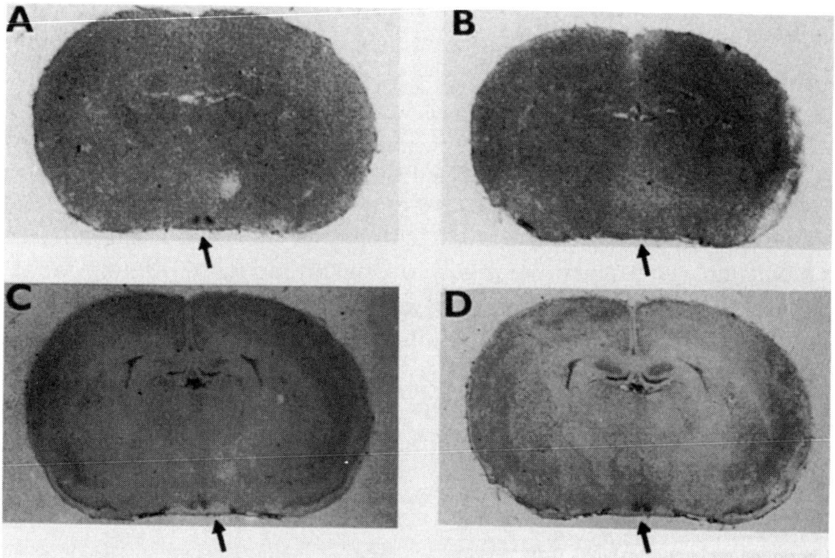

(b)

Fig. 9.1 (a) ^{125}I-2-iodomelatonin. (b) ^{125}I-2-iodomelatonin binds to the suprachiasmatic nucleus in the rat (and many other species). Autoradiography of ^{125}I-melatonin binding in the absence (A) or presence (B) of cold melatonin (1 μM) as a displacer. (C) and (D) illustrate the corresponding brain sections stained with toluidine blue. Arrows point to the suprachiasmatic nuclei (SCN). (Reproduced with permission, from Vanecek J., Pavlik, A. and Illnerova, H.; published by Elsevier, 1987.)

all over the world. ^{125}I-melatonin, in spite of the gross change in molecular weight, is equipotent or more potent than melatonin in the induction of typical melatonin-mediated responses such as dermal melanophore aggregation or short day reproductive responses in hamsters. These observations provide reassurance that this ligand really binds to melatonin receptors.

The molecule is now available at a specific activity of 2000 Ci per mmole. Its use has provoked enormous interest in melatonin 'receptors' with the autoradiographical localization of intense and highly localized binding in the rat SCN initially by Vanecek, Pavlik and Illnerova (1987, Figure 9.1) and then in the sheep pars tuberalis simultaneously by Morgan and coworkers (Morgan *et al.,* 1989c) and Reviers *et al.* (1989) (Figure 9.2). It has identified high- affinity (K_d 25–175 pM), saturable, specific and reversible melatonin binding to cell membranes in the central nervous system, the retina and the pituitary. More recently peripheral sites labeled with ^{125}I-melatonin have also been reported.

Many species have now been investigated for autoradiographical localization of ^{125}I-melatonin binding sites in the brain. Species variation is of course apparent, however in rodents the SCN, the pars tuberalis, the PVN, the paraventricular nucleus of the thalamus

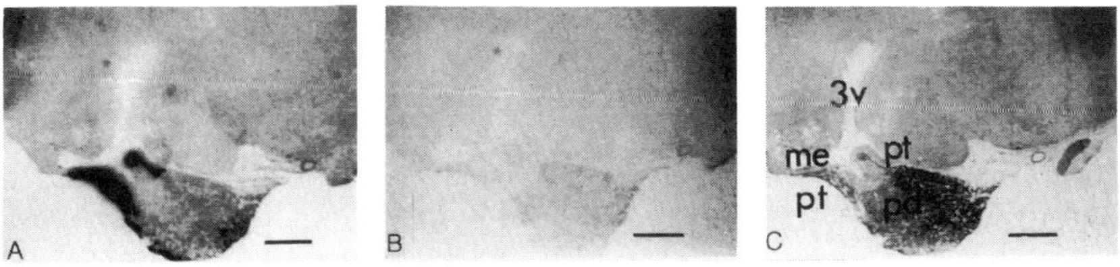

Fig. 9.2 Images produced on X-ray film by 20 μm thick sagittal sections through the hypothalamic-pituitary complex of an adult sheep, incubated with 50 pM [^{125}I] iodomelatonin (a) alone and (b) with 1 μM melatonin. Bar = 2.5 mm. (c) Section which produced image in (a), after staining with toluidine blue revealing the third ventricle (3v), the median eminence (me), the pars distalis (PD) and the pars tuberalis (pt). (Reproduced with permission from Morgan, P.J. *et al.* Melatonin receptors on ovine pars tuberalis; characterisation and autoradiographical localisation. *Journal of Neuroendocrinology*, **1**, 1–4; published by Blackwell Scientific Publications, 1989c.)

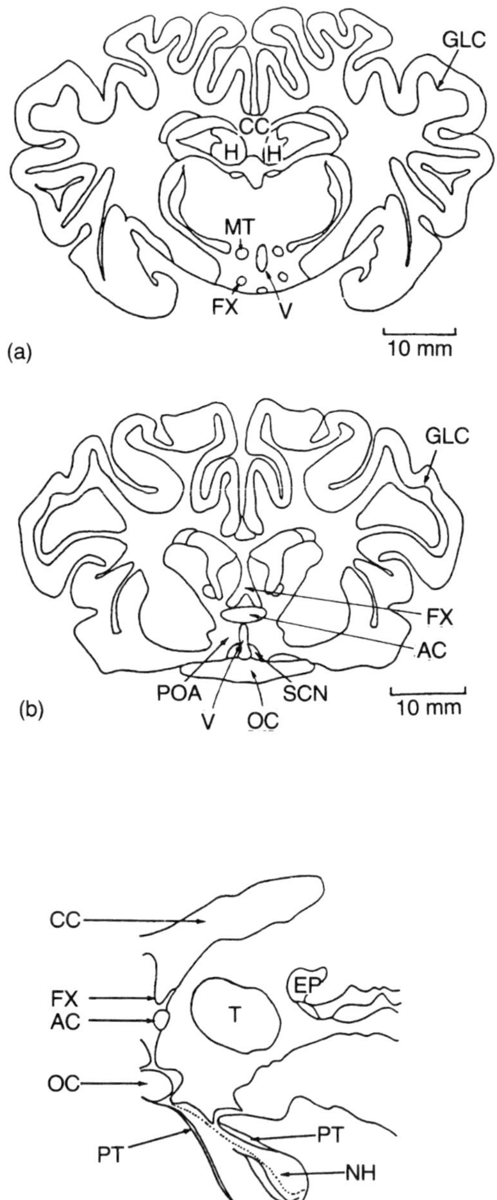
GLC
CC
H
H
MT
FX
V
10 mm
(a)
GLC
FX
AC
POA
SCN
V
OC
10 mm
(b)
CC
FX
AC
OC
T
EP
PT
PT
NH
AH
10 mm
(c)

(PVT), the habenula, the ventromedial nucleus, the pre-optic area, the amygdala and the area postrema have all shown labeling. In sheep binding is found in the pars tuberalis, in septal and pre-optic areas, in the inner and outer molecular layers of the hippocampus and the stratum lacunosum, the enterorhinal cortex, subiculum and apical interpeduncular nucleus. The SCN is variously reported as free of binding or with detectable binding in sheep. Goats show patterns similar to sheep with diffuse binding in the granular layer of the cortex (Figure 9.3). All the anterior hypothalamus together with the hippocampus, the cortex and the pars tuberalis have binding sites in the rabbit. The SCN shows clear binding in human post-mortem tissue (Figure 9.4). Interestingly, the pars tuberalis appears to be lacking in binding sites in most humans. Only a small number have been studied however.

Curiously adult ferrets, a classical photoperiodic species, appear to have no binding in the brain but it is present in the pars tuberalis. However comparison of photoperiodic hamsters and marginally photoperiodic rats and house mice shows few differences in binding and likewise photosensitive and photorefractory hamsters in different photoperiods do not differ in binding patterns. Further examples can be found in reviews (Cassone, 1990; Krause and Dubocovich, 1990; Morgan and Williams, 1989).

Evidently some of these binding sites (anterior hypothalamus, pre-optic area, SCN) represent potential targets for the effects of melatonin on seasonal and circadian rhythms and the inhibition of dopamine release from the hippocampus. By far the most consistent

Fig. 9.3 Diagram of sections of the goat brain showing major areas where ^{125}I-melatonin binding (indicated by an asterisk below) was present by *in vitro* autoradiography. (a) Transverse section of the goat brain at 22 mm rostral to the vertical interaural plane. *GLC = granular layer of the cerebral cortex; *CC = corpus callosum; *H = hippocampus; MT = fasciculus mamillothalamicus, *FX = fornix. (b) Transverse section of the goat brain at 30 mm rostral to the vertical interaural plane. *AC = anterior commissure; V = third ventricle; OC = optic chiasm; *POA = pre-optic area; *SCN = suprachiasmatic nucleus. (c) Mid-sagittal section through the goat brain. *PT = pars tuberalis; AH = adenohypophysis; NH = neurohypophysis; EP = epiphysis (pineal), T = thalamus. Bars indicate the scale in man. The distribution is similar to that of sheep although in this species SCN binding may not be present. (Reproduced with permission from Deveson, S., Thesis, The effects of photoperiod and melatonin on seasonal breeding in goats. University of Surrey, 1990.)

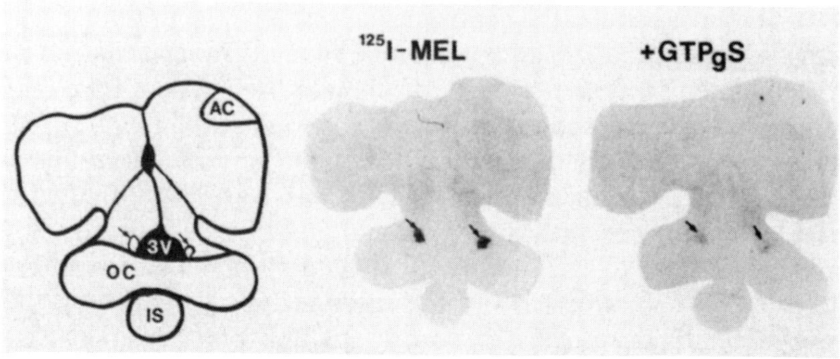

Fig. 9.4 Autoradiographic localization of ^{125}I-melatonin binding in the human SCN. The line drawing (left) represents a coronal section through the human hypothalamus, illustrating the location of the SCN (arrows). The centre image is the autoradiographic image produced by a section incubated in ^{125}I-melatonin (40 pM; total binding). Note the intense ^{125}I-melatonin binding in the SCN. The image on the right (+GTP gamma S) is from an autoradiogram generated by an adjacent section incubated with ^{125}I-melatonin (40 pM) in the presence of GTP gamma S (100 μM). The intensity of binding in the SCN was markedly reduced by the guanine nucleotide. 3V = Third ventricle; AC = anterior commissure; IS = infundibular (pituitary) stalk; OC = optic chiasm. Magnification × 2.4. (Reproduced with permission from Weaver, D.R. *et al.* Melatonin receptors in human hypothalamus and pituitary: implications for circadian and reproductive responses to melatonin. *J. Clin. Endocrinol. Metab.* **76**, 295–301, 1993 © The Endocrine Society.)

binding site from species to species is the pars tuberalis, whose function is unknown at the time of writing. The rise of the pars tuberalis as a primary candidate for a melatonin target tissue has stimulated work on function. Gonadotrophes constitute part of the cell population. *In vitro* co-culture with median eminence from the rat suggests that melatonin-provoked LH release from the pars tuberalis inhibits LHRH release from the median eminence (see Chapter 5). Most of the ovine pars tuberalis is constituted of agranular unidentified cells which, according to Morgan, Lawson and Davidson (1991), are largely responsible for melatonin-induced effects (Figure 9.5).They have proposed that they secrete an entirely new hormone which subsequently mediates the physiological effects of melatonin – an exciting prospect.

It is of particular interest that humans do not consistently show melatonin binding sites in the pars tuberalis. This suggests that

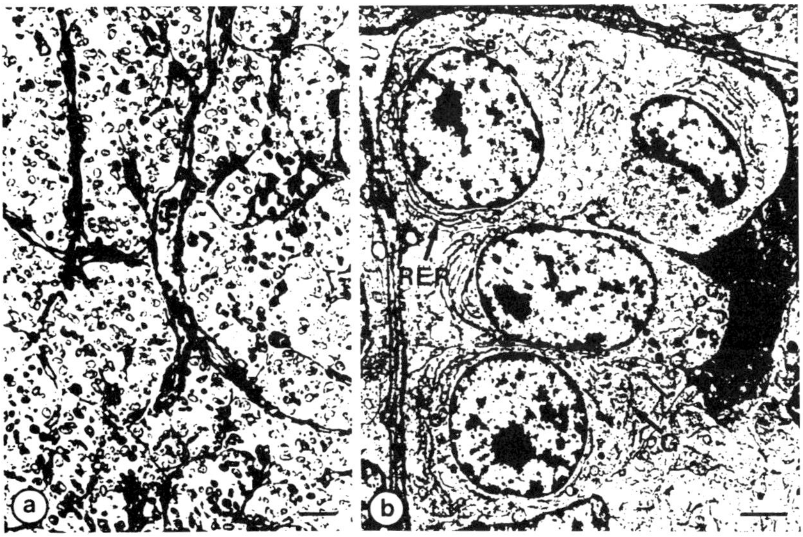

Fig. 9.5 (a) Light micrograph of the ovine pars tuberalis showing the columnar arrangement of cells. (b) Electron micrograph of agranular secretory cells of the ovine PT thought to be melatonin target tissue. RER = Rough endoplasmic reticulum; G = Golgi apparatus. (a) × 310. Bar: 25 μm; (b) × 5000. Bar: 2 μm. (Reproduced with permission from Morgan, P.J. *et al.*; published by Elsevier, 1991.)

human neuroendocrine responses to melatonin may be different from other mammals, even from primates, where binding in the pars tuberalis is evident. Moreover it is not inconceivable that the presence of pars tuberalis receptors in humans may signify an excessive sensitivity to seasonal changes such as is found in seasonal affective disorder.

Much more extensive binding can be seen in the brain of birds, especially in retinorecipient areas. Similarly widespread intense binding is present in lampreys, trout, and the African clawed toad with possibly the greatest density in lizard brain. The latter observation has provoked Rivkees, Conron and Reppert (1990) to initiate isolation of this lizard 'receptor' which appears to have a molecular weight of around 110 000.

There are changes in detectable binding with age, for example in fetal rats the first appearance of binding is in the pituitary, pars distalis and pars tuberalis with SCN labeling appearing in later gestation. Pars distalis binding is absent in adult rats but persists after birth in the neonate. This suggests that binding may indeed underlie function as melatonin will inhibit LHRH-induced pituitary LH release in prepuberty but not in adults.

There is some evidence that melatonin receptors are located on neurons in the brain. In general, however, the cell types and connectivity have not been identified conclusively as yet.

In parallel with these brain studies, Dubocovich (1985) has demonstrated a functional melatonin receptor initially in rabbit and chicken retina (inhibition of calcium-dependent dopamine release, Figure 9.6) which is localized in the inner plexiform layer containing dopamine amacrine cells in rabbits. Receptors may also be present in the outer and inner segments in mice and possibly the pigmented layer in some mammals.

9.6 HIGH- AND LOW-AFFINITY BINDING SITES

Using tritium or ^{14}C-labeled melatonin, low-affinity binding in the high nanomolar range was observed in the brain of rodents. Using high-specific-activity ^{125}I-melatonin, autoradiographic studies are considered to identify high-affinity sites. However a range of low nanomolar to low picomolar binding to membranes has also been demonstrated using this ligand. Dubocovich has called high-affinity and low-affinity sites ML1 and ML2 respectively and has reported different pharmacological profiles for the two hypothetical sites.

The problem with low-affinity binding is that the best guess as to the endogenous concentration of melatonin at brain target sites is picomolar, although there are suggestions that higher concentrations can be encountered in the hypothalamus, and in some species in the CSF. In most mammals melatonin is thought to be secreted primarily into the blood, subsequently entering the brain, and blood concentrations are in the pico- to femtomolar range. For such concentrations to be physiologically relevant (and clearly they are,

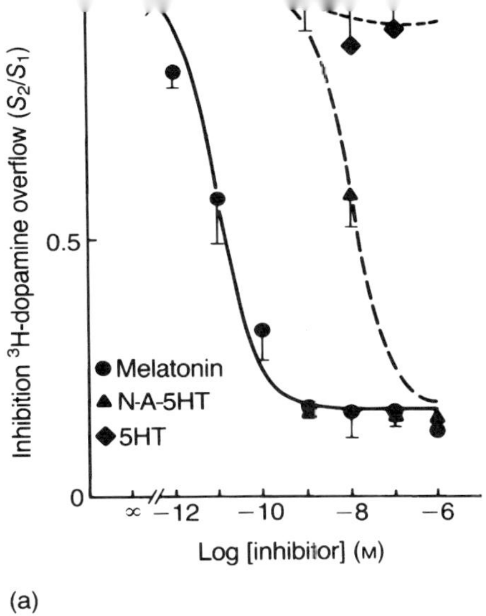

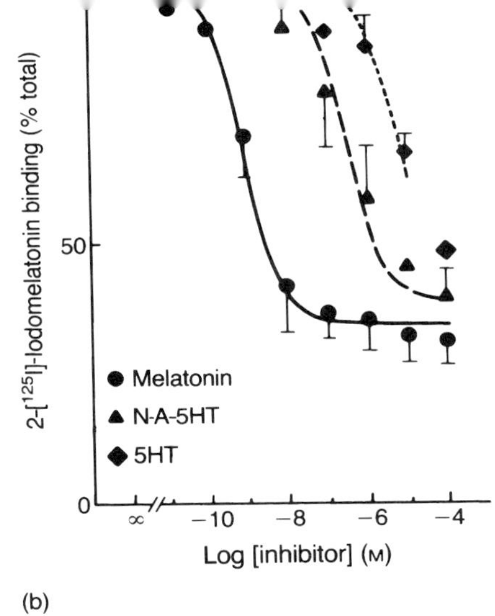

Fig. 9.6 (a) Effect of melatonin, *N*-acetylserotonin and serotonin on the calcium-dependent release of ^{3}H-dopamine from rabbit retina. Retinas from albino rabbits were dissected, labeled with ^{3}H-dopamine, and superfused. The results are expressed as ^{3}H-dopamine overflow as the percentage of total radioactivity released by field stimulation above spontaneous levels of release. The ratio between the first and second stimulation periods (with and without inhibitors) in the same experiment is given. The calcium-dependent release of ^{3}H-dopamine was elicited by electrical stimulation at 3 Hz in periods of 1 min (20 mA, 2 ms). Abscissa: molar concentrations of melatonin (●), (IC_{50}=9 pM), *N*-acetylserotonin (N-A-5HT) (▲), (IC_{50}=8.6 nM), and serotonin (5HT), (◆). Mean values ± SEM of three to eight experiments per group are shown. (Reproduced with permission from Dubocovich, M.; published by *The FASEB Journal*, 1988.) (b) Competition curves for inhibition of ^{125}I-2-iodomelatonin binding by melatonin, *N*-acetylserotonin and serotonin in rabbit retinal membranes. Washed rabbit retinal membrane was incubated with ^{125}I-2-iodomelatonin (50–100 pM) and various concentrations of melatonin ((●), K_i=1.03 nM), *N*-acetylserotonin ((▲), K_i=2.6 nM) and serotonin ((◆), K_i=3.7 μM). Values are means of at least three independent determinations. (Reproduced with permission from Dubocovich, M.; published by *The FASEB Journal*, 1988.)

as shown *inter alia* by infusion studies in hamsters and sheep, Chapter 5), receptor affinity must be in this range. Melatonin has many effects if administered in quantities such as to generate nanomolar concentrations or above. They may or may not be related to endogenous effects. They are nevertheless of interest and some are exploitable therapeutically. If these responses function through nanomolar affinity receptors there is every reason to investigate the properties of these receptors.

In some but not all cases both high- and low-affinity receptors have been shown to vary systematically with physiological status. For example the amount of low-affinity binding in rodent hypothalamus, hippocampus and medulla pons increases from morning to evening, and this difference can be modified by pinealectomy, ganglionectomy or constant light. Aging attenuated diurnal variations. Ovariectomy and castration decreased binding in the rat which was reversible by administration of steroids. Seasonal and diurnal variations have recently been described in high-affinity binding in the hamster (Figure 9.7) and in this species (and no doubt others) melatonin administration may 'down regulate' binding.

In general, the number of binding sites appears to peak in the early evening, close to the time of onset of melatonin secretion. In one study high-affinity binding was maximal in the very early morning. It has been suggested that the affinity of a single class of receptors can be shifted by the actions of cations and guanine nucleotides. Recent work by Sugden and Chong (1991) strongly supports the concept of a single class of sites with a similar pharmacological profile in different species and it is possible that the distinction between low- and high-affinity binding sites will prove to be artefactual.

The increase in high-affinity sites at the end of the photophase correlates with the time course of melatonin induced 2-deoxyglucose uptake by the SCN and the window for inducing phase advances of circadian rhythms.

9.7 MELATONIN RECEPTORS

The first functional melatonin receptor to be described was that of the amphibian melanophore. Cultured melanophores exhibit

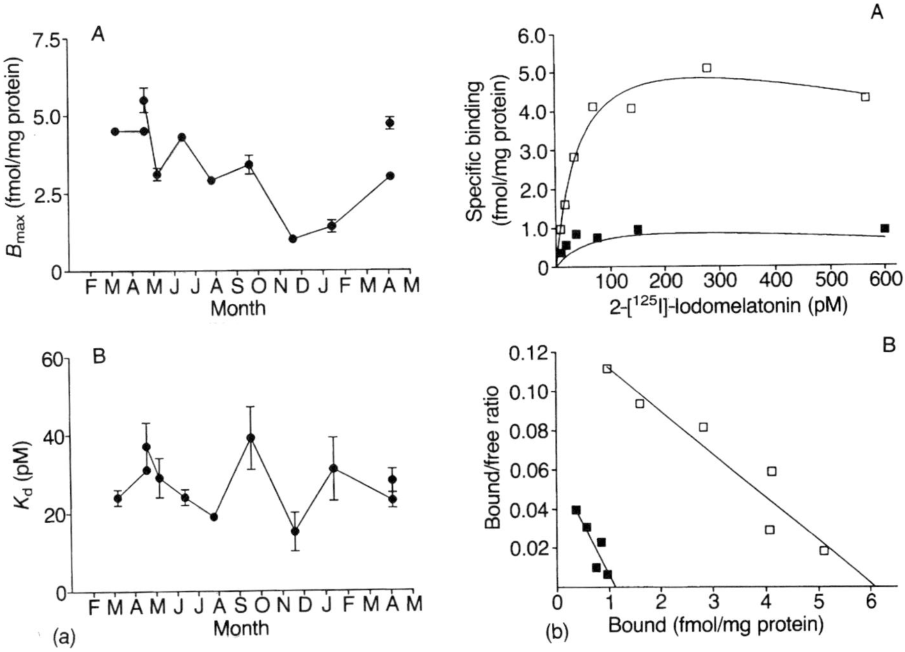

Fig. 9.7 (a) Changes in ^{125}I-2-iodomelatonin binding sites in the pars tuberalis of European hamsters housed under natural environmental conditions, and killed at different times of the year. (A) Number (B_{max}) and (B) affinity (K_d) of binding sites. Each data point ± bar represents the estimate ± SEM of the estimate obtained by LIGAND analysis of the saturation study. Where SEM bars are not shown they are encompassed by the symbol. The breaks in the graph indicate a freshly caught batch of hamsters. (Reproduced with permission from Skene, D.J., Masson-Pevet, M. and Pevet, P. Seasonal changes in melatonin binding sites in the par tuberalis of male European hamsters and the effect of testosterone manipulation. *Endocrinology*, **132**, 1682–6, 1993 © The Endocrine Society.) (b) (A) Specific ^{125}I-2-iodomelatonin binding in the pars tuberalis of European hamsters housed under natural environmental conditions and killed in mid-April (□) or in mid-November (■). Points are means of duplicate samples. (B) Scatchard plots of the data. K_d values of 37 ± 6 pM (estimate ± SEM) and 15 ± 5 pM and B_{max} values of 5.5 ± 0.4 fmol/mg protein and 1.0 ± 0.1 fmol/mg protein were obtained for the mid-April and mid-November group, respectively. (Reproduced with permission from Skene, D.J., Masson-Pevet, M. and Pevet, P. Seasonal changes in melatonin binding sites in the par tuberalis of male European hamsters and the effect of testosterone manipulation. *Endocrinology*, **132**, 1682–6, 1993 © The Endocrine Society.)

extreme sensitivity to melatonin with some cells responding by pigment migration to 10^{-16} M melatonin). An initial pharmacological profile was derived with dermal preparations and subsequently with cultured cells. It indicated that the melatonin receptor possessed a unique response to various analogs and neurotransmitter agonists and antagonists. Rollag and coworkers (White, Sekura and Rollag, 1987) have subsequently demonstrated that melatonin-induced pigment aggregation is a pertussis toxin sensitive system and that melatonin inhibits forskolin-activated cAMP formation.

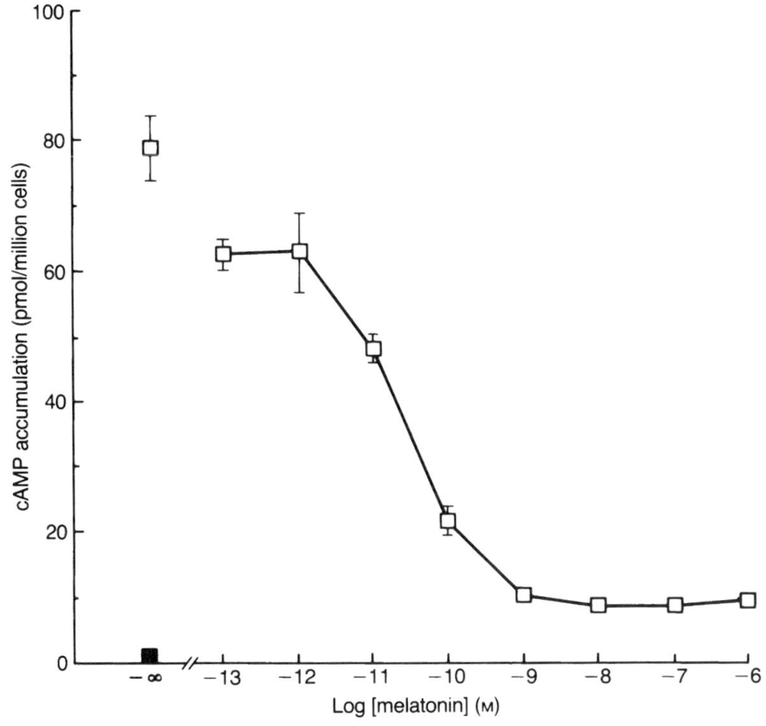

Fig. 9.8 Dose-dependent inhibition of forskolin-stimulated cAMP production in ovine pars tuberalis cells. Cells were incubated with forskolin (1 μM) and melatonin as indicated (open symbols). The levels of accumulation in unstimulated cells are shown for comparison (solid symbol). Data are mean ± SEM for a single typical experiment. Where errors do not appear they lie within the symbol. (Unpublished data kindly provided by Dr David G. Hazlerigg after Morgan *et al.*, 1989a,b, and Morgan *et al.*, 1991.)

Inhibition of cAMP production may be a general feature of melatonin receptors. Intensive investigation of the properties of the pars tuberalis binding site has revealed that physiological doses of melatonin inhibit forskolin-activated cAMP production *in vitro* in a time and dose-related manner (Figure 9.8). There is extensive evidence that the binding sites are coupled to G-proteins. GTP analogs which interfere with the regeneration of G1-coupled receptors decrease the affinity and sometimes the capacity of ^{125}I-melatonin binding in reptiles, birds and mammals (Figure 9.9). Vanecek and Klein (1992), investigating the ability of melatonin to inhibit LHRH-stimulated cAMP and cGMP in the neonatal rat pituitary consider that an increase of intracellular calcium is a *sine qua non* for activity.

A possible link between receptor pharmacology and the physiological effects of melatonin has emerged recently. Hazlerigg *et al.*

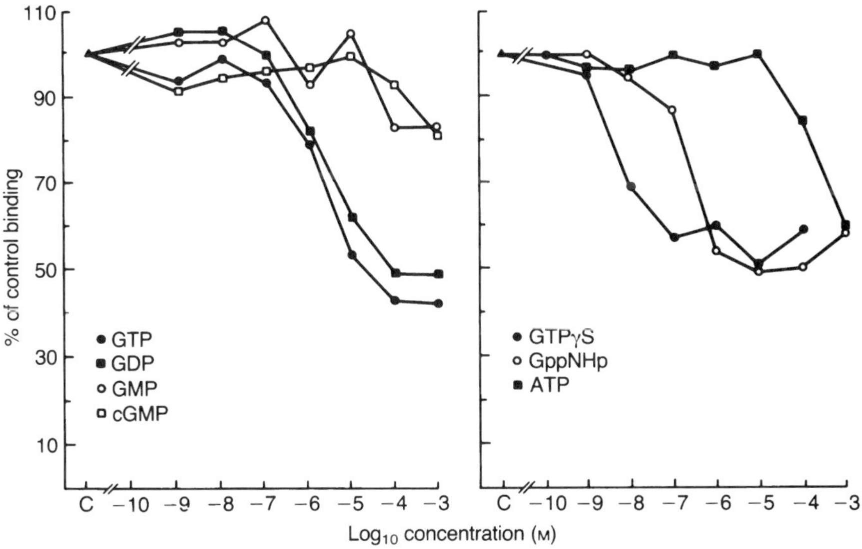

Fig. 9.9(a) Effect of guanine nucleotides and ATP on the specific binding of ^{125}I-2-iodomelatonin to membranes from the ovine pars tuberalis. Specific binding was measured using a fixed concentration of ^{125}I-2-iodomelatonin (80 pM), in the presence and absence of 1 μM melatonin. Results are expressed as the % of specific binding in the absence of a nucleotide (control). Each point is the mean from three independent experiments, where duplicate measurements were made. Error bars omitted for clarity but the average SEM was 4.96 %. (Reproduced with permission from Morgan, P.J. *et al.*; published by Karger, Basel, 1989b.)

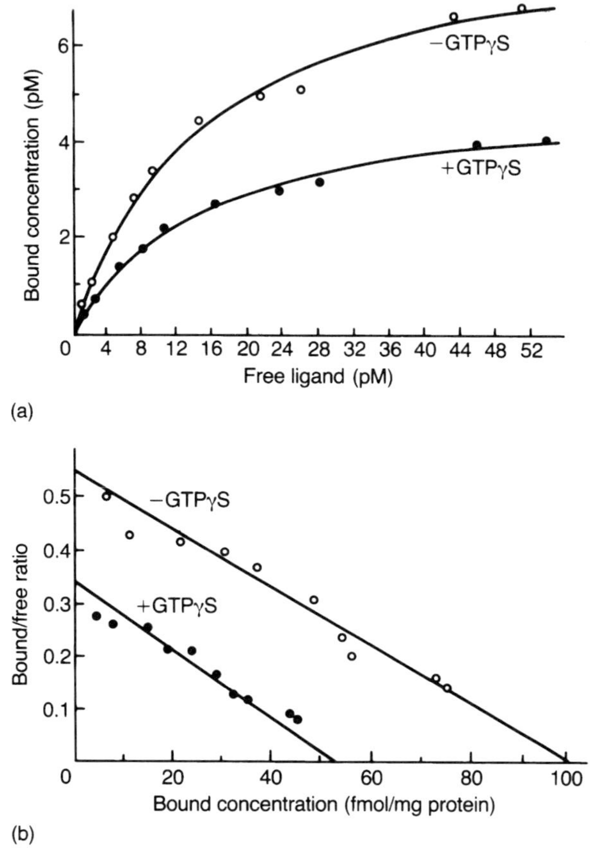

Fig. 9.9(b) (a) Saturation isotherms of specific binding of ^{125}I-2-iodomelatonin to ovine pars tuberalis in the presence and absence of 1 μM GTP gamma S. Specific binding was measured as in Figure 9.9(a). (b) Scatchard plots of saturation data as in (a). The data are taken from a typical experiment where measurements were made in triplicate. (Reproduced with permission from Morgan, P.J. *et al.*; published by Karger, Basel, 1989b.)

(1993) have described the sensitization of the pars tuberalis response to forskolin by prolonged pretreatment with melatonin. Maximum sensitization occurred with 16-h pretreatment and parallels with short day effects of 8L 16D are evident (Figure 9.10). Such pretreatment also induces an increase in the IC_{50} for the acute inhibition of forskolin-

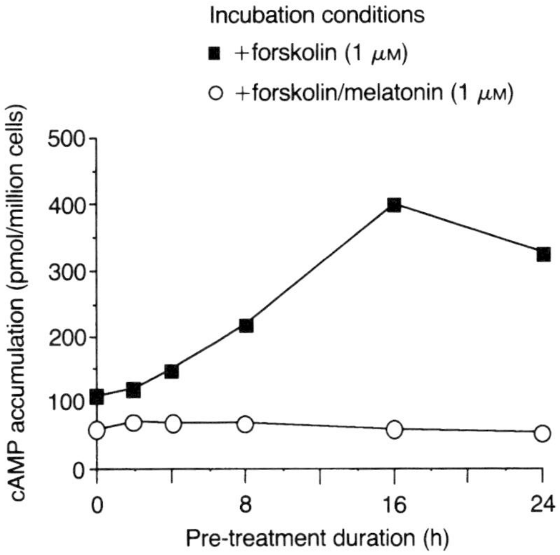

Fig. 9.10 Duration-dependent sensitization of ovine pars tuberalis cells to forskolin stimulation by treatment with melatonin (100 pM). Cells were pretreated with melatonin for the duration indicated prior to washing and challenge with forskolin (1 μM) with (open symbols) or without (solid symbols) melatonin (1 μM) and assayed for the accumulation of cyclic-AMP. Data are mean ± SEM of a single typical experiment, errors lie within symbols. Note that the maximal inhibitory effects of melatonin are unaffected by pretreatment with the hormone. (Unpublished data kindly provided by Dr David G. Hazlerigg after Hazlerigg *et al.* 1993.)

stimulated cAMP by melatonin, and prolonged treatment also reduces the total ^{125}I-2-iodomelatonin binding to pars tuberalis cells.

A major problem concerning the link between binding capacity, cAMP response and target organ response at least in seasonal terms is the slow time course of events. Each step has to be separately identified. To date the evidence indicates that ^{125}I-melatonin binding sites are genuine receptors although by no means have all binding sites been investigated. So far, however, there is no conclusive direct link to a physiological response such as onset or offset of seasonal reproductive activity. We may anticipate that such a link will emerge in the very near future.

9.8 MELATONIN ANTAGONISTS AND AGONISTS

There are a number of published reports of synthetic analogs of melatonin.

The 6- and 2-substituted halogenated melatonins are good agonists both *in vivo* and *in vitro*. Presumably in both cases but particularly the 6-substituted molecule there is protection against metabolic degradation and hence a lower dose requirement for the induction of a given effect.

A series of agonists have been developed from naphthalene derivatives recently (Figure 9.11). They show a range of affinity for the pars tuberalis melatonin receptor, some being of much higher affinity than melatonin. The most interesting will entrain free-running circadian rhythms in rats, induce short day effects in photostimulated hamsters when administered in such a way as to gener-

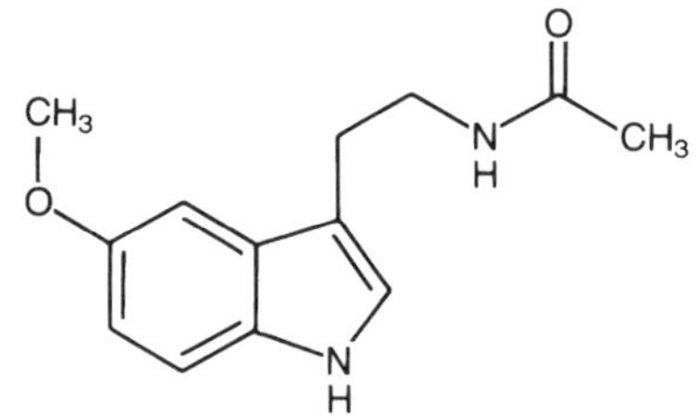

(a) Melatonin (*N*-Acetyl-5-methoxytryptamine)

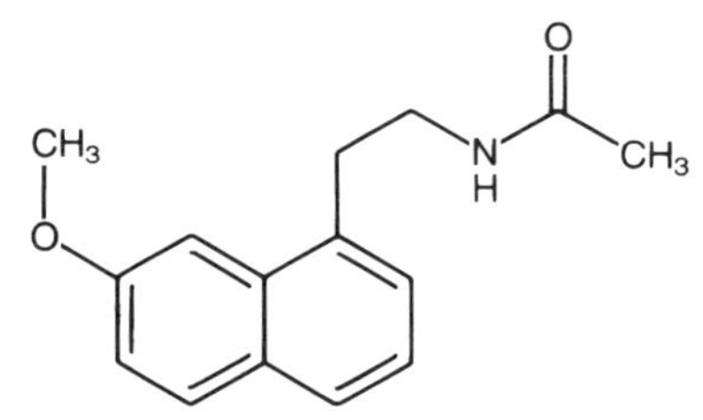

(b) MNEA (*N*-[2-(7-Methoxy-1-naphthalenyl)ethyl]acetamide)

Fig. 9.11 A naphthalenic melatonin analog (b) with similar receptor pharmacology and physiological effects to melatonin (a).

ate a long duration profile, and hasten entrainment of the endogenous circadian rhythm of aMT6s in rats.

Only two antagonists have been reportedly investigated in detail. Luzindole, a melatonin molecule with a benzyl group substituted in the 2-position clearly inhibits the effects of melatonin on calcium

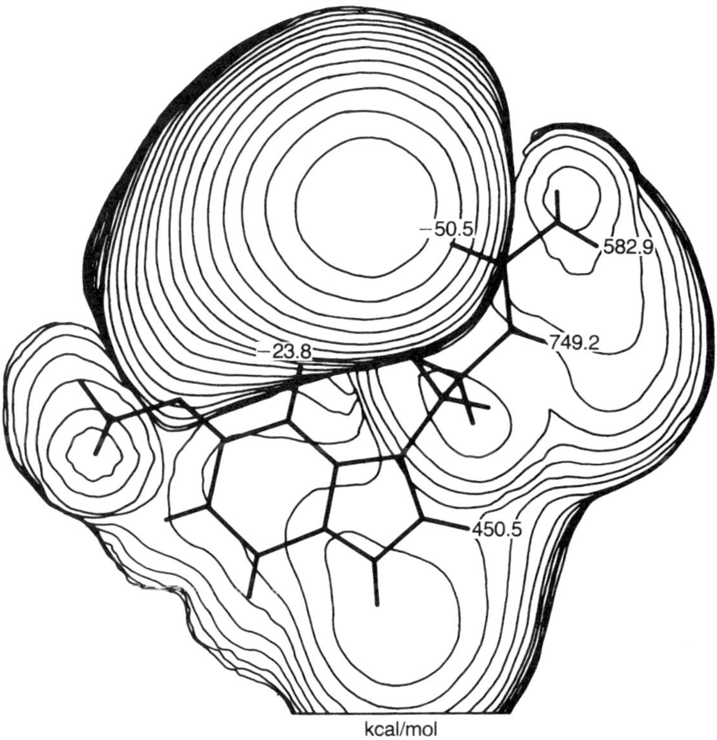

Fig. 9.12 The molecular electrostatic potential energy contour surface of melatonin. The upper area indicates the region of negative potential energy whereas the lower area indicates positive potential energy. Contours are in steps of 5 kcal/mol: maxima and minima are shown in their corresponding regions. Using this information it is possible to make predictions as to the structure of melatonin binding sites and to devise competitive analogs. (Reproduced with permission from Lewis, D.F.V., Arendt, J. and English, J. Quantitative structure-activity relationships within a series of melatonin analogues and related indole alkylamines. *Journal of Pharmacology and Experimental Therapeutics*, **252**, 370–3, © Williams and Wilkins, 1990.)

dependent dopamine release in the retina *in vitro* but has not proved very potent *in vivo* with respect to melatonin dependent reproductive changes or circadian entrainment in rodents, probably due to a lower affinity than melatonin.

ML23, *N*-(2,4-dinitrophenyl)-5-methoxytryptamine is an *in vitro* inhibitor of low-affinity melatonin binding and was reported to counteract melatonin-induced delay of puberty and the effects of melatonin on the constant estrous anovulatory syndrome in rats. It does not however appear to counter reproductive effects of melatonin in hamsters or sheep and has low displacement activity with respect to the high affinity melatonin receptor.

9.9 STRUCTURE OF THE MELATONIN RECEPTOR

Apart from the molecular weight information concerning the lizard receptor little information is available. Very recently the melanophore receptor has been cloned (Ebisawa *et al.*, 1994) and cloning of the mammalian receptor(s) is eagerly awaited .

Lewis, Arendt and English (1990), using structure activity relationships and available information on the structure of the serotonin receptor, have constructed a model of the electrostatic potential energy profile of the melatonin molecule (Figure 9.12) and have used this to predict an amino-acid sequence for the melatonin binding site. Only the future will tell whether this is accurate or not.

REFERENCES

Cardinali, D.P., Hyyppa, M.T. and Wurtman, R.J. (1973) Fate of intracisternally injected melatonin in the rat brain. *Neuroendocrinology,* **12**, 30–40.

Cassone, V. (1990) Effects of melatonin on vertebrate circadian systems. *Trends in Neurosciences,* **13**, 457–63.

De Reviers, M.M., Ravault, J.P., Tillet, Y. and Pelletier, J. (1989) Melatonin binding sites in the sheep pars tuberalis. *Neuroscience Letters*, **100**, 89–93.

Deveson, S.L. (1990) Thesis, The effects of photoperiod and melatonin on seasonal breeding in goats. University of Surrey.

Dubocovich, M. (1988) Pharmacology and function of melatonin receptors. *Journal of FASEB,* **2**, 2765–73.

Ebisawa, T., Karne, S., Lerner, M.R. and Reppert, S.M. (1994) Expression cloning of a high affinity melatonin receptor from *Xenopus* dermal melanophores. *Proceedings of the National Academy of Sciences, USA*, **91**, 6133–7.

Hazlerigg, D.G., Gonzalez-Brito, A., Lawson, W. *et al.* (1993) Prolonged exposure to melatonin leads to time dependent sensitization of adenylate cyclase and down regulates melatonin receptors in pars tuberalis cells from ovine pituitary. *Endocrinology,* **132**, 285–92.

Joss, J.M.P. (1981) Autoradiographic localisation of sites of uptake of [^{3}H]-melatonin in the brains of the lizard and the lamprey, in *Pineal Function* (eds C.D. Matthews and R.F. Seamark) Elsevier/North Holland Press, Amsterdam, pp. 223–34.

Krause, D.N. and Dubocovich, M.L. (1990) Regulatory sites in the melatonin system of mammals. *Trends in Neurosciences,* **13**, 464–70.

Laudon, M. and Zisapel, N. (1986) Characterization of central melatonin receptors using I^{125}-melatonin. *FEBS Letters,* **197**, 9–12.

Lewis, D.F.V., Arendt, J. and English, J. (1990) Quantitative structure-activity relationships within a series of melatonin analogues and related indolealkylamines. *Journal of Pharmacology and Experimental Therapeutics,* **252**, 370–3.

Morgan, P.J., King, T.P., Lawson W. *et al.* (1991) Ultrastructure of melatonin responsive cells in the ovine pars tuberalis. *Cell and Tissue Research,* **263**, 529–34.

Morgan, P.J., Lawson, W. and Davidson, G. (1991) Interaction of forskolin and melatonin on cyclic AMP generation in pars tuberalis from ovine pituitary. *Journal of Neuroendocrinology,* **3**, 497–501.

Morgan, P.J., Lawson, W., Davidson, G. *et al.* (1989a) Melatonin inhibits cyclic AMP production in cultured ovine pars tuberalis cells. *Journal of Molecular Endocrinology,* **3**, R5–R8.

Morgan, P.J., Lawson, W., Davidson, G *et al.* (1989b) Guanine nucleotides regulate the affinity of melatonin receptors on the ovine pars tuberalis. *Neuroendocrinology,* **50**, 359–62.

Morgan, P.J., Williams, L.M., Davidson, G. *et al.* (1989c) Melatonin receptors on ovine pars tuberalis: characterisation and autoradiographical localisation. *Journal of Neuroendocrinology,* **1**, 1–4.

Reviers, M.M.de, Ravault, J.P., Tillet, Y. *et al.* (1989) *Neuroscience Letters,* **100**, 89–93.

Rivkees, S.A., Conron, R.W. and Reppert, S.M. (1990) Solubilisation and purification of melatonin receptors from lizard brain. *Endocrinology,* **127**, 1206–14.

Skene, D.J., Masson-Pevet, P. and Pevet P. (1993) Seasonal changes in melatonin binding sites in the par tuberalis of male European hamsters and the effect of testosterone manipulation. *Endocrinology,* **132**, 1682–6.

Sugden, D. and Chong, N.W.S. (1991) Pharmacological identity of 2-(125-I)iodomelatonin binding sites in chicken brain and sheep pars tuberalis. *Brain Research,* **539**, 151–4.

Vakkuri, O., Rintamaki, H. and Leppaluoto, J. (1985) Presence of immunoreactive melatonin in different tissues of the pigeon (*Columba livia*). *General and Comparative Endocrinology,* **58**, 69–75.

Vanecek, J. and Klein, D. (1992) Melatonin inhibits gonadotropin-releasing hormone-induced elevation of intracellular Ca^{2+} in neonatal rat pituitary cells. *Endocrinology*, **130**, 701–707.

Vanecek, J., Pavlik, A. and Illnerova, H. (1987) Hypothalamic melatonin receptor sites revealed by autoradiography. *Brain Research*, **453**, 359–62.

Weaver, D.R., Stehle, J.H., Stopa, E.G. *et al.* (1993) Melatonin receptors in human hypothalamus and pituitary: implications for circadian and reproductive responses to melatonin. *Journal of Clinical Endocrinology and Metabolism*, **76**, 295–301.

White, B.H., Sekura, R.D. and Rollag, M.D. (1987) Pertussis toxin blocks melatonin-induced aggregation in Xenopus dermal melanophores. *Journal of Comparative Physiolology*, B, **157**, 153–9.

FURTHER READING

Anton-Tay, F., Chous, C., Anton, S. *et al.* (1968) Brain serotonin concentration: Elevation following intraperitoneal administration of melatonin. *Science,* **162**, 277.

Bittman, E.L. (1993) The sites and consequences of melatonin binding in mammals. *American Zoologist,* **33**, 200–211.

Brzezinski, A., Schenker, J., Fibich, T. *et al.* (1992). Effects of melatonin on progesterone production by human granulosa lutein cells in culture. *Fertility and Sterilisation,* **58**, 526–529.

Cardinali, D.P., Ritta, M.N., Pereyra, E. et al. (1982) Role ofprostaglandins in rat pineal neuroeffector junction. Changes in melatonin and norepinephrine release *in vitro. Endocrinology,* **111**, 530–4.

Cardinali, D.P., Vacas, M.I. and Boyer, E.E. (1979) Specific binding of melatonin in bovine brain. *Endocrinology,* **105**, 437–41.

Deveson, S., Howarth, J., Arendt, J. *et al.* (1992) *In-vitro* autoradiographical localisation of melatonin binding sites in the caprine brain. *Journal of Pineal Research*, **13**, 6–12.

Dubocovich, M.I. (1985) Characterisation of a retinal melatonin receptor. *Journal of Pharmacology and Experimental Therapeutics,* **234**, 395–401.

Dubovich, M.L. and Takahashi, J.S. (1987) Use of 2-(125I)-iodomelatonin to characterize melatonin binding sites in chicken retina. *Proceedings of the National Academy of Sciences, USA,* **84**, 3916–20.

Franchi, A.M., Gimeno, M.F., Cardinali, D.P. *et al.* (1987) Melatonin, 5-methoxytryptamine and some of their analogues as cyclo-oxygenase inhibitors in rat medial basal hypothalamus. *Brain Research,* **405**, 384–8.

Fraschini, F.O., Mess, B. and Martini, L. (1968) Pineal gland, melatonin and the control of luteinizing hormone secretion. *Endocrinology,* **82**, 919–24.

Glass, J.D. (1988) Neuroendocrine regulation of seasonal reproduction by the pineal and melatonin. *Pineal Research Review,* **6**, 219–59.

Hastings, M., Herbert, J., Martenz, N.D. *et al.* (1985) Melatonin and the brain in photoperiodic mammals, in *Photoperiodism, Melatonin and the Pineal,* Ciba Symposium Foundation, Pitman, London, **117**, pp. 57–70.

Hastings, M.H., Maywood, E.S. and Ebling, F.J.P. (1991) Sites and mechanism of action of melatonin in the photoperiodic control of reproduction, in *Advances in Pineal Research* (eds J. Arendt and P. Pevet), **5**, 147–57.

Hastings, M.H., Roberts, A.C. and Herbert, J. (1985) Neurotoxic lesions of the anterior hypothalamus disrupt the photoperiodic but not the circadian system of the Syrian hamster. *Neuroendocrinology,* **40**, 316.

Herbert, J. (1989) Neural systems underlying photoperiodic time measurement: A blueprint. *Experientia,* **45**, 965–97.

Laudon, M., Yaron, Z. and Zisapel, N. (1988) *N*-(3,5-dinitrophenyl)-5-methoxytryptamine, a novel melatonin antagonist: effects on sexual maturation of the male and female rat. *Journal of Endocrinology,* **116**, 43–53.

Le Bars, D., Thivolle, P., Vitte, P.A. *et al.* (1991) PET and plasma pharmacokinetic studies after bolus intravenous administration of ^{14}C-melatonin in humans. *Nuclear Medicine and Biology,* **18**, 357–62.

Lincoln, G. and Maeda, K. (1992) Effects of placing micro-implants of melatonin in the mediobasal hypothalamus and preoptic area on the secretion of prolactin and β -endorphin in rams. *Journal of Endocrinology,* **134**, 437–48.

Liu, Z. and Pang, S. (1992) (125I)-Labelled iodomelatonin binding sites in the duck bursa of Fabricius: binding characteristics and diurnal variation. *Neuroscience Letters,* **146**, 163–6.

Martin, J.E. and Klein, D.C. (1975) Melatonin inhibition of neonatal pituitary response to luteinising hormone-releasing factor. *Science,* **191**, 301–2.

Morgan, P.J., Barrett, P., Davidson, G. *et al.* (1992) Melatonin regulates the synthesis and secretion of several proteins by pars tuberalis cells of the ovine pituitary. *Journal of Neuroendocrinology,* **4**, 557–63.

Morgan, P.J., King, T., Lawson, W. *et al.* (1991) Ultrastructure of melatonin-responsive cells in the ovine pars tuberalis. *Cell and Tissue Research,* **263**, 529–34.

Morgan, P.J. and Williams, L.M. (1989) Central melatonin receptors; implications for a mode of action. *Experientia,* **45**, 955–65.

Nakazawa, K., Marubayashi, U. and McCann, S.M. (1991) Mediation of the short-loop feedback of luteinising hormone (LH) on LH-releasing hormone release by melatonin-induced inhibition of LH release from the pars tuberalis. *Proceedings of the National Academy of Sciences, USA,* **88**, 7576–9.

Pang, C. and Pang, S. (1992) High affinity specific binding of 2-(125I) iodomelatonin by spleen membrane preparations of chicken. *Journal of Pineal Research,* **12**, 167–173.

Rémé, C.E., Wirz-Justice, A. and Terman, M. (1991) The visual input stage of the mammalian circadian pacemaking system: 1. Is there a clock in the mammalian eye? *Journal of Biological Rhythms,* **6**, 5–29.

Reppert, S.M., Weaver, D.R., Rivkees, S.A. *et al.* (1988) Putative melatonin receptors in a human biological clock. *Science,* **242**, 78–81.

Rosenstein, R.E. and Cardinali, D.P. (1986) Melatonin increases *in vivo* GABA accumulation in rat hypothalamus, cerebellum, cerebral cortex and pineal gland. *Brain Research,* **398**, 403–6.

Steinlechner, S. (1991) In search of a physiological role for retinal melatonin, in *Advances in Pineal Research* (eds J. Arendt and P. Pevet), Suppl. 5, pp. 123–128.

Sugden, D. (1983) Psychopharmacological effects of melatonin in mouse and rat. *Journal of Pharmacology and Experimental Therapeutics,* **227**, 587–91.

Vitte, P.A., Harthe, C., Lestage, P., Claustrat, B. *et al.* (1988) Plasma, cerebrospinal fluid, and brain distribution of ^{14}C-melatonin in the rat: A biochemical and autoradiographic study. *Journal of Pineal Research*, **5**, 437–53.

Vriend, J. (1983) Evidence for pineal gland modulation of the neuroendocrine-thyroid axis. *Neuroendocrinology*, **36**, 68–78.

Wang, Z., Cheng, K., Brown, G. *et al.* (1992) Characterization of 2-(125I) iodomelatonin binding sites in quail testes at mid-light and mid-dark. *Neuroscience Letters*, **146**, 195–8.

Weaver, D.R., Namboodiri, M.A.A. and Reppert, S.M. (1988) Iodinated melatonin mimics melatonin action and reveals discrete binding sites in the fetal brain. *FEBS Letters*, **288**, 123–7.

Weaver, D.R. and Reppert, S.M. (1990) Melatonin receptors are present in the ferret pars tuberalis and pars distalis, but not the brain. *Endocrinology*, **127**, 2607–9.

Williams, L.M., Martinoli, M.G., Titchener, L.T. *et al.* (1991) The ontogeny of central melatonin binding sites in the rat. *Endocrinology*, **128**, 2083–90.

Yous, S., Andrieux, J., Howell, H.E. *et al.* (1992) Novel naphthalenic ligands with a high affinity for the melatonin receptor. *Journal of Medical Chemistry*, **35**, 1484–6.

Zisapel, N., Oaknin, S. and Anis, Y. (1991) Melatonin receptors in discrete areas of the rat and Syrian hamster brain: modulation by melatonin, pinealectomy, testosterone and the photoperiod, in *Advances in Pineal Research* (eds J. Arendt and P. Pevet), **5**, pp. 175–181.

Appendix: Glossary

Simple and understandable definitions of terms in order to help those who are not familiar with the chronobiology jargon are provided. More accurate definitions, while differing among authors, can be found in the set of books listed.

Acrophase, Φ, is the estimated span of time to reach the peak of the validated rhythm for the period (tau = τ) under consideration. When using the Cosinor method, Φ represents the peak time of the best-fitting mathematical function approximating the data. It is expressed as an interval from a designated phase reference (Φ_0).

Amplitude, A, is the amount of variability due to a given rhythm. When the Cosinor method is used, it is numerically equal to one-half of the extent of rhythmic changes for a particular τ. In other words, $2A$ is the crest-to-trough difference.

Chronogram is an individual or averaged display of data as a function of time, e.g. raw data of a time series.

Chronopharmacokinetics designates biological time-related changes in the pharmacokinetics of an agent quantified by parameters of one or several curve patterns (models).

Chronopharmacology designates rhythmic (predictable-in-time) differences in the pharmacologic effects of a substance.

Circa is the Latin prefix meaning about. Many curious terms have been coined to designate bioperiodicities. Among them are:

Circadian: about 24 hours

Circaseptan: about 7 days

Circatrigintan: about 30 days

Circamensual: about 30 days

Circannual: about 1 year

Cosinor is an analytical procedure used to detect and quantify periodicities in terms of 'period', 'mesor', 'amplitude' and 'acrophase' (defined elsewhere). Best fitting (cosine) function approximating all data (least squares method).

Mesor, M, is the rhythm-adjusted mean. When the interval of time between data sampling (Δt) is constant, M equals the arithmetic mean (X).

Period, **tau or** τ, is the duration of one complete cycle of a rhythmic variation. It is customarily expressed in units of time (e.g. s, min, h, day, or year).

Photoperiod is strictly the length of the light phase of a particular light–dark cycle, but may be used to describe the whole light–dark cycle.

Scotoperiod is the length of the dark phase of a particular light–dark cycle.

Skeleton photoperiod is used to describe one or more light pulses of given intensity and duration which mimic a full photoperiod in the induction of seasonal or circadian responses, e.g. two spaced pulses interpreted as dawn and dusk.

Spectral analysis is a technique developed by physicists and used by chronobiologists to determine periodicities in data collected longitudinally overtime.

Synchronizer, **time cue** or **zeitgeber** refers to an environmental periodicity capable of determining the temporal staging, with respect to clock hour or calendar date, of a given endogenous rhythmicity. Currently, it is the environmental signal(s) (e.g. dawn and/or dusk) used by the organism to reset its biological clocks.

Chronobiologists rely upon certain abbreviations to convey pertinent information about synchronizer type and schedule. LD : 12 12 indicates the synchronizer to be the Light (L)–Dark (D) cycle with L and D each having a 12 h duration. LD : 8 16 is the light–dark cycle with the duration of the former being 8 h and the latter 16 h. Time or clock hour is expressed using the international designation, e.g. 1:00 pm is referred to as 1300. HALO means Hours After Light Onset.

T refers to the period of an imposed light–dark cycle of given photoperiod and light intensity.

Ultradian or pulsatile is the biological variation with a period shorter than 20 h, e.g. 12, 8, 6, 3, 1.5, 1 h, etc. The word **circhoral** (about one hour) is used to designate these periodicities dealing mainly with hormonal secretions such as that of LHRH.

FURTHER READING

Mills, J.N. (1973) *Biologic Aspects of Circadian Rhythms*. Plenum Press, London, NY.

Moore-Ede, M., Sulzman, F.M. and Fuller, C.A. (1982) *The Clocks that Time Us*. Harvard University Press, Cambridge, Mass.

Halberg, F., Carandente, F., Cornelissen, G. and Katinas, G. (1977) Glossary of chronobiology. *Chronobiologia*, **4**, suppl. 1.

Reinberg, A. and Smolensky, M. (1983) *Biological Rhythms and Medicine*. Springer-Verlag, New York.

Index

Page numbers appearing in **bold** refer to figures and page numbers appearing in *italic* refer to tables.